For Joseph
5/20/13

[signature]

stillpoint

⊕

stillpoint
the geometry of consciousness

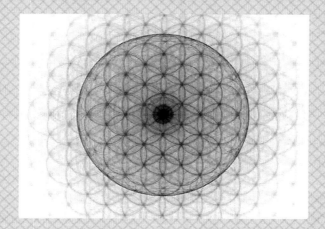

james ross godbe

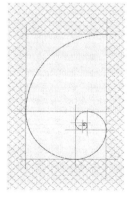

for my Mother
Marion Frances Ross

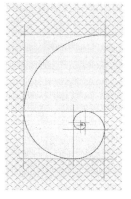

Acknowledgment

The following people are responsible for just about everything that you're about to read. I am grateful for their dedication to what they love. Without their work, and some for their personal support, I would never have been able to put the following together, nor have this moment happen. Some are personal friends and relatives, some are acquaintances, some have created brilliant books and films with invaluable information who I've never met and some are living and some are not.

Many thanks to family and friends John August, Bill Elwell, Foster Gamble, Alexandra Godbe, Roberta Godbe-Tipp, William S. Godbe, Dugan Hammock, Dylan Kaufman-Obstler, the incredible Normans, Susan Powell, Beth Shaffer, Jeff Tipp and Marcel Vogel.

I also have so much gratitude for the work and intangible contributions of Greg Braden, Alan Butler, Pierre Teilhard de Chardin, Rodney Collin, Crazy Horse, Keith Critchlow, Gyorgy Doczi, Masaru Emoto, Buckminster Fuller, James Gardner, George Gurdjieff, Stanislav Grof, John Major Jenkins, Christopher Knight, John Martineau, Hans Jenny, Abraham Lincoln, John Michell, Van Morrison, Jeremy Narby, Jill Purce, Wilhelm Reich, Red Cloud, Mouni Sadu, Hehaka Sapa, Michael S. Schneider, Freddy Silva, Bill Schul, Rupert Sheldrake, Richard Tarnas, Alexander Thom, Arthur Young . . . and many more.

I am especially grateful to Michael Godbe, without whose help in so many ways, so much of what follows would never have happened, as well as to Karen Szybalski who flew in like an angel and took a hard look at everything being said and gave inspired advice.

And ultimate gratitude to the Spirit that lives within and around everything and everywhere in this place called Inyo – in Paiute, 'a dwelling place of the Great Spirit'.

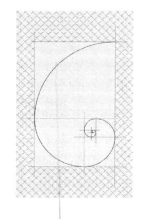

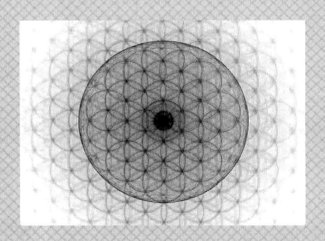

TABLE OF CONTENTS

'Geometry is an 'exact' science. It leaves nothing to chance. Except for its axioms, it can prove everything it teaches. It is precise. It is definite. By it we buy and sell our land, navigate our ships upon the pathless ocean, foretell eclipses, and measure time. All science rests upon mathematics, and mathematics is first and last, geometry, whether we call its extension 'trigonometry' or 'differential calculus' or any other name. Geometry is the ultimate fact we have won out of a puzzling universe . . . There are no ultimate facts of which the human mind can take cognizance which are more certain, more fundamental, than the facts of geometry.'

Foreign Countries (1925) Carl H. Claudy

'There are two ways to be fooled.
One is to believe what isn't true;
the other is to refuse to believe what is true.'

Soren Kierkegaard

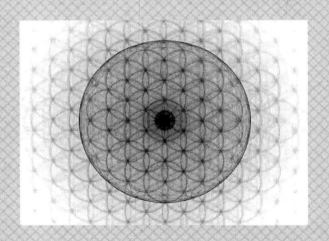

PREFACE

I live at the edge of a tiny community in the midst of the high desert of eastern California. I live with two cats – one of whom has a great sense of humor that helps to get me through the day. Each morning I have coffee and a cigarette in a chair facing north, just outside my bedroom door, watching the Sun come up, mountain shadows retreating towards the east, flooding this huge valley with the magic of light – seeing it all 'coming alive." Then I go back inside and turn on the computer and the depth of Abraham Lincoln's eyes stare back at me from my desktop. Soon I'm assaulted by the daily atrocities. I don't turn my eyes away. Often, I break down in tears. My heart was broken long ago.

I am now making information public that in my view is world-changing – and making this public is perhaps the last thing I want to do. This has a lot to do with the fact that everything that follows is a threat to the powers that be, as well as to thousands of years of organized religious belief and hundreds of years of materialistic scientific belief. While it's all too easy to get lost in the endless conversation, the book is about the courageous action now required. Anyway, little is private these days . . . and apparently I can no longer avoid the conversation. I wonder who you are, reading this note in a bottle, cast out into the infinite in hope of reaching those who will understand – and be of a mind and capability to do something about the idiotically precarious situation we now find ourselves in.

My only connection to you is through the miraculous extension of our consciousness called the Internet as it exists today – and now through these pages. I am sharing the following because the endless unnecessary suffering imposed upon us by ourselves is now beyond unconscionable - and because we are now threatening all life on Earth. It's clear to any thinking, sensitive person that dramatic change as never seen before is now required. The existence of this suffering world-wide, side by side with obscene greed and wealth, is perhaps the most damning truth regarding our *general level of consciousness* – and it is this suffering that drives me to share this.

As these pages are finally being sent to the printer, I discover this morning that wildlife have disappeared by more than half in the last 40 years . . . as our population has nearly doubled.[12] The daily assault. If you're about to read this, you care about this. If you don't, put the book down.

There are almost 7 and a half billion of us now . . . almost *three* times more people than when I was born - a blink of an eye in our thousands-year-old history. Our numbers didn't make much difference in the past . . . but now they do. We are ignorant locusts on a tiny blue miracle of a planet, devouring all before us – our evolution to this point necessarily dependent primarily upon a built-in survival mechanism based upon fear as it slowly evolves towards a more expanded, aware, compassionate way of being. The 'lucky' ones devour the most, made obese by the glut offered them for their obedience, while billions live in unspeakable poverty. The very few who control the masses are psychopathic in the drive towards the shallow rewards their greed demands, as they spin their stories and seduce the multitudes with their mesmerizing trinkets and tokens. They will not be swayed, and our destinies are already written without some kind of miracle.

Our Universe emerged out of the Great Mystery around 14 billion years ago. A bit over 4 ½ billion years ago our Sun and planets coalesced out of the cosmic vapor, creating the solar system that is our home in this infinite vastness. Shortly after that, Universally speaking, a rogue celestial body of some kind crashed into the budding Earth, creating the Moon, stabilizing the Earth's rotation and making life possible here.[3] A billion or so years later the miracle of life appeared on this still furious Earth . . . evolving through its myriad stages and seemingly endless array of life-forms until *we* emerged, perhaps a couple of hundred thousand years ago – *we*, the only life-form here with self-awareness; awareness capable of true compassion, and the power of self-determination. Civilization grew in ways we still have no complete knowing about, eventually informed with the ability to craft and move stones, unliftable even today, towards monumental architecture, until it was swept away by fire and flood around 11,500 years ago.[4] Since that time, as we have regrouped and rebuilt and reinvented ourselves, civilizations once again capable of monolithic building, as well as Shakespeares, Beethovens, Einsteins, Gandhis, Kings and Buddhas, have emerged from this darkness . . . along with terrifying, 'ignorant armies that clash by night,'[5] murdering hundreds of millions and torturing other millions through the devastation they impose, each armed with their own particular ideology – all too often a religious one - their *reason* for justifying what they do . . . and driven by those who, unbeknownst to them, own them.

Long ago, we sailed and rowed across the endless water to land on the luxurious shores of what we called Rapa Nui - what we now call Easter Island . . . only to consume everything there, doomed to isolation and extinction when there were no more trees left to craft the canoes so necessary for survival. Easter Island is a speck in the vastness of the Pacific Ocean, the humans stranded there long ago by their own means, the island ravaged, their choices gone – one of many such examples.

We now find ourselves in just that situation . . . stranded on a finite planet of dwindling abundance in the infinity of the Universe, devouring everything in our way, refusing to curb our appetites, oblivious to the dark future we are certainly creating. We have not changed much where it counts the most.

There are people who believe that the technology of science is our canoe, ready to jump outwards toward the stars as our dying world can no longer support us . . . or ready to merge with the world of artificial intelligence – certainly *that* will save us from ourselves! Some believe that a *natural* (Earth generated) purge of these billions will save 'us' (but of these with whom I've talked - always those who love the Earth and all its beauty, not those with any agenda of such - it is never *them* that will be a part of that purge). Others believe that of the many social, political, economic, environmental or religious revisionings that have temporarily moved us forward in the past will do so again – the hope that if we change the *system* we're trapped in – capitalism say . . . that all will be OK. Others believe that a 'free' energy that breaks the chains of our addiction to the burning oil that is annihilating our environment will make the difference – hoping that making 'free' energy available to the multitudes will fix what's wrong.

What remains, if any of those revisionings were to happen, is our general level of consciousness that has, at the very least, an 11,500 year record of consistency. None of these temporary 'solutions' would address the fundamental, underlying problem – who we presently *are*.

Given the direction we're headed, it is only understandable that we would find ourselves living in myriad forms of denial – seeking it even. And that denial is provided to us daily by the mainstream media, which so many depend upon for the information that informs their lives. While I can understand this, seeking such denial is not what the following is about – we can no longer permit our consciousness to be controlled by the powers that be, blind to the fact that we are being led down a path not of our choosing:

> 'The conscious and intelligent manipulation of the organized habits and opinions of the masses is an important element in democratic society. Those who manipulate this unseen mechanism of society constitute an invisible government which is the true ruling power of our country . . . We are governed, our minds are molded, our tastes formed, our ideas suggested, largely by men we have never heard of. This is a logical result of the way in which our democratic society is organized. Vast numbers of human beings must cooperate in this manner if they are to live together as a

smoothly functioning society. In almost every act of our daily lives, whether in the sphere of politics or business, in our social conduct or our ethical thinking, we are dominated by the relatively small number of persons . . . who understand the mental processes and social patterns of the masses. It is they who pull the wires which control the public mind.' *Edward Bernays – Propaganda – 1928*

The billions now living in poverty - and much worse - are invisible to the other billions . . . who themselves don't have the time, furiously trapped on the wheel created for them, doing whatever they can to support themselves and their families, lost in the illusion of needing the latest widget offered, mowing their lawns or mesmerized by the corporate entertainment supplied by the Trojan horse sitting in the middle of their living room or at the foot of their bed or both. And frankly, too many simply do not care – and caring is what is missing – a reflection of our collective level of consciousness.

The care that is required now is beyond the capabilities of this paradigm. It is beyond the compassion necessary to feel deeply the fact that there are now over 65 million refugees – 65 *million* men, women and children who have been brutally forced out of their homes and onto the road; beyond the compassion necessary to feel deeply the fact that 50,000 people, 85% of them *children*, die of starvation *each day*; beyond the compassion necessary to feel deeply the quality of life for billions of the poor and hopeless; beyond the compassion necessary to understand the suffering of billions of animals trapped in the concentration camps of the global food industry or care deeply for what we have done to the world's forests and oceans . . . and this list goes on and on and on and on. What is required now is a level of empathy where we *are* those people and animals – *they* are *us*. I am *that*. *We* are One. How do we get from here to there? What can any of us do? *Very little* that will make any difference towards what is coming.

To those of you who are doing all you can – and I know there are countless people all over the world doing everything possible to make a difference, making every effort they know of in their particular field - keep working, do not give up, we can not give up . . . and hopefully someday our efforts will be aided by the only chance that is left to us. If some kind of critical mass is reached and we break through to a higher general level of consciousness, all the good efforts from all the good people throughout time will not have been in vain.

We have been *given* this incomparably beautiful jewel of a planet such that the evolution of consciousness can proceed . . . but we are at the most critical juncture in our thousands-of-years-old collective attempt at civilization. Greed and fear dominate and now threaten the very existence of life on Earth. *It doesn't have to be this way.* Whatever you may *believe*, it is time for a critical mass of us to move beyond those beliefs into a new paradigm, a new worldview, a new way of being.

But *how?*

I remember vividly, years ago now, sitting up all night in a tee-pee with twenty others, all of us surrounding the fire at the center. I was the only 'white' face in the group . . . all Native Americans and most with faces lined and scarred by the cruelties of life, abuses I'd never had to endure, life experiences I couldn't even begin to imagine. I was humbled far beyond words as we all ingested peyote and chanted to the constant drum and prayed and prayed and prayed and prayed . . . an indescribable combination of Native mysticism and Christianity – all the while chanting/praying/drumming for *mercy*. *'Have mercy on us Grandfather. Have mercy have mercy have mercy.'* I knew that 'mercy' would never come . . . and perhaps they did too . . . but what else could they do? What else *is* there to do? If there is very little that we can do, and if 'God' is not going to part the Red Sea for us, what is left?

The only possibility we now have to alter our destiny is a shift in the fundamental nature of who *we* now *are* . . . our *collective* level of consciousness . . . and we are quickly running out of time. *A shift in global human consciousness is all that will save us now* – and this is what the following information is entirely about. I am aware how this idea sits with most. Too aware. The information presented will never be a matter of consensus, and because of this I've been looking for that small group of people - 'Never doubt the power of a small group of people to change the world. Nothing else ever has.'[6] My experience has shown me that this information represents a new worldview and requires thinking from a new paradigm. This information is now *yours* . . . and we shall see.

I do believe though, that if we were to achieve a level of consciousness that lives in the awareness of our Oneness, our intrinsic connectedness, and the level of compassion and empathy inherent in that kind of awareness, that the beauty of who we truly are will finally become evident – and the world as we know it will change for the good, the darkness losing its power to seduce, and we will move into our healthy future. But how to get there? The startling new empirical data and its implications presented in the following pages suggest a possibility filled with hope.

Because the heart of all that's about to be said has to do with phenomena that in past understanding has been caused either by 'God' or the randomness of materialistic science, I feel that I must make it clear that this has *nothing* to do with the 'Intelligent Design' movement of fundamentalist Christianity, based upon old and worn out religious beliefs, nor science's myopic concept of the evolution of life.

This is about the *evolution of consciousness*. It is about hope in a world looking into the abyss.

Ultimately what follows is about *action* . . . finally releasing us from the endless, tired conversations of a thousands-of-years-old paradigm whose death-throes we are now experiencing. The significance of the information lies in our decision to do something with it.

I believe that the information that follows, once forged into action, is a bridge . . . a connection . . . a completion . . . a possibility in a world where few possibilities now exist. It is an attempt to create an opening to that other, higher world with the intention of reaching the critical mass needed for a shift in global consciousness.

It is this - a dramatic shift in *who we are* - that will express itself in the arising of compassion and the alleviation of the endless unnecessary suffering.

That is the deepest aim of this prayer.

James Ross Godbe
Owens Valley, California
Summer, 2017

OWENS VALLEY SKY SPIRIT
(Imagine this flaming orange at sunset)

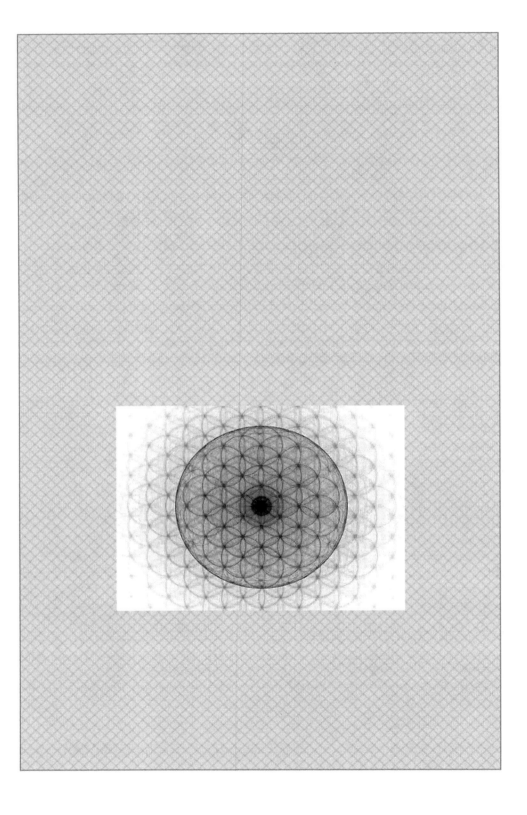

INTRODUCTION

What follows is the here-to-fore unknown discovery of *empirical* information that is worldview shattering: *the heart of sacred geometry - **a geometry that does not manifest while being the source of all that does**[7] . . . the geometry of consciousness, the Stillpoint geometry -* has been found in the dimensions of the Earth, Moon and Sun. This is 'impossible' according to any orthodox scientific understanding. This, along with more recent discoveries in other disciplines, is a *message* to us not only proving the existence of a consciousness infinitely greater than our own, but which suggests bodhisattvic intention that offers guidance towards a way through the interval that life on Earth is now experiencing.

Most of the following pages were written after I'd had an intimate experience with this *Stillpoint* - leading directly to the discovery. This was the most transcendent and important experience of my life and will be discussed in more detail shortly. Because of this discovery, I was thrown into all kinds of exhilarating corroborative research of all related subjects that ultimately contradicted thousands-of-years of organized religious belief in a fixed, a priori concept of 'God', as well as hundreds of years of materialistic science's misguided belief in the randomness and purposelessness of existence. That is, the research led to a sublime synthesis of the two - proving the existence of an *evolved* higher consciousness most would consider to be 'God', while also providing a much more expanded and encompassing understanding of evolution - the evolution of *consciousness* in the Universe.

When I finally decided to make the heart of this research public, I began the excruciating process of editing. As this process wore on, an unrelenting inner voice appeared within me, questioning every assumption I'd made - on the surface a healthy thing. In the Preface I mention seeing the Sun come up every morning . . . and immediately the voice arrogantly chimed in saying: 'The Sun doesn't 'come up' . . . it *appears* as the Earth rotates towards the East.' The inner response, from the part of me that I trust the most, was a sarcastic 'Thanks for sharing.'

On the other hand, I know how critically important it is that comments are not casually thrown out there that can easily be questioned by established scientific consensus and I have done my best to make sure this does not happen. The truth is that the following pages focus on areas that orthodox science simply does not recognize - and that is the danger of the myopic, scientifically trained and funded mind. On that level, the most important event in my life, the one single event that threw me into all this discovery, was – at the scientific level – hearsay. I came to see that much of what I talk about here is just that – if 'hearsay' is defined by the strict vocabulary of materialistic science.

That kind of science understands *consciousness* – a concept I use repeatedly – as something generated in the brain itself. It's not. Consciousness is a *non-physical field* of awareness that, in fact, evolves . . . our brain connected to, and a part of it, in a non-local, or universal, way: an *opinion* that the book goes to great length to justify. In one sense, it is the difference between hooking up a skull cap full of electrodes to a meditating Buddha in an inane attempt to discover the nature of consciousness . . . and the expanded sense of Self that is actually being experienced.

While I trust and respect the remarkable discipline and rigor of the scientific approach, and am grateful for the convenience it has brought – the miracle of being able to type these words into a computer and send them in an instant across thousands of miles, an irrigation system I installed in my yard that is on timers, the stove I use to cook my food, the telephone that connects me to my son, and, oh my, Google Earth . . . the list is endless, I am also aware of the inconceivable damage it's done to our world and recognize that the corporations and governments that subsidize the arms/intelligence/industrial juggernaut uses the dark side of science to make their atrocities possible. So, there's science and then there's science and then there's the 'science' that's not yet considered science - work often done by the most creative minds.

When I was in school, it was all liberal arts for me – that is where my own magnetic center pulled me – until I majored in geology in my last years. But even here I was attracted not because of any scientific *use* that could be turned into wages and salary and career – I learned that others in my classes were there because they wanted to work for the oil industry. I was there because I was able to learn about the Earth and how it came to be this way. My best memories from school are from a five-week field course in the White Mountains of Eastern California, where I walked by myself in deep silence and complete ecstasy across wild and untamed land, mapping the geologic formations and trying to figure out what happened and when. My lasting memory of this time was lying on my back somewhere by myself out in this vast wilderness looking up at the cumulus clouds ever so slowly moving through the sky while their shadows slid over the land. And it was *this* impression that helped me to understand the immensity of geologic time for the first time – reminding me of when I was five and sleeping in my family's back yard and looked up at the Milky Way and *felt* the immensity of the Universe for the first time.

It is from *these* experiences that I became connected to that which is far far beyond the ken of the scientific mind. No matter how many quarks or leptons or bosons science finds, the result will never match what is available to a mind unconstrained by the purely material world. It is in the world of Spirit that the truest answers will always be found.

I am *not* a scientist. In fact, if I am any*thing* at all I am a designer entranced and in love with the essence of Beauty – that which is True but not necessarily manifest – and with beauty as it expresses itself in form. That is, I am an expert in none of the fields I address in the following pages and because of this include many quotes from people who are - especially in the more purely scientific areas. In the other areas that are not in strict compliance with that double-blind, falsifiable world, I quote people who inspire me and who describe things I have an intuition or partial knowing about . . . and in many instances, simply because they say what I'd like to say in a much more beautiful and informed way.

I am aware that in some of these instances there is controversy surrounding those quoted, and I'm also aware that because of the nature of these profound, other-side-of-the-veil concepts, it must by definition be an *opinion* of the writer. In fact, all of what I've come to believe about all of this, bolstered by discoveries in many related areas, is based on the *opinion* I came across over 50 years ago by the Indian spiritual teacher Meher Baba about the evolution of consciousness. In some sections, I also refer to subjects and opinions that are controversial to many . . . especially to those who are educated by the mainstream.

Try not to let beliefs you may hold about any of these subjects deflect you from the heart of what is being presented. I often use the phrase 'I believe' because, while the evidence strongly points in the direction of certainty, we're talking about unknown territory and I'm just being honest. I am also aware that I repeat myself – mostly, this is done intentionally with the hope that it sinks in, so please have patience.

All of the above is an attempt at an explanation to you, the reader, to ward off the vampire of my critical, patriarchal, academic inner voice that is eating my soul. So, let me use the worn out analogy of the smoking gun . . . there is a *whole lot of smoke here!* So much of what is spoken of pushes the boundaries of what science is currently able to accept – or understand - but it's also clear that the most expanded in that arena are coming along. At its core, this is about Spirit and about Spirit's interface with our reality. I *know* what is there . . . and I know that those of you who'll resonate with the following know this too. So, I'm not going to worry so much about that pedantic inner voice for now.

Ironically I suppose, and perhaps a little gleefully, what the entire idea that follows is based upon is a totally transcendent and *subjective* experience of what I consider to be the Truth, that led *directly* to the discovery of the most inconceivable *empirical* information that I ever could have imagined – had I even been able to. This information alone should alter 'scientific' understanding of cosmology forever – at least as regards our own solar system – and theology certainly. It also establishes the critical importance of the evolution of consciousness - a consciousness that materialistic science does not deign to recognize as critical to much of anything, except begrudgingly, as it seems to muddle up the experiments of which it's so fond.

In Melville's remarkable book, written in the mid-19th century, extensive detail about the great whale and about whales in general is laboriously discussed. It covers the extent of Man's knowledge of this incomparable animal at the time. At one point the author speaks in Shakespearian terms regarding the absence of any cetacean voice . . . that is, *academic* Man had never heard the mystical, haunting song of the whale, hence it did not exist. This made me think of two things. The first is that this misinformation didn't detract from the standing or magnificence of the story one jot. The second is that the following story describes an even deeper and subtler voice that Man – as far as I can tell - has not yet heard . . . nor imagined.

Melville also goes to great length to inform the reader all there is to know about whales and whaling including information that seems not directly related to the story itself. But without the mention of the strength of the whale's tail, and that whales can't see in front of them because their eyes are on opposite sides of their gigantic heads, or that Man hunted and murdered them relentlessly so that 'civilization' could light its lamp at night . . . the story would have little of the depth it has. And so it is with much of the information presented here. It may seem to the reader that areas are covered that needn't be, or do not engage as perhaps they should . . . but they are included to help tell the deeper story.

Let's just say that it covers a lot of territory. Find the subjects that catch your attention or interest the most, and turn the page when they don't. As you get deeper into the story, you may return to those missed pages as they take on more meaning.

I realize too that the world is being flooded with an endless onslaught of information and that it is becoming more and more difficult to be able to know what is true and what isn't – and perhaps worse . . . that fewer and fewer people are interested in or have the time to read a book whose pages do not turn themselves. Along with all this, I am only too aware that the world does not need another book . . . but something much much more profound – which is what this book is really about.

Having said that, please do not mistake this as simply another story amongst millions of stories. It's not.

⊕

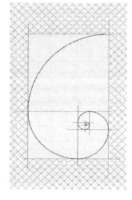

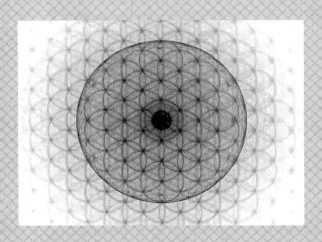

Earth Moon Sun

Eleven and a half thousand years ago, something catastrophic happened to this miraculous Earth that is our home. An 'unknown cosmic assailant[8] careening through our solar system ventured too close, wreaking havoc all over the globe. Coming out of an ice age, it was a planet thrown into upheaval . . . a time of earthquakes, volcanic eruptions turning day into night, torrential rain and global flooding. Whatever our budding humanity had been up to this moment in our collective history, it was no more. Those of us who could, fled to higher ground or sought protection underground until the dust settled and the waters receded. This event was recorded in the unwritten myths of antiquity from all around the world, and came to be known as the Great Flood in the Western world.

Life here was hard, but we began to rebuild. It was all tooth and claw. We gathered together for protection in this valley or the next, and with our larger brains, opposable thumbs, and primitive tools we prospered. We had no competition . . . except from ourselves. Whenever one tribe of us bumped into another, conflict was often inevitable given 'human nature,' but we worked it out. After all, there was always another valley and land and trees and game and water galore – the resources seemingly endless - and our numbers grew as we expanded our territories to the next valley, and the next. Over the course of these intervening years - all eleven and a half thousand of them – it took us until only a couple of *hundred* years ago to reach a global population of 1 billion people.

But in the last two hundred years out of many thousands, with the advent of the Industrial Revolution, advances in medicine and sanitation, and a myriad of other factors, we have multiplied ourselves by over 7 times – *way* faster than our ability to take care of the world that has given us so much. And there are no more valleys.

Today, we are experiencing another kind of upheaval. *Humanity* is threatening all complex life on Earth. *We* are threatening all life. 'We' are collective *human consciousness* at war with itself, and we no longer have the luxury of living peacefully in our secluded valley - in the original Garden of Eden we were given. The provincial crises have only recently become global. The Holocene Epoch has ended and the Anthropocene (human influenced) has begun – but it will be short-lived unless something almost miraculous happens - now. For the first time in our global, collective history for at least the last 11,500 years - and, in fact, far beyond human history to a time 65 million years ago - all complex life is threatened. This time it is not due to an asteroid striking Earth. This time the cause is human consciousness.

I

When we mow down the rainforests, those primordial forces of evolution that transform carbon dioxide into the oxygen we breath, to plant genetically modified crops to feed the animals we make into hamburger, and kill the bees that make everything bloom, and destroy the oceans and the fish that live in them . . . we are also destroying ourselves. In fact, all of this *is* Ourself.

Besides the rain forests – the planet's lung and transformer of CO2 into oxygen – and through our refusal to ween ourselves from the comforts that burning fossil fuels bring, we are also killing the other primordial producer of the oxygen we breath - the phytoplankton in the oceans. A new study suggests that if global warming continues at this pace, at a certain point - around the year 2100 – *there will not be enough oxygen to support human life.*[9] Hello?

We've assembled enough nuclear weapons alone to destroy all life on Earth. We've murdered at least 160 *million* of *ourselves* through war in the last century alone. Half of us live in a kind of poverty driven hopelessness. We are watched by a 21st century panopticon that tracks our every action, stored for future recall – when needed. In every way, we are approaching the dystopian worlds described in Orwell's *1984* or Huxley's *Brave New World*. This litany goes on and on and on and on.

Journalists, authors, filmmakers, thinkers and speakers from all around the world are embroiled in an endless conversation of what the problem is and what can be done about the insanely precarious situation we now find ourselves in . . . whether it be a political, social, economic, environmental or religious stab at an answer. *All* of these conversations exist in the old paradigm of the last 11,500 years. Solutions that may have worked on a smaller scale, for a time and in the past, will no longer work now. And we're not going to meditate our way out of this. While we've learned many things in all these years and made progress in many areas, we are essentially the very same people we were long ago when it comes to the survival-based, fear-based, ignorant way we view and interact with the world. We are killing the biosphere we live within . . . totally unaware that *we are that biosphere*. We are One Being who mistakenly thinks it is many separate beings.

Our crisis is entirely one of *consciousness* . . . or, rather, the lack of it. If we do not somehow transform our global consciousness towards an awareness that all of this is an interconnected evolving being that is dependent upon itself to move forward – we will not make it. While billions of us are locked in poverty and hopelessness, much of the other billions are just too busy . . . stuck on the hamster wheel that life so often is, and overwhelmed by the daily assault of the incomprehensible breakdown of countless aspects of our environment all around the world. It is all just too much, and we cling to our trivia and old ideologies.

To realize that only a transformation of global consciousness has the potential to save us from ourselves can be a very grim thought - after all, none of us has ever witnessed anything like that, nor is there any accepted evidence of it ever happening.[10] It is almost impossible to even imagine. Surely it rings of fantasy.

Still, we find ourselves in an almost impossible situation, looking into the abyss while the problems that face us grow exponentially out of control. Something must be done to shift the global human consciousness responsible for so much damage. And we are running out of time.

What *is* this consciousness that needs to be transformed? What *is* consciousness? It is defined in many ways – here are a few dictionary versions:[11]

> 'The state of being conscious; awareness of one's own existence, sensations, thoughts, surroundings. The thoughts and feelings, collectively, of an individual or of an aggregate of people. Awareness of something for what it is; internal knowledge: consciousness of wrongdoing. Concern, interest, or acute awareness: Philosophy. The mind or the mental faculties as characterized by thought, feelings, and volition. And finally, to raise one's consciousness - to increase one's awareness and understanding of one's own needs, behavior, attitudes, etc., especially as a member of a particular social or political group.'

This last description – 'understanding one's own needs,' is telling. The kind of consciousness I'm speaking of is going beyond this to be aware of the needs of 'other.' Another common understanding of this term is that of being aware of *information* . . . but 'we are buried beneath the weight of information, which is being confused with knowledge.'[12] This is *not*, ultimately, about absorbing more 'information.'

The prevailing materialistic scientific worldview understands consciousness as something generated in the brain itself. It's not. Consciousness is a non-physical *field* of awareness that evolves, our individual awareness intimately connected to and a part of it.

The late Terrance McKenna, the brilliant ethnobotanist, mystic and author famous for the promotion of the use of psychedelics, points out a critical aspect of consciousness:

> 'Consciousness is the generalized word that we use for this coordination of complex perception to create a world that draws from the past and builds a model of the future and then suspends

perceiving organism in this magical moment called the 'now,' where the past is coordinated for the purpose of navigating the future.'

I use the word to mean *the kind of consciousness that grows to understand at the deepest levels the interconnectedness of all that is, and gradually comes to experience the **compassion** inherent to this state of awareness.*

How do we get from here to there . . . from our present, survival-based, me-or-you awareness, to an awareness that we are all part of an evolving wholeness, dependent upon the whole to progress?

The key to this is the field-like nature of consciousness. If we are to survive, our *field* of consciousness needs to shift, affecting each of us individually at our own stage of evolution. How to we do this? How do we effect this 'shift'?

My goal in writing this book is to share a startling new discovery, the heart of which is empirical, scientific, objective information whose implications are unprecedented. It is the only possibility I know of that has the potential to shift our global consciousness. Some will believe what I have to say and some will not. But for those of you who yearn for meaning in a world that has gone hopelessly insane, as well as those who are willing to open to a possibility that brings hope, all that follows will be welcome news. But make no mistake, any number of the facts or theories that follow will likely challenge beliefs long held. It has to be this way given what is happening. I would never make these claims were it not for the power of what is about to be presented. I am sharing the following with you in the hope that something can be done before it is too late.

And I'm afraid that it may already be too late . . . all one has to do is witness the daily onslaught of world news to find proof of the systemic madness tearing our planet apart. Incomprehensibly, we are in the process of committing planetary genocide. Besides the forests and the oceans and the human casualties already mentioned, we torture and kill 70 *billion* animals *every year*. Does anyone think Chernobyl and Fukishima are aberrations? After 70 years, we *still* have found *no place on Earth* to safely store the thousands of metric tons of nuclear waste produced *every year*. Insanity? Surely. Unless we awaken from this collective nightmare and transform our global consciousness towards an awareness that we are all an interconnected evolving being – we will eliminate ourselves at the top of the food chain and pull everything else down with us. I know that no one wants to hear this.

I suspect what some of you are thinking . . . 'I don't have time for this - living in the modern world is hard enough. It is all just too much.' I do understand and it's the same for me. Most of the more fortunate don't have the time or inclination to care about the big picture, perhaps because of its immensity and perhaps because it's *them*, not *me*, while the rest of us scramble for daily survival. Far too many of us are desensitized and simply do not care . . . and *global compassionate action* is missing.

This is all a reflection of our current state of consciousness. So . . . how is it possible to shift our global consciousness?

As mentioned, a recent discovery has been made that has the only potential I know of that can trigger a shift in global consciouness. Similar, also recent, discoveries support it. All of this new information is based upon scientific, objective, empirical data. I dare to hope. To my mind, there has never been anything as paradigm changing as the following, and I honestly do not know of anything more important . . . *ever.*

Here I must ask the reader for patience. The information that follows, while it is everything I say above, has nothing to do with flipping a newly discovered cosmic switch that will disappear all that ails us in a twinkling, propelling us into an era of Light. In fact, as mentioned, it will likely challenge every belief you've ever held. But this is where the story must begin . . . we need to open to this new information and what it implies and go beyond what we've come to believe about our world. If we don't, it will die stillborn and we will fail. If we are to succeed, we will not only have to discard old ideas and old belief systems – in itself an almost impossible task – but must take action that will involve spiritual courage.

Further, the essence or heart of what is about to be shared has little to do with the day-to-day world we're all familiar with, and has nothing to do with any of the symptoms of unevolved consciounessness mentioned above that we often mistake as causes – the endless conversations regarding social, economic, technological, political, environmental or religious 'causes.' In fact, this has to do with who we are and why we're here, and will address these ancient questions in a dynamically new way. And, certainly, it is new territory for most of us. It has to do with the very heart of so-called 'sacred' geometry – and the very *point* of the genesis of Universal existence.

I know that this is a push for most of us . . . the unfamiliar and seemingly irrelevant world of arcane geometry that has *nothing* to do – we think – with our own lives. But this is not about the dry geometry – or trigonometry or calculus - we met in school and never looked back. What follows concerns only circles, squares, triangles and their three-dimensional counterparts, along with the proportional magic of π and $\emptyset$.

All familiar primal symbols that help to tell the greater story.

To begin, know that *whatever* you may believe regarding how we all got here . . . whether it be a God in Heaven, or the random chance combined with necessity of materialistic science, or any metaphysical combination of the two, *sacred geometry* is the primal reality upon which all of manifestation is based, regardless. I use the term 'sacred' because, 'on every scale, every natural pattern of growth or movement conforms inevitably to one or more geometric shapes.'[13] *Everything* begins here. It is that important. I am certainly not the first to suggest this. While the following goes beyond what most have come to believe about why we're all here, awareness of the geometric underpinnings of the Universe has been a thoroughly documented subject by thinkers including Plato, Da Vinci, Kepler and many others.

What *is* so-called *sacred* geometry?

> 'The strands of our DNA, the cornea of our eye, snow flakes, pine cones, flower petals, diamond crystals, the branching of trees, a nautilus shell, the star we spin around, the galaxy we spiral within, the air we breathe, and all life forms as we know them emerge out of timeless geometric codes. The designs of exalted holy places from the prehistoric monuments at Stonehenge and the Pyramid of Khufu at Giza, to the world's great cathedrals, mosques, and temples are based on these same principles of sacred geometry.'[14]

Everything in manifestation is based upon this 'sacred' geometry.

Certainly, you may be asking, 'What has this to do with my life, or anyone's life today for that matter?' Please know that nothing matters more, for you, for me, for us . . . *today*.

There is one geometrical description that stands apart from all other such expressions. In fact, *all* of the geometry upon which all of manifestation is based, is generated from this primal geometry. It is the *heart* of sacred geometry and – like consciousness itself - is the genesis of all this is . . . and this is why I refer to it as the *geometry of consciousness*.

It is called the Stillpoint or Vector Equilibrium or Flower of Life.

Critically it is a geometry *that does not manifest* anywhere in the known or observed Universe *- except that it has now been discovered at the heart of our solar system: in the dimensions of the Earth, Moon and Sun.*

While this is 'impossible' within any scientific understanding, the fact that it is true remains. The implications are worldview-shattering. The following will take you through a journey of discovery and helps to piece together a profound cosmic riddle that affects us all – ultimately a journey *leading back to you and me and our future.*

The implications of the following *empirical* information not only irrefutably prove the existence of an almost infinitely evolved consciousness relative to our own, but that this is a *communication* to *us,* and an invitation to *act* - towards the purpose of accelerating the evolution of consciousness on this planet at the very time we need it the most. Other recent discoveries affirm this. Most essentially, what follows regards a possible global shift in human consciousness – a shift ultra-critical to our future. This chapter introduces the *evidence*, its *implications*, and a possible *response.*

Let's begin at the beginning . . .

⊕ THE EVIDENCE:

I came across the following diagram for the first time in 1987:

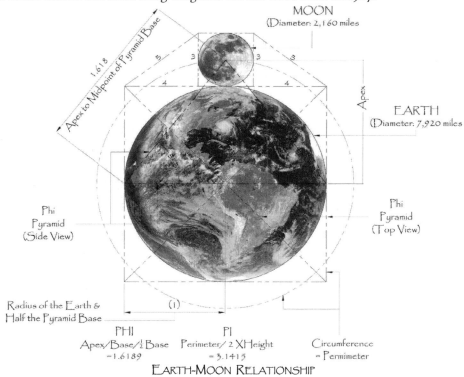

EARTH-MOON RELATIONSHIP

Remove the Earth and Moon from the image above and what is left is pure sacred geometry that includes squaring the circle ('marrying Heaven and Earth'), the Great Pyramid (integrating the mystical numbers of proportion Phi (Ø) and Pi (π)), and the prime Pythagorean triangle (the 3-4-5 triangle, sides whole numbers) – each shape inter-connected, the one naturally flowing into the next.

What may not be so obvious is that the Earth and the Moon have no rational reason for being a part of this geometry - there is *no* scientific explanation for this . . . not physics, not geometry, not mathematics, not chance, not necessary for life. But, 'impossibly,' the Earth and Moon fit perfectly into this profound expression of sacred geometry. Why?

And then there's this:

TOTAL ECLIPSE OF THE SUN

The total eclipse of the Sun by the Moon. *Perfect.* Conveniently making possible Arthur Eddington's verification of Einstein's Theory of Relativity - gravity bends light . . . and the discovery of the composition of stars. The study of the cosmos exploded.

The Sun and Moon the exact same size in the sky. How is this possible? The practical answer is that the Moon's diameter is 400 times smaller than the Sun's while the Moon is 400 times closer to the Earth than the Sun at the time of total eclipse.[15]

The common . . . and 'scientific' . . . explanation for all of this is *coincidence.* *Quite* a coincidence it turns out, as none of the other major 65 moons in our solar system come anywhere close to this kind of precision. This is often the easiest way to begin to engage with this information. Honestly, what are the odds of this happening by accident?

In 1987 I knew that the Earth/Moon diagram was not a coincidence, but had no idea what it ultimately meant or where to go with it.

This general summary, as well as the pages that follow, would never have been possible had I not had an experience in late 1999 with what some consider the most powerful hallucinogen in the world. I am mentioning this with full awareness of the general population's prejudice and fear of hallucinogens, and I'm mentioning it at the beginning because I believe it is critical to everything that these pages are about – there are other dimensions or levels of reality on the other side of the veil, in the invisible and implicate world, that the prevailing scientific mindset cannot or will not acknowledge - and it is long past time that this *most* real of worlds is recognized for the incomparable importance it holds – whether it be accessed by hallucinogens, meditation, breathwork, epiphany, revelation or whatever. In any case, this experience led directly to the discovery of the empirical information that I'm about to share.

This was a direct experience with the still, single, non-existent *point* that is the opening to the reality from whence all the manifested Universe emerged – what the Kabbalah calls Kether, *The Point of Creation* at the top of the *Tree of Life* and the interface between the Kabbalah's *Ayn* or the Egyptian *Nun* – the Limitless or God - with all of manifestation. The very *point* of connection.

I'd come to have a fairly deep *understanding* of what this 'Stillpoint' was through many years of studying sacred geometry, but it wasn't until November 1st, 1999, when I smoked the synthesized venom of the Bufo alvarius toad – 5-MeO-DMT – that I *knew* what I'd come to understand to be *true*. Before this experience, I was unaware that this substance even existed. Then, three months later, in the book *Maya Cosmogenisis 2012* by John Major Jenkins, I discovered that it was used by Mayan shamans to journey to the 'cosmic *center*.' This is exactly what happened to me. The indescribable, ancient smoke had given me something very important . . . a moment of complete, pure awakening to the eternal timeless moment where past, present and future exist simultaneously, outside of/including all time, infinite in all directions - the essential Oneness of all that is and isn't.

We are One.

But most importantly and, ultimately - for *me* - this was the experience of one . . . still . . . *point* . . . the cosmic center . . . an experience of one, still, timeless . . . *point*. This *Stillpoint* is the doorway to what the Lakota call *Wakan Tanka* – or Great Spirit . . . the Mystery.

While I'd come to *understand* this heart of sacred geometry . . . I now deeply *knew* all this to be absolutely true. I am aware that I cannot give the reader, or anyone else, an authentic experience of what is essentially inexpressible,[16] but it is too powerful and too important not to acknowledge this experience as the *source* to the incomparably important discovery that follows - so please bear with me.

9

In the Rig Veda, the most ancient scripture of India, the Universe was born and developed 'from a core, central point.'[17] The unequaled significance of the geometry that describes this point - the 'Maha-Bindu', the point representing simultaneously the source of creation and the transcendence of all polarities and final integration at the end of the spiritual journey"[18] - is defined geometrically by the Vector Equilibrium, twelve equally spaced points defining a sphere equidistant to one central point, and described by Buckminster Fuller below in his inimitable way (hold on to your hat):

> '[the Vector Equilibrium is the] zero phase . . . inexpressible inter-relationship of all universal events . . . the inherently invisible Vector Equilibrium self starts and ever regenerates life . . . . *Zero pulsation in the Vector Equilibrium* is the nearest approach we will ever know to eternity and *God:* . . . . the magical shape of the Vector Equilibrium transforms the octahedron into the square into the circle into the tetrahedron and so on forever . . . . [and] is the anywhere, anywhen, eternally regenerative event inceptioning any evolutionary accommodation and **will never be seen by man in physical appearance.** *It represents the stillpoint.* Yet it is the *frame* of the evolvement. *It is not in rotation. It is sizeless and timeless. The Vector Equilibrium is a condition in which nature never allows herself to tarry.* Ever pulsive and impulsive, nature never pauses her cycling at equilibrium – she refuses to get caught irrecoverably at the zero phase of energy. Everything that we know as reality has to be either a positive or a negative aspect of the omnipulsative physical Universe. The whole of physical Universe experience is a consequence of our not seeing instantly, which introduces time. As a result of the gamut of relative recall time-lags, the physical is always the imperfect experience, *but tantalizingly always ratio-equated with the innate eternal sense of perfection – thus the mind induces human consciousness of evolutionary participation to seek cosmic zero.*'

This is the most powerful and *sacred* and pure geometry there is . . . the geometry of the Stillpoint from which all creation emerged. It is the First Word. The primal blueprint. Thoth's Flower of Life. I believe it is the Philosopher's Stone. I call it the geometry of consciousness partly because, just as consciousness, it is by ancient tradition the genesis of all that manifests and partly because of my direct, personal experience.

Another understanding of this geometry points towards its incomparable significance . . . the joining and perfect balance of archetypal polarities: Spirit and Matter, Yin and Yang, Feminine and Masculine, Sphere and Octahedron. The

Sphere below is tangent (touching) to the midpoints of the *edges* of the Octahedron – the 12 points of intersection equally spaced around the central point and equidistant from the center – the Vector Equilibrium.

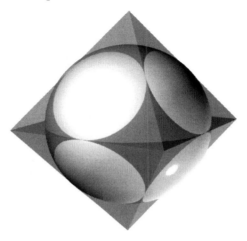

SPHERE-OCTAHEDRON-VECTOR EQUILIBRIUM
(Drawn by John August)

Because of the force and depth of this transcendent experience, as well as the verification of the incomparable importance of the Stillpoint, I wondered if the enigma of the Earth/Moon diagram could possibly be explained through the Stillpoint geometry . . . could it, in fact, be an expression of that geometry that *does not manifest*, impossible as that may seem? If this were so, the *reason* for its expression on this neon billboard in the sky would now be evident – it would then clearly be a *message* having to do with consciousness.

I went looking for it in the geometrical relationships of the Earth and Moon . . . and there it was. Yes . . . *and* the Sun . . . the dimensions of the Earth, Moon and Sun, the three celestial bodies absolutely critical to our very existence, 'impossibly' and perfectly express this geometry of the Stillness – *a geometry that does not manifest*. I do not wish to encumber this summary any more than necessary, and am including the geometric proof for this at the end for reference, *but this is the reason all of this has been written, so please verify this for yourself in Appendix A* . . . and for a 3d animation of this complicated explanation, see www.stillpointdesign.org. Further, this phenomenon is not of the quantum or micro manifestation one might expect. It is a grand, *macro* manifestation. It is an intentional, *symbolic* expression – a *communication - pointing* to the significance of the geometry itself. Please think about this.

This is the non-existent point called Kether, or Point of Creation, through which the *Ayn* ('The Limitless') of the Kabbalah, the *Great Mystery* of the Lakota, the *mind of God,* or the *cosmic plenum of the Metaverse* explodes into existence, beginning the long evolutionary process described by Arthur Young (a brilliant mathematician at Princeton in the 30's, the man who made the helicopter fly, and the author of an amazing scientific and philosophical book on the evolution of consciousness in the Universe called *The Reflexive Universe: Evolution of Consciousness)* such that '. . . Creation comes at last to recognize itself' through the *evolution of consciousness.* This is defined in Kabbalic wisdom by the ten Sephirot of the *Tree of Life* ('the 10 attributes or emanations through which God reveals himself'). It is the point that science, in its reductionist vision, attempts to quantify and refers to as the 'point of singularity,' one-billionth the size of a proton and of infinite density, from which the entire Universe exploded. This is as intrinsically different as is movement from eternal stillness – yet here it is, the unmanifest, dimensionless Stillpoint expressed in the proportions of the Earth, Moon and Sun.

This is what everything that follows is about.

The expression of this particular geometry that **does not manifest** in the dimensions of the Earth, Moon and Sun is 'impossible' in any kind of purely 'scientific' explanation one wants to consider. What follows is a summary of information that came later that reinforces this discovery, as well as what I believe are its implications.

By early 2000, I began a serious attempt not only to express this emerging worldview in writing while I looked for more 'coincidence' within our solar system. What I found astounded me. This research began with another diagram I'd also seen for the first time in 1987 - a diagram representing the discovery published in 1596 by Johannes Kepler regarding the mean orbits of the planets and the Platonic solids. The following quote is from physicist Lee Smolin's *The Trouble with Physics*, speaking about this discovery of Kepler's in the late 1500's:

> 'The cube is a perfect kind of solid, for each side is the same as every other side, and each edge is the same length as all the other edges. Such solids are called Platonic solids. How many are there? Exactly five: besides the cube, there is the tetrahedron, the octahedron, the dodecahedron, and the icosahedron. It didn't take Kepler long to make an amazing discovery. Embed the orbit of Earth in a sphere. Fit a dodecahedron around the sphere. Put a sphere over that. The orbit of Mars fits on that sphere. Put the tetrahedron around that sphere, and another sphere around the tetrahedron. Jupiter fits on that sphere. Around Jupiter's orbit is the cube, with Saturn beyond. Inside Earth's orbit, Kepler placed the icosahedron,

about which Venus orbited, and with Venus's orbit was the octahedron, for Mercury.'

Kepler based his model on circular orbits centered on the Sun – perfectly circular orbits were a religious certainty at the time due to the flawlessness of God's plan. He thought that he had discovered a geometric blueprint for the entire Universe designed by this commonly accepted 'Creator-God.' After he had discovered this incredible relationship between the planetary orbits and the Platonic solids, published in his 1596 *Mysterium Cosmographicum*, he continued to work towards more and more precise measurements that would make the discovery all the more credible.

What he discovered . . . that the orbits of the planets were, in fact, ellipses, as well as the fact that the centers of their almost circular orbits were not centered precisely on the Sun . . . only served to undermine his discovery – at least to the scientifically inclined mind obsessed with a certain understanding of precision and accuracy. In this process he discovered three very precise laws of planetary motion – the discoveries that made him famous.

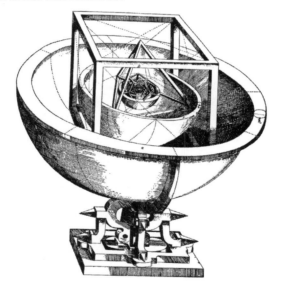

NESTING OF THE PLATONIC SOLIDS IN THE ORBITS OF THE PLANETS
(From Kepler's *Mysterium Cosmographicum* (1596))

Through the work of Steve Wilmoth, animated by Dugan Hammock, we now know that 'all of the Platonic Solids are nested inside each other, with the Golden Ratio exhibited in the proportions between the edge lengths of the polyhedra.'[19]

In fact, the orbits of the planets are not precisely circular . . . but extremely close – most whose eccentricity is within *thousandths* of being a perfect circle - which is why I will use the term 'mean' orbits. Still, as incredible and 'unexplainable' as these phenomena are, science moved on because it could make no falsifiable predictions[20] based upon the information. In 1609, Galileo discovered the moons of Jupiter and they did not conform to Kepler's theory based upon the Platonic solids.[21] Science completely ignored the phenomena because it couldn't explain it within its own reality. Still, Kepler's discovery remains.

But Kepler never abandoned his original theory. In fact, twenty five years later, in 1621, he published a more expanded version of his original work, where he – rightfully – placed the more accurate measurements of the orbits in *footnotes* – where they should be. Science to this day dismisses his Platonic/orbit theory . . . only because the orbits are eccentric by thousandths from the ideal circle. But think about it! What is left, thousandths of a percentage point from the ideal, is the fact that if the spheres were perfect (which they almost are), the *five* Platonic Solids (not one or two or three, etc.) are *all* represented . . . whether inscribed within a particular planet's orbit or circumscribed around that orbit to create the next. This is incredible! . . . yet science dismisses it, obsessed with experimental precision of reducing the whole to its parts and afraid of the implications. So, I repeat: there is nothing in physics or in fact all of science – no 'law' that explains this phenomenon. Nor could this be explained by any stretch of chance or accident. It is also not necessary for life or the evolution of life. Something else is happening here.

It's also worth noting that Kepler, a devote Lutheran, saw all this as a pure expression and proof of the Creator-God so common to so many religious beliefs. In his *Mysterium Cosmographicaum* he makes this clear:

> 'God himself was too kind to remain idle, and began to play the
> game of signatures, signing his likeness into the world; therefore I
> chance to think that all nature and the graceful sky are symbolized in
> the art of geometry.'

I include this quote because it alludes to something I've come to believe about all this 'coincidental' phenomena – that the expression of this geometrical precision in *this* solar system *is*, in fact, an intentional 'signature,' or more importantly, a *message* whose authorship is clearly of a vastly higher order than ourselves – from a 'God' that was too kind to remain idle.' Kepler understood the phenomena through the lens of his own belief system and understandably thought that what he observed within the solar system was universal[22] . . . it's not. It is my conviction that it can be understood accurately through neither the purely scientific *nor* religious view – but through the lens of the much more expanded view of the *evolution of consciousness*.

I continued looking for 'coincidental' phenomena occurring in our corner of the cosmos, especially as it related to the Stillpoint, or unity - discussed in depth later - and began to see a very clear picture forming. I also found a great deal of information regarding the particular conditions that permit complex life to exist . . . the distance from the home planet to its parent star, the size of the satellite or moon orbiting the planet, the tilt of the planet creating the seasons and so on. Then, in 2005, I found an entire compilation of this scientifically 'coincidental' information in the 2004 book/film called *The Privileged Planet*. Given the overwhelming empirical evidence, the authors understandably default to the Judea-Christian model . . . the old Creator-God. That is, if nothing in known science can explain this phenomena, the only other alternatives are either an as yet undiscovered law or - 'God' did it.

But the facts presented stand on their own and make an astounding case for not only the incredible uniqueness of our planet, but of its ideal environment not simply for complex life forms, but also for scientific discovery having nothing to do with life only, but with knowledge . . . awareness . . . consciousness – *and* it provides the context for the information I've just shared that is beyond even the scope of that book/film. Certainly, all this suggested *intention – but from a different source than what has always been assumed.*

All this pointed, in my mind, to an explanation filled with exciting and hopeful potential . . . that our solar system was created intentionally by an *evolved* consciousness for a profound purpose. What follows will explain why I feel this way and why I feel it is so important.

To date, as I understand, there are some *twenty* factors that have been discovered to be essential for the existence of complex, carbon based life forms.[23] Besides the conditions mentioned above, these factors include the existence of liquid water, that the planet has an oxygen/nitrogen rich atmosphere, the kind of star necessary, that the planet has a nearly circular orbit . . . etc. Apparently, all of these twenty conditions (*see Appendix C*) have to exist *simultaneously* for complex life to exist . . . *and they are not directly related to each other*. Regarding the possibility of a planet like ours existing in our galaxy, if a very conservative, in the generous sense, 10% chance is given to only thirteen of factors required,[24] and each one in ten chance is multiplied by the others, and then multiplied by the 1 billion stars in our galaxy, the number resulting from this *conservative* calculation, representing the likelihood of a planet like ours existing, is 0.01 – meaning that its very unlikely that even *one* such planet exists in our galaxy. So, yes, a planet like ours is apparently very rare, even when seen through the lens of the random creation of materialistic science. But, as we will see (*See Appendices A & B*), there are other factors that science cannot possibly explain – making our solar system literally 'impossible' in scientific terms.

My present understanding is that 3,500 exoplanets (planets that are a part of another solar system) have been discovered, due mostly to the launch in 2009 of the Kepler Space Telescope.[25]   Yet almost all of these planets, for various reasons, have no chance of supporting life as we know it. Very recently, the Spitzer Space Telescope discovered a star 40 light years away from us that supports *seven* Earth-like planets . . . three of them situated in the so-called Goldilocks Zone – the area surrounding a star that permits the existence of liquid water . . . and life[26] *(see What are the Chances that Our Universe Could Exist Just as it Is, Permitting Life as we Know it to Exist, page 86)*. But these planets have few of the other requirements for life.

Even if one considers the fact that there are a whole lot of stars in the many billions of galaxies in the Universe - perhaps as many as 10^{22} - and many more planets – the *rarity* of complex life still holds. But here *we* are . . . along with ample evidence of extra-terrestrial life. Hmmmm. There is no possible 'law' that could encompass all this '*coincidence,*' leaving only intentional creation or chance. So what could the reason be? Organized religion defaults to the Creator-God, while science scrambles for ways to explain this away in terms of infinite multi-universes within which all possibilities are present - it is the science's only answer to all that follows and is addressed later in more depth.

I continued looking for other 'coincidental' facts about our corner of the cosmos, and was once again amazed at what I found. It turns out that beyond the 20 conditions mentioned above, our own solar system displays other extraordinary phenomena that have nothing to do with the requirements of complex life. If in fact the Universe is infinite in size (and the *observable* Universe is close enough for this purpose), it is inevitable that there will exist planets that can support life. What *this* information regards goes far beyond the statistical certainty, although still a scientific rarity, that other such planets exist. *This* information has to do with precise, geometrical, *intentional* relationship that is not universal. The twenty factors, regardless of the rarity of all of them happening simultaneously, are all universal conditions. The usual assumption is that what we experience here is universal. It's *not* . . . and I'd like to now make the critical distinction between how universal principles and laws manifest in the cosmos as a whole and how we witness their occurrence in our very special place within it – our own solar system.

Beyond the infinitesimal unlikelihood of a planet like ours even existing in scientific terms, I realized how important it would be to find out, if possible, if what was occurring here is witnessed anywhere else . . . that is, is what we experience here *universal*? I *knew* that it wasn't coincidental, but it would be useful to see if there was any real evidence that this was so, beyond statistical improbability - that the facts I'd compiled were, in fact, *unique* to our corner of the cosmos. I wrote to two internationally respected astrophysicist/astronomers, Dr. Geoffrey Marcy at Berkeley and Dr. Brad Carter in Queensland, who specialize in the discovery of new solar systems. I shared with each of them a few unlikely facts regarding our own solar system, asking them if they thought the phenomena happened by chance and whether or not any such phenomena had been observed in the 200 or more solar systems already discovered by that time.

Here are the facts I presented to them: If one can envision the galactic plane, as well as the plane of ecliptic, as two separate plates you could hold in your hands, one can see that there is a literal infinity of possible orientations between the one and the other. It so happens that of these *infinite* possibilities, the plane of the ecliptic happens to be aligned *perfectly* with the center of the galaxy - making possible the Winter Solstice alignment with the Earth, the Sun, and this galactic center.[27] It is also true that of the *infinite* possibilities of the *angle* between the ecliptic and the plane of the galaxy, that angle is almost precisely 60°, the principle angle belonging to the geometry of the Stillpoint, the Vector Equilibrium. It also happens that while our solar system is located very close to 26,000 light years from the center of the galaxy, it also takes the wobble of the axis of the Earth to transit 25,920 years in one revolution . . . the Precession of the Equinoxes called the Great Year . . . the light emitted near the center of the galaxy reaches us exactly *one* Great Year later. Imagine this: the center of the galaxy is 26,000 *light-years* away from us - the *time* it takes *light* traveling at the *speed* of light – *186,000 miles* a *second* - to reach us. Imagine looking at the center of the galaxy, and then *slowly* turning around in a circle until you were once again looking at the center of the galaxy . . . and this took you 25,920 *years* to complete . . . point to point. This is the extraordinary timing and precision that is but one aspect of what follows.

Why is this?, I asked each of them. I asked if perhaps it was the enormous amount of gravitational pull from the black hole at the center of the galaxy that *entrained* the plane of the solar system, the ecliptic, to align so perfectly with the center of the galaxy. I knew that I was already pressing against their scientific comfort zone and so did not include Kepler's remarkable discovery regarding the mean orbits of the planets and their relation to the Platonic solids (phenomena long since dismissed by science), or the many other 'coincidental phenomena' I'd come across, nor the discovery of the Stillpoint geometry at the heart of the solar system.

Dr. Marcy responded with all the hubris of the purely scientific view, letting me know that all the phenomena I'd mentioned had *'of course'* happened entirely by coincidence, and that, further, all observed orientations of new solar systems was also random . . . *and no such similar coincidences have been noted.* Dr. Carter very kindly told me the same thing, explaining in more scientific terms, regarding the possible entrainment by the center of the galaxy, that the distance was so great to the center that any gravitational pull was irrelevant. This was the information that I was expecting – that is, affirmation by the orthodox scientific community that everything happens randomly, regardless of the empirical data presented to it. It is one thing to verify the laws of physics operating randomly in newly discovered solar systems, and quite another to dismiss the unexplained 'coincidences' in our solar system off-hand. This is an example of assessing factual results through the lens of cherished beliefs. What we are witnessing *here*, in our solar system, is a uniqueness completely ignored or dismissed by the orthodox scientific community.

Sometime after this I came across a wonderful and beautiful book, John Martineau's 2001 *A Little Book of Coincidence in Our Solar System*. Using a computer, he mapped the orbits of the planets and their relationships to each other - and revealed the beautiful and elegant - and unexplained - geometric relationships between all the celestial bodies in our solar system . . . very much the same kind of phenomena as Kepler had come across without the aid of a computer. Later, this led to my reading Richard Tarnas's *Cosmos and Psyche* which includes many examples of the more scientific aspects of astrology – and the idea that the celestial bodies in our solar system embody primal archetypes of the focused collective unconscious, the *Akashic* record of all we've learned in our eons-long evolution.

A Little Book of Coincidence presents a tour de force of empirical evidence depicting previously unimaginable geometric relationships between the planets themselves, as well as between the planetary orbits, that is almost overwhelming. I can't possibly represent all of the information that Mr. Martineau uncovers, but will mention a few examples to give the reader a sense of the scope of the geometric precision evident in *this* solar system.

Below are two images the author created graphically displaying the relationship between the orbits of the Earth and Venus, and Earth and Mercury. In the first is a compilation of lines drawn between the two planets – Earth and Venus - as they orbit the Sun. Because Venus orbits the Sun more quickly than the Earth, the 5-fold image shown represents 8 Earth years to Venus's 13 – all Fibonacci numbers. Also, if one inscribes a pentagram or an octagram within the circle of Earth, the planet Mercury is defined. Further, 'Earth's and Saturn's relative orbits *and* sizes are both given by a 15-pointed star' and 'if one inscribes a square within the mean orbit of Jupiter and then draws four circles tangent to each other at the square's corners, the resulting smaller sphere at the center, drawn tangent to the four circles, is the diameter of Mars.'[128] The examples of planetary/orbital relationships seem endless and truly need the beautiful graphics of *A Little Book* to make them clear, but I hope that the reader will get the idea.

LEFT: THE CYCLIC RELATIONSHIPS BETWEEN EARTH AND VENUS
RIGHT: THE CYCLIC RELATIONSHIPS BETWEEN EARTH AND MERCURY
(From *A Little Book of Coincidence in the Solar System*)

Some years ago, I wrote to John and asked him what he thought was the cause of all the remarkable 'coincidences' displayed so beautifully in his book. He kindly responded:

> 'The reason I called my little book *A Little book of Coincidence* is simply because I believe coincidence, a term for unexplained resonant structures, has a role to play in the universe. The coincidences surrounding our home planet are very great, yet there is no science to support them as anything other than random. So we have this axis along which various thinkers arrange themselves.
>
> My own belief is that there are universal non-physical laws for conscious biological life which, as Keith Critchlow [perhaps the world's preeminent sacred geometer] would say, are fundamentally Platonic, which is to say they prefigure any manifestation. I would also tentatively suggest that they involve the golden section. From the evidence around our own planet, aside from the habitable zone science [has] well-developed today, these other stranger laws may require certain macro manifestations of the Fibonacci series or phi to imbue the host planet with conscious life.'

I believe that his thinking is changing regarding all this, but I include his response because it represents so thoughtfully the view of that small faction of the scientifically inclined who have moved beyond the limitations of orthodox science. The solar system, as well as its creation, obeys the laws of physics – but the 'coincidences' within it are beyond staggering and cannot be dismissed. This kind of geometric precision is not normal by any stretch of the imagination. It appears that, due to the overwhelming statistical evidence that what happens *here* does not happen out *there*, that this is not an expression on a cosmological level of an as yet unexplained coherent *macroscopic* quantum system . . . or theoretical 'non-physical laws for conscious biological life.'

That 'these other stranger laws *may* require certain macro manifestations of the Fibonacci series or phi to imbue the host planet with conscious life' expresses a best hunch, based, possibly, upon hundreds of years of scientific reaction to what organized religion did to the divine. That is, there *must* be *some* explanation for this, *something* we haven't yet discovered in the a priori laws of the universe . . . *something*. From *Coincidence:* 'The reason for this harmony is still unknown . . . there *must* be a reason for this beautiful fit between the ideal and the manifest, but none is yet known.'[29]

There is now. If we apply Occam's Razor - 'among competing hypotheses, the one with the fewest assumptions should be selected' – we can see that while *many* separate, as yet undiscovered, 'laws' are required to explain away all the endless 'coincidental' phenomena, one by one, only *one* answer is required if we are willing to venture beyond the limited view of materialistic science – a possibility that holds vast potential. The elephant in the middle of the room is being avoided – the one science shudders at the thought of. For now we'll call it 'coincidence,' and wait for science to discover a law that explains it. Perhaps the answer to this is not to be found in a new, complicated array of fixed 'laws,' but in the *process of the evolution of consciousness itself.*

I'd like now to return to Kepler's amazing discovery of 1596, having to do with the uncanny correspondence between the orbits of the planets and the five Platonic Solids. The two astronomer/astrophysicists responded to the information I'd given them straight from the orthodox scientific view . . . as was expected. Everything observed is random. I recently found the same view expressed by physicist and Nobel laureate Frank Wilczek in his 2015 book *A Beautiful Question:*

> 'It is the question . . . that inspired Kepler: What determines the size and shape of our Solar System? To Kepler's question, the modern answer is, basically, 'It is an accident. No fundamental principles fix the size and shape of our Solar System.' There are many possible

ways that matter can condense into a star surrounded by planets and moons, just as there are many possible poker hands. Which one you get is the luck of the draw. Indeed, astronomers are now exploring systems of planets around other stars than our Sun, and finding that they are arranged in many different ways. All these systems evolve according to the laws of physics. But those laws are dynamical. They do not fix the starting point. Newton's dynamical worldview wins out over Kepler's aspiration for the geometrically ideal.'

There you have it: it's all random . . . accidental. Besides the threat to organized religious beliefs that this information presents, is the threat to long held scientific beliefs. As the story continues, this backbone of the materialistic/reductionist view – as well as the orthodox religious view - will continue to be challenged.

Similar to the opportunity given to science regarding the discoveries made possible by the total eclipse of the Sun by the Moon[30] - a phenomenon that does not happen 'coincidentally,' nor does it conform to any 'law' - it also appears that this incredibly narrow window of circumstance that allows us to exist also provides us with the best cosmological setting for making scientific discovery. Apparently there are at least a dozen such discoveries in various fields of research, each significant to its field, that are permitted only because of the unique, 'impossible,' conditions we find here on Earth . . . conditions *not* necessary for *life* . . . but essential for the advancement of knowledge – an aspect of awareness, an aspect of consciousness.[31]

So, here we have all this remarkable 'coincidence' that is, at the same time, empirical, *unexplained* fact ignored by the scientific worldview:

- The phenomena mentioned to the astrophysicist /astronomers (and all the rest not mentioned addressed later).
- The twenty rare conditions that have to occur simultaneously for us to even *exist*.
- The dozen similar phenomena that permit scientific discovery completely unrelated to the requirements of life yet essential to the evolution of consciousness.

Random? Happy coincidence? I think not.

Meanwhile, in the background of all this exploration since 2000, was Martin Rees's 1999 book *Just Six Numbers*. Rees – a cosmologist, astrophysicist and Britain's Astronomer Royal - argues that six numbers underlie the fundamental physical properties of the Universe, and that each is the *precise* value needed to permit life . . . and, in fact, if any *one* of the numbers were different 'even to the *tiniest* degree, there would be no stars, no complex elements, no life.'

One can assign the same conservative likelihood as above (one in ten) that any of these numbers, or values, would be so precise, to each of the numbers (and there are more than six - perhaps as many as 36 as Ervin Laszlo mentions in *Akashic Field*) - when each likelihood is multiplied with the next to establish the likelihood of *all* of them existing at once, permitting our Universe - the infinitesimal possibility that even our *Universe* would be here at all is glaring. Anyone other than an orthodox scientist or a committed atheist would jump to the logical conclusion that this implied a Designer . . . and, understandably so, this would be the old Creator-God prototype. I couldn't explain this and, given the overwhelming evidence for the intentional creation of our solar system by an *evolved* consciousness (to be explained in 'Implications'), lived in a kind of denial about the Universal, preexisting, constants that implied a fully complete, a priori . . . Designer.

And then I had an epiphany. I'd come upon a great deal of evidence from scientific and spiritual disciplines that made it clear, at least to me, that *everything* was an expression of the *evolution of consciousness. Everything*? If 'as above, so below' is a fundamental universal principle, and if our Universe sprang from the Great Mystery of the Lakota or the Ayn of the Kabbalah, through the still/timeless Point of Creation – than *all* universes, and particularly all *previous* universes, also obeyed this universal law. This meant to me that our Universe had *learned* how to express itself this way because evolving consciousness had learned what did and did not work in other attempts it had made at universal expression and the will to evolve – to become aware. Within this Universe it had finally learned/created the physical laws needed to provide for higher and more complex and highly evolved life forms . . . *us* and *beyond*. It was euphoric in some way for me. Everything now fit, except that the heart of all this information was almost impossible to communicate.

And then, while reading *Tryptamine Palace*, by James Oroc – a book about his experience with 5-MeO-DMT - I came across an astounding quote by Ervin Laszlo, a brilliant and prolific philosopher of science, systems theorist and integral theorist:

> 'Laszlo's view of the history of the universe is of a series of universes that rise and fall, but are each 'in-formed' by the existence of the previous one. In Laszlo's mind, the universe is becoming more and more in-formed, and within the physical universe, matter (which is the crystallization of intersecting pressure waves moving through the zero-point field) is becoming increasingly in-formed and evolving toward higher forms of consciousness and realization.'

While I speak more in terms of the 'Great Mystery' and Laszlo speaks more in terms of the 'Metaverse's quantum vacuum,' I feel we are attempting to articulate the same inexpressible 'trans-empirical' reality – something beyond the veil.

> '. . . the quantum vacuum, the subtle energy and in-formation sea that underlies all 'matter' in the universe, did not originate with the Bang that produced our universe, and it will not vanish when the particles created by that explosion fall back into it . . . . the subtle energies and the active in-formation that underlie this universe were there before its particles appeared and will be there after they disappear. The deeper reality is the quantum vacuum, the enduring in-formation and energy sea that pulsates, producing periodic explosions that give rise to local universes.'   Ervin Laszlo

Whatever has been learned in all the past progression of universes *must* be *available* to the present one in a very real way. Laszlo: 'The vacuum of the Metaverse was not only such that one universe could arise in it, but such that an entire series of universes could. This could hardly have been a lucky fluke. In some way, the primordial vacuum must have been already in-formed. There must have been an original creative act, an act of metaversal Design [I would say *Intention*].'

Long ago now, when I was nineteen, I found a profoundly deep and beautiful suggestion for what this initial creative act may have been, written by Meher Baba, the Indian spiritual master who spent so many years in silence, in his 1957 book *Listen Humanity*, and include it for the first time below . . . inherently 'trans-empirical' as it certainly is:

> 'Before the beginning of all beginnings, the infinite ocean of God was completely self-forgetful. The utter and unrelieved oblivion of the self-forgetful, Infinite Ocean of God in the beyond-beyond state was broken in order that God should consciously know his own fullness of divinity. It was for this sole purpose that consciousness proceeded to evolve. Consciousness itself was latent in the beyond-beyond state of God. Also latent in this same beyond-beyond state of God, was the original whim (lahar) to become conscious. It was this original whim that brought latent consciousness into manifestation (form) for the first time. Slowly and tediously consciousness approaches its apex in the human form, which is the goal of the evolutionary process, and thereby an individual mind gradually differentiates itself from the sea of oblivion . . .'

This initial spark has evolved into self-determining, conscious, *compassionate* creation far *beyond* the human form . . . and it is this compassion that holds so much potential for us at this critical moment in our collective history.

Returning now to the uniqueness of our own solar system within this vast Universe, I finally read *Who Built the Moon?* by Christopher Knight and Alan Butler, which came out in 2005. I'd resisted the flashy title, something I've regretted. In this book, the authors reached the same conclusion regarding the intentional creation of the solar system - but by an entirely different path. There is a truly astounding amount of information in the *Moon* book (as well as their previous book *Civilization One*) proving that the Earth, Moon and Sun did not come into being in the normal way. The evidence the authors present occurs in the form of uncanny, 'impossible,' proportional relationships between the three celestial bodies based upon the discovery of a unit of measurement, common to *each* of them, used some 5,000 years ago – the Megalithic Yard. It was as clear to the authors as it was to me when I discovered the Stillpoint phenomena embedded in the proportions of the Earth/Moon/Sun, that a *message* was being communicated.

I have included their list of 'coincidences' *(see Appendix B, page 324)* that are listed at the end of the *Moon* book that summarize this message called 'The Message in Detail.' Again, for the serious reader, please make sure these 'coincidences' – this *message* – is taken in in full. Note too that there are many more details to this message, one of them being that Earth's orbital speed just happens to be *1/10,000th* of the speed of *light*, while the orbital speed of the Moon is almost precisely 3 times the speed of *sound*[32] *(see Harmonograph Drawings, page 190)* . . . each beautiful whole number proportions. But, just as I realized when I first saw the Earth/Moon diagram, Knight and Butler knew that they didn't know what the message meant: *'It seems certain that we have only identified the first 'introduction' aspect of the message from the UCA* [unknown creative agency]. *The details of the message are likely to hold the key to the next phase of human development: information that will change our destiny forever.'* Yes.

To review some of the information in the *Moon* book: Alexander Thom, a Scottish engineer and surveyor, had done extensive and precise work in the 1950's measuring Megalithic sites from the British Isles and France and had discovered a unit of measurement common to all . . . what he named the Megalithic Yard. But he had no idea what it had been derived from. Knight and Butler solved this mystery many years later by demonstrating that it was derived by correlating the swing of a pendulum to the 366 day orbit of the Earth around the Sun (an idea Thomas Jefferson also had) - a phenomenon that has remained constant for a *very* long time - to a unit of measurement that is geodetic – that is, proportional to the Earth itself.

In their previous book, *Civilization One*, they meticulously made the point, based upon Thom's work, that the ancient builders who appeared seemingly out of nowhere around 5,000 years ago, used standards of measurement that were related to each other (from the Megalithic Yard, to the Egyptian *Remen*, the Sumerian second, the Minoan foot, etc.) – as though the myths were correct and a highly developed civilization from somewhere other than Earth had shared this information with the world's budding civilizations. Because of this, and because this ancient geodetic unit of measurement had been preserved in stone structures throughout Britain, Thom was able, through careful measurement, to re-discover it for today's world.

I cannot compliment the author's of the *Moon* book enough for the meticulous work they did, connecting all of Thom's own precise work to the deep reasons why Thom found what he did. But they found *much* more. I'd like now to let the authors speak for themselves. The following is how *Civilization One* was concluded. It's wonderful to read because one can feel the humility and lack of pretension as they share their own moment of revelation:

> 'The ancient builders used a pendulum to beat 366 times during Venus' movement across a 366th part of the horizon, and produced a pendulum length that corresponded precisely to the units that describe the circumference of the Earth, *Moon and Sun*. This surely verges on the magical . . . . nevertheless, no matter how bright [they] might have been, there are some issues arising here that transcend any human power . . . . as though the very blueprint of our solar system was saying, 'Recognize that all this has been designed for you . . . .'
>
> Add these observations to the other well-known improbabilities of the Moon's very existence and perhaps we should be speaking to a scientifically-minded theologian!
>
> *But how clever is it possible for any human to be?*
>
> We have agonized over this for many months. We are not religious people . . . . we have resisted the idea for as long as we can but we have had to concede that it does look as though *our place* in the universe *has been designed*, and that *the designer has laid down some very deliberate clues so that we will recognize the plan.*
>
> But the message became jumbled and almost lost as modern humankind began to believe in its own importance as the pinnacle of all intellectual achievement. Perhaps fortunately, the essence of the message has never been totally lost. There is infinitely more to all this that we haven't touched upon as yet . . . *the real work begins now.*'

But they, too, default to a pre-existing 'Great Architect of the Universe,' implying the omniscient, omnipotent Creator-god that was always just so. What they did not and could not touch upon is that the Stillpoint geometry is the key to all of this. They also make it clear in their book that this is all a matter of *timing*[33] –

while the Moon has always been the same size, it has been gradually moving away from the Earth. We are living within that miraculous window where a *perfect* total eclipse[34] of the Sun is possible. According to a 2012 Scientific American article, this window is around 100 million years . . . or slightly over 2% of the 4.6 billion years the Earth and the Moon have existed.[35]

I would say also that the timing was not based simply upon the time it would take evolving consciousness and intelligence to be advanced enough to be able to read and understand the message woven into the heart of the solar system, nor the timing related to the very gradual movement of the Moon away from the Earth, but that it was timed *precisely* when the Winter Solstice alignment with the Earth and Sun was transiting the center of the galaxy . . . the 'cosmic center' that the Mayan shamans considered the source of their spiritual wisdom, our local Stillpoint . . . between 1987 and 2018.[36] This is the window that we are now living within and the time span that contains the explosion of information of all kinds . . . but particularly the different sources mentioned above that apply to the very 'magical' information demonstrating that this corner of the universe was intentionally designed.

The late John Michell was one of the world's preeminent masters of sacred geometry and ancient wisdom. In his beautiful 2009 book *How the World is Made: the Story of Creation According to Sacred Geometry*, Michell lays out an overpowering litany of geometric and numeric 'impossibilities' concerning the Earth and Moon, which he attributes either to 'a barely credible coincidence or by an inexplicable action of conscious design.' While the book is a tour de force of the beauty and depth of sacred geometry, he glances over this most pregnant of statements by simply setting up the 'God as geometer' myth as one of many myths about 'the Creator.' But this is an easy fix and doesn't touch what is truly happening.

Coincidence? No way. Conscious design? Yes . . . but the assumption here, again, is the old Creator-God model, conveniently co-opted to support the God-as-geometer myth to explain the 'impossible' geometric evidence. The 'myth' I have in mind is radically different and holds vast potential for evolution and . . . hope.

I believe that the Stillpoint information embedded within the proportions of the Earth, Moon and Sun is the Rosetta Stone of all of the information from all of the above researchers, demonstrating the intentional creation of this solar system and that it does, in fact, present us with the information that *can* change our destiny forever. This is *exactly* what the Stillpoint message is . . . if we can act.

In *How the World is Made*, Michell speaks about Plato's idea of whole societies being based upon the sacred order of cosmological geometry:

'The code of knowledge behind these societies, said Plato, was divine revelation, a lighting-up of minds' and 'that revelations of the heavenly order occur at different times and always have done. Yet the content of these revelations, and their numerical core, is always the same. Sometimes it must happen that the revelation strikes someone who is in a position to institute it – a young, initiated ruler perhaps who can reform his people on cosmological principles. Far more often, of course, it is to individuals whose power extends little farther than their own minds.'

The experience I had that led to all this discovery fits squarely in the second category . . . except through the note in a bottle that this book surely is. So . . . there's really no way around it – none of this would have happened (these last eighteen years of obsessive research and discovery, writing and designing, reading and self-education) had I not had the awakening experience, journeying to the other side of the veil already mentioned. It was a 'revelation' . . . have no doubt. Ultimately this experience led to the discovery of unknown, scientifically 'impossible' information that was later affirmed by other more recent discoveries.

After thousands-of-years of belief regarding why we're here and what this is all about, Copernicus turned the world upside down and initiated an entirely new way to understand existence through scientific inquiry. Earth was not the center of the Universe. Now, around 500 years later, new empirical discoveries prove an even more world-view shattering discovery. Evidence suggests that our solar system was intentionally created by an *evolved* consciousness. This vastly higher consciousness relative to our own has embedded the most important information it has to share – the Flower of Life/Vector Equilibrium geometry, the geometry of consciousness – in the dimensions of the Earth, Moon and Sun as a message to *us* regarding a possible global shift in consciousness. That it has been discovered when our world is on the brink of the abyss is not an accident. Please put this book down for a moment and think about this and let it sink in . . . *this* is what the following is entirely about.

⊕ THE IMPLICATIONS:

Thus far it's clear how inconceivably 'fortunate' and rare our Earth, and the solar system that contains it, is. Many of the realities expressed by these phenomena are not explained by physics or, in fact, all of science, nor by sacred geometry, are not necessary for life, relate only to the expansion of knowledge or consciousness, and are far outside any consideration of chance or coincidence. The sum of all this points dramatically to a Designer . . . but not that which is normally assumed. From Stanislav Grof's *When the Impossible Happens:*

'The connections revealed by astrology are so complex, intricate, creative, and highly imaginative that they leave no doubt of their divine origin. They provide convincing evidence for a deep, meaningful order underlying creation *and for a superior cosmic intelligence that engendered it.* This raises a very interesting question: is there a comprehensive worldview that could accommodate astrology and assimilate its findings?'

I believe these pages do just that. But 'science' doesn't recognize any validity concerning astrology. Below is a prescient quote by Arthur Young:

'One could then ask, since the objection of science to astrology is theoretical (it could not be true because there is no explanation for it), it must eventually follow that the facts of astrology would win out and make it necessary for science to revise its theories, as has happened so often before – and properly too, since science got its start, and has made its progress, by attention to facts.

Now I realize that science would insist at this point that the facts of astrology have not been proved; but since I have, to the best of my ability, verified many of the claims of astrology I would have to answer science as Newton answered Halley when the latter criticized him for his study of astrology. 'I have studied the matter, you have not,' was Newton's reply.'

Further, Laszlo says in *Akashic Field:*

'A paradigm-shift is driven by the accumulation of observations that do not fit the accepted theories and cannot be made to fit by the mere extension of those theories. The stage is set for a new and more adequate scientific paradigm. The challenge is to find the fundamental, and fundamentally new, concepts that form the substance of the new paradigm. A theory based on [this new paradigm] must enable scientists to explain all the findings covered by the previous theory and must also explain the anomalous observation. It must integrate all the relevant facts in a simpler and yet more encompassing and powerful concept.'

Albert Einstein said this in a similar way:

'The supreme goal of all theory is to make the irreducible basic elements as simple and as few as possible without having to surrender the adequate representation of a single datum of experience.'

The 'more encompassing and powerful concept' is that this 'Designer' was not a fully conscious, omnipotent, omniscient, fully awakened Creator-God, as the world's organized religions believe. While this *presence* has always existed - It . . . *we* . . . have been in the *process* of awakening. And within this long process of the evolution of consciousness, a very highly *evolved* consciousness *created* the remarkable solar system of planetary, archetypal movement to *accelerate* the evolution of consciousness through its awakening – a more perfect synthesis of spirituality and scientific inquiry I cannot imagine.

While this distinction between the Creator-God versus an *evolved* higher consciousness that passes as our common understanding of 'God' may seem relatively unimportant, it is critical. As the last 11,500 or so years have gradually faded into the past, people everywhere throughout this period have felt the presence of something far far greater than themselves and have developed increasingly more complicated and articulated *mis*understandings of this presence. In Barbara Hand Clow's 2001 book *Catastrophobia,* the author lays out the theory that much of the misunderstanding during these years was a reaction to the horror that befell the Earth during the catastrophe that caused the earthquakes, volcanic eruptions and finally the Great Flood some 11,500 years ago – and that the trauma of that time is imprinted on our collective unconscious.

Understandably, people began searching for the answers to *why* this devastation would have happened . . . and continues to happen on all levels of human existence as human-made catastrophes now add to natural ones . . . and all sorts of religions emerged to explain away the incomprehensible. Almost invariably, the common misunderstanding centered upon a god that had always existed and had created the Earth and all of manifestation. Superstition reigned.

Simultaneously, the need to appease this often seemingly angry god became prevalent. Divine help was invoked to ward off plagues and crop failures, volcanic eruptions and other natural disasters that were taken personally. Sacrificial hearts were cut out of victims by the thousands (certainly qualifying for one of today's Darwin Awards), crusades were waged in the name of 'God,' the persecution of entire races was justified by self-serving translations of verses in the Bible, wars were initiated with 'God on *our* side' where millions died, and in yet another twist on the theme came the Inquisition to make sure people were toeing the line, trapped in the iron-clad belief system of the vengeful god . . . and so on. More recently, the Pastor John Hagge, leader of a 20,000 member mega-church in Texas, blamed hurricane Katrina and all that happened there on this angry god punishing the gay population of New Orleans[37] - the same condemnation recently hurled at the victims in the Orlando nightclub.

While it's undeniable that the organized religions of the world, based upon this old idea of God . . . the pre-existing, always existing, inviolate Father in Heaven and all versions thereof . . . have provided temporary solace and needed answers (incomplete as they certainly are) to billions of struggling people, the darker side of this projection has had seemingly endless poisonousness affects for all these years. At its core, this is a belief in an omniscient and omnipotent Creator, *separate* from *His* creation, that has always existed just so and must be appealed to and appeased such that life as we'd like it could proceed. Just as essential to this core is the belief that whatever happens is this Creator-God's *will*. There are many versions of this, including the Deist idea that, while God created everything, once all that was finished, She sat back and let it all proceed however it worked itself out.

If all of these anthropomorphic projections . . . so understandably created in response to the horror show that life so often is . . . could be stripped away, what is left is a cosmic benign *indifference*. The evolution of consciousness marches on, bumping into itself and its limits and doing its best to get to the next level by whatever path works at the time. There is a pervasive quality of the impersonal which we twist into being personal. It has been - from its beginning in *this* universal expression . . . and undoubtedly before this particular expression - a chaotic and volatile experience. Our Universe matches precisely the energy of this infinite creative awakening force . . . the initial Big Bang from either *nothing* (science's most popular present understanding), where all that *is* exploded out of a point one billionth the size of a proton, or as a spin-off from another universal expression through a similar Bang, endless nuclear explosions at the center of stars, massive black holes in the centers of galaxies, intense heat and intense cold throughout relative to our own comfort zone, particles traveling at the speed of light crashing into each other – slowly and gradually evolving into this tooth and claw level of expression we are experiencing here on Earth at this time in our collective history.

Consciousness has evolved past this more chaotic and explosive stage and has evolved sentient beings with the capability of free will and of creation itself . . . and *this* is the crossroads we now find ourselves staring at – evolve into our healthy future based upon the truth of the connectedness of all things, or succumb to dark vortex of the belief in the reality of separateness. Unlike the grand indifference . . . even though always awakening-oriented . . . of the eons-long process of evolution of consciousness, the awakening process is clearly becoming aware of itself and its true nature in the human expression on Earth. *Compassion* is found in this human form for the first time here. It has *evolved* to become this.

And this is the critical distinction between the old, fixed, Creator-God that must be appeased and fought for, and this new, compassionate, evolved god who is in fact *us*, that has come to know that we are all One and has reached back in bodhisattvic love to offer us a possible way out of the quagmire that we're stuck in.

Presently though, we *are* mired in this lethal misunderstanding of what 'God' is – often leading to a total rejection of the idea of divine presence itself. Most essentially, this is a belief in a 'God' that is separate from us . . . the god of duality. An essential aspect of the power of this belief system is that it has morphed into an agenda of world dominance partially through the suppression of critical information proving the existence of extraterrestrial, extremely advanced life - or consciousness - that flies in the face of the geocentric Garden of Eden of the Bible.

Steven Greer, a retired American medical doctor and ufologist who founded the Center for the Study of Extraterrestrial Intelligence (CSETI) and The Disclosure Project, which seeks the disclosure of suppressed UFO information,[38] discusses this in precise terms in a talk he gave in London, on September 17th of 2015. When discussing the fact that heads of state, defense secretaries, generals, admirals and other top level administrators of powerful nations all around the world were completely kept in the dark regarding what is known about the presence of UFOs by a world core-group of the controlling 'elite' – the need-to-know based entirely upon the individual's willingness to go along with the secrecy, or cover-up, he said:

'It is because this subject [the presence of advanced extraterrestrial civilizations] . . . once disclosed properly, and in a hopeful and productive way, would leave no aspect of life on this planet unchanged, up to and including the complete rearrangement of our economic system, our energy system, political system, social system, and ultimately major religious systems and thought. That, of course, is enough to keep the brakes on anything – because, as a very interesting guy at the jet propulsion labs once told me: 'We can't disclose that to the public.' I said 'Why not? He said *Because it would show the connection between ancient civilizations and the development of intelligent life on Earth and that would collapse the foundations of every fundamentalist orthodox belief system of every religion on Earth . . . this big of a change they do not want to see happen.*' So you have that aspect of this, let's call it the paradigm shift, on a deep, philosophical, spiritual level. Then you have the shift that would be on the macro-economic system which we all live off of – the 1940's nuclear power . . . or, late 1800's fossil fuels. We basically have a system that is running on a 70 to 120 year old paradigm of energy and an economic system that goes with it, so, if you change that, you have to be prepared for the biggest change in the history of the human race.'

So, there it is. This makes clear how critical this distinction between the old, fundamentalist idea of a separate God, versus the abundant evidence of evolved consciousness is, as well as the powerful and world-changing potential of *this* new information: proving not only the existence of advanced consciousness beyond the UFO level, but an evolved consciousness responsible for creating the solar system for the acceleration of the evolution of consciousness in the Universe. The old Creator-God is no more, and we have the possibility of freeing ourselves from the 11,500-year-old paradigm based upon the ignorance of separation and the damage it has brought.

This critical distinction is also addressed by Rupert Sheldrake when discussing the Intelligent Design movement, a recent hybrid where fundamentalist Christianity attempts to merge with the irrefutable discoveries of science . . . while attempting to keep the externalized Designer in place. From Sheldrake's *Science Set Free*:

> 'The problem with the design argument is that the metaphor of a designer presupposes an *external* mind. Humans design machines, buildings and works of art. In a similar way the God of mechanistic theology, or the Intelligent Designer, is supposed to have designed the details of living organisms. Yet we are not forced to choose between chance and and external intelligence. There is another possibility. Living organisms may have an internal creativity, as we do ourselves. When we have a new idea or find a new way of doing something, we do not design the idea first, and then put it into our own minds. New ideas just happen, and no one knows how or why.
>
> Humans have an inherent creativity; and all living organisms may also have an inherent creativity that is expressed in larger or smaller ways. Machines require external designers; organisms do not.'

The Stillpoint phenomena proves the existence of a God that is evolving, gradually becoming aware of Itself, gradually emerging out of the eons-long process of relatively unconsciously purpose-driven evolution to become, ultimately, fully enlightened awareness, at one with the Unity and connectedness of all things. This evolved consciousness is an advanced version of who we now are. A far cry from the belief in the externalization of a God responsible for so much devastation.

WE are God waking up. It is up to us.

This corner of the cosmos has been designed and created for consciousness to evolve. The solar system is a precisely designed mechanism to facilitate this process. The effect that the planets and their rhythms have on our lives - the alignments, transits, and cycles of the geometry of astrology – are in fact a conscious, intentionally created construct to move humanity through its cycles of, as Arthur Young would say, 'learning how things work,' similar to Ken Wilbur's image of working through an almost infinite array of 'Atman Projects'[39] - journey's of the soul to reach Unity. The entire system of astrology is based upon this beautiful and precise geometric dance as our evolutionary ancestors are moved through all of their stages over the billions of years of pre-human evolution and - most relevant for us – through the last 25,920 Great Year of human evolution with the 2,160 year rhythms of the ages set in place – through Taurian, Arian, Piscean, etc., ages, learning the lessons required for later evolution.

According to Young, the initial stages of this awakening are driven by initiating purpose alone, through eons of 'time,' beginning with light itself. Max Planck, one of the early pillars of modern physics said, regarding purpose and consciousness:

> 'The photons which constitute a ray of light behave like intelligent human beings: out of all possible curves they always select the one which will take them most quickly to their goal' and 'I regard consciousness as fundamental. I regard matter [energy] as derivative from consciousness. We cannot get behind consciousness. Everything that we talk about, everything that we regard as existing, postulates consciousness.'

For the purpose of simplification, and perhaps poetry, I will be considering the photon - 'a particle representing a quantum of light . . . that has zero rest mass' – as the first, initiating particle. But there were many others discovered after Max Planck was alive . . . the quark, gluon, bosons, gravitons and *eventually* (all this happening between 10^{-43} second and *one* second) hadrons, protons and neutrons developing soon after . . . etc. Photons were likely the very beginning, so let's just say 'let there be light' and leave it at that. Close enough.

Young's seven-stage process theory of evolution in the Universe sees the photon of light as the first stage . . . the monad . . . then morphing into the particles mentioned as the proto-Universe cooled – then the protons, neutrons and electrons that assemble themselves into atoms - and atoms into molecules - eventually into the DNA molecule that *creates the possibility of the cell and life itself* - plants and then animals. But it is only at the seventh stage of this process – the Dominion Kingdom of which the human expression is a part - that *self-determination* enters the equation. Free-will enters the picture.

Evidence suggests that this solar system is a grand-scale consciousness-evolution free-will zone. It was created *intentionally* . . . precisely as Young describes this level of evolution, in a totally *voluntary, controlled,* and *nonrandom* manner. The rhythms of the planets expressed so beautifully in Mr. Martineau's book are *not necessary* for *life* - but I believe they *are* necessary for the acceleration of the evolution of consciousness . . . and we have reached the stage where our destiny is in our own hands.

The possibility that our solar system was created intentionally, *within* the laws of physics, but not *required* by the laws of physics, and astronomically outside of chance or coincidence, for the purpose of teaching us the way things are through dynamic, orbiting Archetype, is *not* the most incredible aspect of the information just shared.

As amazing as this all is, the pure geometric harmony found in the rhythm of the planets is a Platonic expression of the *movement* – while the geometry found at the very heart of our solar system is a description of the *Stillness* from which it emerged. This is . . . essentially . . . different. The *specific* information woven into the proportions of the Earth, Moon and Sun has only to do with the essence of fully awakened consciousness itself – the still, eternal, present, aware, infinite *moment*. It is *this* information that is by far the most important for us . . . *now.*

Each of Young's seven stages is also divided into seven sub-stages. He says of his process theory that the 6th sub-stage of the *Dominion Kingdom* (of which humanity is a part) is 'literally out of this world, for these entities would not be physical . . . and *hence become a beacon or guiding light for the stages below it.* But here is where the mind betrays us . . . man balks at the idea that there are entities higher than himself – is it because there is no tangible evidence?' This is the stage of development just beyond the *genius*, or 5th sub-stage, where ego is still evident but the energy experienced is uncontainable and must be shared – individuals such as Shakespeare, Beethoven, and Einstein belong here.

The 6th sub-stage is the stage of the immortals with no need to physically manifest except for 'special conditions and provisions' and he places of the likes of Buddha and Christ in this 6th sub-stage. Because of the lack of 'tangible evidence' of such beings, the scientifically oriented mind of Modern Man dismisses their existence. But Young's beautiful process theory predicts the existence of these . . .

> '. . . immortal and god-like entities. We would suspect that there must have been many instances of these immortals living on Earth – not only because of the sheer power required to launch a whole civilization or set the pattern for a whole age, but because the account of early ages of all peoples state unequivocally that certain gods came down and taught the people.'[40]

In his 7th and last sub-stage of the last stage, the Dominion Kingdom, he places one of only two question marks in his entire theory. Young simply says that this stage 'is, by definition, ineffable.' Relative to our own level of consciousness this well may be so . . . but the discovery that our solar system was created intentionally is evidence of just such an evolved consciousness that can weave energy and matter . . . and time . . . at will.[41] Refusing to acknowledge this evidence would qualify as 'preventing the emerging paradigm' to 'support the attempted transformation of human vision.' We are witnessing evolving consciousness finally re-emerging through the Omega point at the end of the awakening process, fully able to manifest space and time at will . . . having gained during the long, hard journey the ultimate wisdom that we are One and the infinite bodhisattvic compassion that accompanies that wisdom. From Laszlo's *Akashic Field*:

> 'In the course of innumerable universes, the pulsating Metaverse realizes all that the primeval plenum held in potential. The plenum is no longer formless: its surface is of unimaginable complexity and coherence; its depth is fully in-formed. The cosmic proto-consciousness that endowed the primeval plenum with its universe-creative potentials becomes a fully articulate cosmic consciousness – it becomes, and thenceforth eternally is, THE SELF REALIZED MIND OF GOD.'

Yes.

This is an idea from a new paradigm and because of this all the temptation and inertia and blindness inherent in the old paradigm is at work to refute it. From *Cosmos and Psyche*:

> 'As it now stands, our cosmic context does not support the attempted transformation of human vision. No genuine synthesis seems possible. This enormous contradiction that invisibly encompasses the emerging paradigm is precisely what is preventing that paradigm from constituting a coherent and effective worldview.'

While indifference aptly describes the present universal expression, compassion is the domain of evolved consciousness . . . and a fully 'self realized mind of God' would exhibit this compassion, perhaps in a way beyond our understanding. It is clear to me that the expression of the Stillpoint geometry found at the heart of our solar system, given its incomparable significance, was done *for* us, to be discovered somehow when we needed it the most . . . and is a supremely bodhisattvic, compassionate act.

35

I consider this consciousness to be an *evolved* consciousness, precisely as Young's theory predicts, because we do not observe this kind of macro-geometric-astrological precision happening *anywhere* except here . . . and because of the *compassion* I believe is being demonstrated by the embedding of the most sacred principle known, having only to do with consciousness, within the proportions of the three celestial bodies most critical to our existence.

'Compassion' because it seems clear that it is a sign, a guidepost, a message - put there *for* us.

But *why?*

I suggest that this rarest of all incubators of life, this Earth and the solar system which it inhabits, with Earth's *extremely* rare liquid water in the midst of such infinite vastness of intense heat and cold, was created much the same as a walled garden perfectly designed for the *acceleration* of the evolution of consciousness . . . *but evolution now on its own terms and by its own will*. But, even if we accept that it was all created intentionally for the evolution of *life*, it could just as easily have been done without weaving into it the profound geometry of our origin and our destiny . . . the geometry of the *stillness* – and herein lies the key to *why?*

Many essential components conspired to create this miraculous island upon which higher life forms could exist and by natural selection and morphic resonance/formative causation evolve to more complex expressions . . . but the geometry of the stillness is not one of them – *it is not necessary for life* and has *only* to do with *fully awakened awareness*. This appears to be the answer to the deeper question of *why?* for this *particular* geometry . . . it was intended that this information - specifically regarding the original portal - be *used* when a quantum leap in the *global* evolution of consciousness was needed. Now.

I can only believe that it was done this way for the most profound of reasons. I can only think that who or whatever created this walled garden for 'our' evolution was not simply sharing with us important information . . . but was sharing the *most* important information 'they' know of on the grandest stage imaginable – its significance unavoidable - available to us only when our level of consciousness was ready to understand its implications. I also believe that sharing this most important information in the way it was done, was not done simply for the sharing of information per se (as in, 'Isn't this interesting?'), but was done on this grand stage to invite action, a *response* – *a specific response that uses the information shared*. It also implies a benevolent or bodhisattvic intention from a consciousness that, while it will not or can not do it for us (save us from ourselves), is there to guide, or encourage from a distance.

I can hardly begin to put into words the importance of all this.

I imagine and believe that the consciousness responsible for setting us up in this way - that is, with the perfect planet for our sort of life - is intimately aware of the immense suffering we are now experiencing. I've wondered if 'It' is so advanced, so aware that there is no 'death,' no real suffering at that level of enlightenment, that It/They cannot 'feel' what it's like to suffer here in our relative ignorance. I don't think so. And this is where bodhisattvic compassion comes in – and why this consciousness is an *evolved* consciousness, *essentially* different than the indifference normally experienced.

I imagine that consciousness in anthropomorphic form; a calm, enlightened, meditating Buddha, with a tear running down his cheek. By Law, It cannot interfere. It cannot reach down and part the Red Sea . . . or alter the direction of a stray force of nature like a careening asteroid or Great Flood. But It *can* offer guidance, It can teach, It can leave messages. As above so below - just as, while we cannot live our children's lives for them, we can impart guidance and inspiration.[42] If this is so, then the conundrum haunting civilization for thousands of years . . . that portrayed so memorably by Job . . . will have been solved. There is no terrifying God playing life and death games with Satan at the expense of life on Earth, but an evolved, compassionate awareness which has done all it can within the Law to encourage and guide us and who, in fact, suffers over the tragedy we're creating.

There was a remarkable, circular, binary code crop formation[43] that appeared at Crabtree, England, in 2002, with the decoded message below:

CRABTREE CROP CIRCLE

'Beware of the bearers of false gifts and their broken promises.
Much pain but still time. Believe.[44] There is good out there.
We oppose deception. *Conduit closing.*'

I've shared above what I believe that conduit to be . . . the winter solstice alignment with the center of the galaxy happening now. It has never been more critical to align with the 'good that is out there.' This alignment began in 1987 and is over in 2018. Mid-2001 was the midpoint, after which the conduit began to close.

The time is now.

The Stillpoint phenomena found in the Earth, Moon and Sun is a bodhisattvic gesture of the highest order - a blueprint for a quantum leap in global consciousness waiting for us all this time - once we were ready to understand and use it.

It is, in fact, the endless unnecessary suffering that drives me in all of this – unnecessary in the Gurdjieffian[45] sense – the suffering imposed upon humanity and our environment by Man's 'false personality,' not the suffering that is inherent to life, the suffering that Buddhism addresses so well – loss, death, pain, attachment, aversion. Sometimes when I've only shared the geometric information with someone, the response will be 'That's interesting . . . but so what? How can this make any difference?' I agree . . . it has to make a difference if it means anything at all. I'm interested in intellectual understandings of God only so much as those understandings lead us to the transformation of our consciousness as a species – because it is only here that the compassion arising from such a transformation can alter the way we relate to our world, to each other and to the animals and other species and the forests and the oceans and the air and the Earth itself.

In my reading, I've come across a very remarkable and growing list of maybe two dozen people to date who have each said, in their own way, that given the direction the world is headed, the only thing that can shift or alter that direction is a dimensional shift in *who we are* . . . a global transformation of consciousness. Representative among these people are Howard Zinn (history/sociology), Stanislav Grof (psychology), the Dalai Lama (spirituality), and Albert Einstein (science). Each of these people is unquestionably respected by his peers and has earned the right to a uniquely expanded and respected worldview.

Here are two others from this group who say the same thing in their own way:

'What we now want is closer contact and better understanding between individuals and communities all over the earth, and the elimination of egoism and pride . . . peace can only come as a natural consequence of universal enlightenment.' Nikola Tesla

'An armed Conflict between nations horrifies us. But the economic war is not better than an armed conflict. This is like a surgical operation. An economic war is prolonged torture. And its ravages are no less terrible than those depicted in the literature on war properly so called. We think nothing of the other because we are used to its deadly effects. The movement against war is sound. I pray for its success. *But I cannot help the gnawing fear that the movement will fail if it does not touch the root of all evil – human greed.'*

M.K. Gandhi

While reading Ervin Laszlo's *Quantum Shift,* I came across other such quotes . . . one by Krishnamurti, another by Válclav Havel and finally, the following by Laszlo:

'*This quantum shift in the global brain is humanity's best chance.* Margaret Mead said, 'Never doubt the power of a small group of people to change the world. Nothing else ever has.' Small groups of people with an evolved consciousness will change the world – *if they grow into a critical mass in time.* There could not be a nobler or more important task in our day than to empower this evolution.'

I couldn't agree more. I have nothing left in me – no rationalization, no illusion, no practice, no faith - to buffer me from the suffering which has already come to billions of sentient beings in this world – nor would I wish to.

Richard Tarnas, from *Passion of the Western Mind*:

'. . . the constant movement toward differentiation, the gradual empowerment of the autonomous human intellect, the slow forging of the subjective self, the accompanying disenchantment of the objective world, the suppression and withdrawal of the archetypal, the constellating of the human unconscious, the eventual global alienation, the radical deconstruction, and finally, *perhaps,* the emergence of a dialectically integrated, participatory consciousness reconnected to the universal.'

'*Perhaps*' is the critical word. The darker side of *who we presently are* has caught up with us on a global scale and there's no guarantee, as much as hope would wish, that we will reconnect with the universal, and move to the next octave. Saturn, Pluto and Mars will continue their dance around the Sun and there's nothing we can do about the stress they bring . . . *except for changing the context within which those forces are experienced – our own consciousness.* I do not believe that it will or can happen by itself. If there is any possibility at all of altering the direction we're headed, it will be through a *global transformation of consciousness* generated by *us.*

But *how?*

I can only hope that at this point the reader will sense the potential this information holds and, perhaps, the necessity of action in its regard that may generate this transformation. Many hold the hope that such a transformation may be effected through methods that have worked on a smaller scale in the past, whether it was through meditation or yoga; or social, environmental, political, or radical economic reform; or the advancement of scientific and technical knowledge; or the spread of information through book or film; or the use of psychotropic hallucinogens or holotropic breath-work.

Something fundamentally different is now being asked for. Tarnas says in *Cosmos and Psyche*:

> 'No amount of revisioning philosophy or psychology, science or religion, can forge a new worldview without a radical shift at the cosmological level.'

From Stan Grof's *Psychology of The Future*:

> 'In the last analysis, the current global crisis is basically a psychospiritual crisis; it reflects the level of consciousness evolution of the human species. It is, therefore, hard to imagine that it could be resolved without a radical inner transformation of humanity on a large scale and its rise to a higher level of emotional maturity and spiritual awareness.'

Perhaps it is that a radical shift at the cosmological level (which this information suggests), directly connected to a radical inner transformation of humanity (which this information may provide), is exactly that which produces the new worldview that triggers the shift to a new paradigm.

This is what we're talking about.

I believe Arthur Young, in *The Reflexive Universe*, is indirectly speaking of this when he talks about purpose inviting the future, our own future being *self-determined* purpose, but even more specifically, Buckminster Fuller gets right to the point when he says: '. . . the physical is always the imperfect experience, but tantalizingly always ratio-equated with the innate eternal sense of perfection – thus the mind induces human consciousness of evolutionary participation to seek cosmic zero.'

Cosmic zero is presently outside of our common experience, but it is where transformation at this level will be found. We've come a long way to reach this moment of opportunity and this encoded message at the heart of our solar system is pointing the way. We must consciously add something to the brew from the outside of the paradigm we've been living and growing in for these thousands of years . . . just as each of Young's odd stages in his process theory added something entirely new upon which all future evolution was based - *a seed from the future* – something that *who we were at the time* could not possibly have envisioned, yet was the critical ingredient that permitted us to continue to evolve.

This 'seed,' as I envision it, is equivalent to what theoretical and sometimes experimental biologist Rupert Sheldrake calls the *germ* of a new morphic field . . . 'germ' in the sense of being 'something that initiates development or serves as an origin.'[46]   The ultimate goal here is to create a new morphic field . . . and this can only happen if there is an original 'germ' around which to form this field.

Belgian physicist, chemist and Nobel laureate Ilya Prigogine termed the precise moment a system goes from order to chaos a bifurcation point. As a system approaches bifurcation it only takes a very small and seemingly inconsequential event to create chaos.   From chaos the system reorganizes itself, functioning at another level of resonance.   Whether the resonance is a *higher or lower level than the original system* is determined by what mathematicians call 'strange attractors,' which can be visualized as seeds of the new order sown during the old order.   When the old order disintegrates, the new order reforms around the vibration of these seeds.

This seed is from cosmic zero – the stillness of our origin.

> 'If the positive innovations connect *exponentially* before the massive breakdowns reinforce one another, the system can repattern itself to a higher order of consciousness and freedom without the predicted economic, environmental, or social collapse . . . If the system could go either way, a slight intervention to assist the convergence of the positive can tip the scales of evolution in favor of the enhancement of life on Earth.'   Barbara Marx Hubbard

I am suggesting just such a slight intervention . . . a comic spark to effect the shift and move us into our healthy future.

It has long been held to be true in some spiritual disciplines that there is a *field* in the Universe that holds and conveys information – a record of everything learned in all universal expression.   The Theosophy movement in the late 1800's coined the term *Akashic Record* – from *akasha*, the Sanskrit word for aether.   Edgar Cayce, the famous American psychic, felt that he was accessing this field to obtain information regarding the future.   It has been suggested by some (Laszlo, Sheldrake, physicist David Bohm and others) that the more modern name for this field is what science calls the Zero Point Field.

Science has determined that even in a vacuum, free of any known matter or energy . . . there is still *something* – an as now mysterious interaction between *some* form of energy and virtual particles[47] occurs at absolute zero . . . zero having to do with the fact that this phenomenon occurs at absolute zero: -273° Celsius. It is this most basic and primal field, Zero Point or Akashic, that may be a storage medium which our own consciousness is able to access. My understanding is that this field is the ground state field of energy, which 'constantly interacts with all subatomic matter.' All matter is sustained by this field, with particles continually interacting with it *(see the David Bohm quote, in Science's Last Gasp to Explain Away any Initiating Cause, page 95)*.

Ervin Laszlo feels that it is critical for us, in this most critical of times, to access this *universal information field* – or Akashic Record – as do I. From *Akashic Field*:

> 'At this crucial juncture in the evolution of human civilization it would be of particular importance to cultivate our long-neglected faculty for accessing the in-formation conserved in the A[kashic]-field. We would not only develop closer ties to each other and to nature; we might also gain crucial insights into ways to cope with the problems of our technologically evolved but largely rudderless civilization . . . . the civilizations that met this challenge found ways to achieve a condition of sustainability. What ways did they find? *The answer must be in the A-field.* Accessing it would be to our advantage . . . this could make the crucial difference between bumbling along in a fateful gamble with trial and error, and moving with intuitive wisdom toward the dynamically harmonized sustainable conditions that more mature civilizations have already achieved on their home planets . . . . Consciousness evolution is from the ego-bound to the transpersonal[48] form. If this is so, *it is a source of great hope* . . . [that] could have momentous consequences. *It could produce greater empathy among people, and greater sensitivity to animals, plants, and the entire biosphere . . . when a critical mass of humans evolves to the transpersonal level of consciousness a higher civilization is likely to emerge . . .*'

Accessing this wisdom, this reservoir of higher consciousness (of *all* evolving consciousness throughout all universes) is precisely what the response I propose would attempt to *do* on a global scale. Our time is running out and we need to initiate viable action that could effect change from the deepest source.

Is this possible? I don't know.

Similarly, Steven Greer, in his London talk in September of 2015 addresses this same point: 'The attempt to communicate with advanced civilizations is critical' and goes on to say why. France has put together a paper summarizing the exact protocol for making this contact. Greer:

> 'Why is this important? Because it's a baby step taken by a major country doing something that should have happened in the 1940's, 50,' and 60's, had there not been a complete subversion by covert military interests, which Eisenhower warned us about . . . The first director of the CIA, Admiral Roscoe Hillenkoetter . . . states that the secrecy attending this issue is a grave threat because the folks who have created the secrecy do not have the best interests of the Unites States or the United Kingdom or humanity in their interest. They have the most venal interests at heart, and that is global control and domination through a centralized economic system [that is essentially slavery by another name]. It's just that the American plantation is a little nicer than the one in the Congo, but there is a masters and slaves mentality amongst the uber-elites that control hundreds of trillions of dollars of assets that require a centralized petro-dollar system and a centralized energy system which has been an obsolete system for well on to 100 years now.'

The incomparable importance of making contact with higher consciousness equates to the life or death issue of dealing with the so-called elite – the .1% of us that are in almost total control of life as we know it; the fact that humanity is now . . . right *now* . . . at the crossroads of its collective history: *this* is how critically important the Stillpoint information is to life on Earth at this particular time.

I first wrote to Stan Grof after reading his book *When the Impossible Happens.* I shared with him for the first time with anyone who didn't know me personally, that the source of my *knowing* all that I was about to share with him came from an awakening of consciousness induced by a powerful hallucinogen. I knew that this information, so out-of-bounds for the mainstream, would be immediately understood by him. His book, and others by him, is a compilation of a tiny sampling of the thousands of experiences of expanded consciousness - generated within the realm of their own being - that he and the thousands of people he's guided or witnessed, experienced - as was my own.

The anthropologist Jeremy Narby's book *The Cosmic Serpent: DNA and the Origins of Knowledge* is filled with the experiences of Amazon shamans who travel into other dimensions through the use of the herbal drug *ayahuasca* to access the boundless knowledge they hold in regard to the medicinal properties of the thousands of plants common to their lives. There is also a theory that our ancestors of perhaps 35,000 years ago, after maybe 165,000 years of incremental or no evolutionary evidence in the archaeological record, experienced a quantum leap in their collective consciousness – demonstrated through the revolutionary symbolic expression of extraordinarily sophisticated impressionist cave paintings – due to ingesting hallucinogens that had also evolved on this planet,[49] accelerating the evolution of their (our) consciousness.

My point is only this: awakening happens by many means, expanded dimensions exist, and access to expanded dimensions is real - and the above is only *one* of many paths to awakening.

The evolution of consciousness is the big picture and it is happening at all levels and in a myriad of ways. The crop circle phenomena clearly establish the fact that higher consciousness exists and is interacting with us and that an aspect of that higher consciousness is to teach, inspire, encourage and guide. Countless UFO sightings/experiences also verify the existence of extraterrestrial life of varied expression. I have had two such experiences myself. Ancient myth and archaeological/architectural evidence also supports this reality. Embedded within these phenomena is the apparent fact that while this higher consciousness obviously exists and can offer guidance, it cannot intervene to actively *change* what is happening – except perhaps in certain instances.

One segment of the *Thrive* film by Foster and Kimberly Gamble focuses on a man who saw with his own eyes a UFO systematically dismantle the armed nuclear warhead his plane was carrying. There are other examples of this same interference happening regarding our nuclear proclivities, some provided by the late Edgar Mitchell, the sixth man to walk on the Moon and founder of the Institute of Noetic Science in California. It is clear that there are other life forms in this Universe with us and that a number of these life forms surround and interpenetrate this five-dimensional world that we live in.

The crop circles demonstrate a level of benevolent, highly evolved consciousness along with an understanding of technology far beyond our own - a consciousness that does not interfere with the choices that are only ours to make. The warhead story is another example of this from a slightly different perspective. In this case, it seems obvious that our predilection to detonate nuclear weapons, killing hundreds of thousands and threatening billions of people and all life on Earth, will not be permitted *outside* of the sandbox we live in. To threaten the use of technology like this, outside of our environment and a threat to extra-terrestrial life, is a no-no. In this scenario, intervention is clearly within the 'law' followed by highly evolved consciousness . . . the law of mutual non-violation.

So . . . the above phenomena and experiences are all examples and expressions of the existence of a consciousness more evolved than our own, a consciousness that expresses the compassion earned through the *evolutionary* process – and is willing and has in fact demonstrated that it/they wish to help us - even while they cannot do it for us. I am asking that this line of thought be extended to its logical conclusion . . . a consciousness capable of manifesting space and time at will. Arthur Young called this conclusion ineffable. Relatively speaking, it may well be so. But I can imagine it. How can there be any doubt that the Earth/Moon/Sun phenomena is evidence for its existence?

As was suggested at the beginning of this chapter, just imagine for a moment what this would mean if I'm correct about this. That is, not only does this fully evolved expression of the evolution of consciousness exist, but that it has constructed, at a much higher level of expression than simply visiting Earth to dispense critical information, a cosmic walled garden for the acceleration of the evolution of consciousness – our Earth and solar system - and in bodhisattvic fashion has embedded the information necessary for our further evolution on a neon billboard in the heavens.

This information is specifically about the cosmic opening called the Stillpoint geometry – the geometry of consciousness. It has never been more important for us to create a bridge to this higher consciousness that has left this message and/or to access the Akashic Record of all experience – by, I believe, recreating this archetypal portal. At this point in our history, as it relates to feelings of scarcity of time and impending danger, and accents the incomparable importance of a global transformation of consciousness - this is a possibility that cannot be ignored.

In this section - on the implications of the evidence that our solar system was created intentionally and for a profound purpose, I have tried to make it clear that *all* of this . . . this Universe we experience and grow within . . . is governed by the initial urge to awaken – initiating what I've called the big picture, or larger context, the evolution of consciousness.

An equally important aspect of the Vector Equilibrium, Stillpoint geometry is its expression in its *movement* aspect – the spiraling geometry called the *torus*.

The torus is a possible model not just for our Universe, but for the evolution of consciousness itself (*see Light . . . and the Torus: page 98*) . . . and it may well be that it is the key to 'free energy.'[50] Personally, I feel that the torus model is accurate . . . and includes within it a merging of the static Universe theory (always was and will be) with Big Bang theory (began in the singularity) – that is, a particular universe emerges from the infinity of the Stillpoint at the center of the torus, evolves and expands and then contracts and disappears back into this other-dimensional Stillpoint reality, to emerge again in a new universal expression: it simultaneously expands *and* contracts, no beginning, no end, always was, is and shall be, constantly remembering itself in its journey through the Stillpoint . . . all the while evolving and learning. Meanwhile, during its 'existence' in manifestation, this process happens at the sub-atomic level, with virtual particles appearing from and disappearing back into the world of the Stillpoint reality (*see the David Bohm quote, in Science's Last Gasp to Explain Away any Initiating Cause, page 95*).

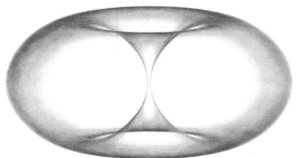

VECTOR-EQUILIBRIUM / TORUS

The infinite in-breathe and out-breathe of the eternally spiraling, evolving, growing awareness, emerging from the Stillpoint at its center, born anew into yet another stage of its evolving and expanding . . . whether another spiraling orbit into the future demarcated by the point of winter solstice, or another universe . . . then making the turn and contracting, returning once again to death and rebirth in its journey into and out of the Stillness at the center, the opening to the Great Mystery of the eternal present moment – 'having been symbiotically united with the nourishing womb . . . the beloved *center* of an all-comprehending supportive world.' This last quote is Stan Grof describing discoveries regarding the as-above-so-below nature of the all-essential perinatal birth experience and how it mirrors what is going on at the cosmological level. From Laszlo's *Quantum Shift*: 'Akasha . . . is the womb from which everything our senses can perceive has emerged and into which everything will ultimately redescend.' From *Akashic Field*:

> 'According to Steinhardt and Turok, [physicists suggesting that space and time may have always existed in an endless cycle of expansion and rebirth] the universe undergoes an endless sequence of cosmic epochs, each of which begins with a 'bang' and ends in a 'crunch.' Each cycle includes a period of gradual and then further accelerating expansion, followed by reversal and the beginning of an epoch of contraction.'

Here's another description of this process and of the universal working of the torus mentioned above from Tarnas's *Passion of the Western Mind*:

> 'For the birth of every new paradigm is also a conception in a new conceptual matrix, which begins the process of gestation, growth, crisis, and revolution all over again. Each paradigm is a stage in an unfolding evolutionary sequence, and when that paradigm has fulfilled its purpose, when it has been developed and exploited to its fullest extent, then it loses it numinosity . . . while the new paradigm that is emerging is felt as a liberating birth into a new, luminously intelligible universe.'

Very powerfully implied by the fact that we are only now being made aware of the Stillpoint phenomena at the center of our solar system at *this most critical point in humanity's history*, is that we are now, in fact, experiencing the death of a perhaps 25,920 year-old paradigm and the shift into a new one. But the fact is, *this* mind continues to think within its box – it is a koan or a riddle that requires being outside of today's mind. The phenomena regarding the geometry of the Stillpoint/torus, is just that.

Evidence strongly suggests that a response is invited whose essence is an appeal to a higher consciousness that has gone to the trouble of showing us the way in the grandest manner . . . but a response that puts the responsibility in our own hands - as I believe it has to be. I have no way of *really* knowing what any of this truly *would look like* . . . only *that we must do something to create an opening to this intelligence*, this consciousness - if possible – making it more available and influential in our world - *in effect uplifting humanity and triggering the natural arising of compassion, alleviating the suffering of all sentient beings and the healing of our environment.* I have been obsessively working on such a response, but have no way to know if it is the correct one. But for now I wish to say that we're talking about an experiment that has never been made, but which must be made . . . and which is essentially just that - an experiment. From *Passion of the Western Mind*:

> 'A threshold must now be crossed, a threshold demanding a courageous act of faith, of imagination, of trust in a larger and more complex reality; a threshold, moreover, demanding an act of unflinching self-discernment . . . [the masculine and feminine] synthesis leads to something beyond itself: It brings an unexpected opening to a larger reality that cannot be grasped before it arrives, because this new reality is itself a creative act.'

This statement points towards something essential - ultimately this is not about ideas, it is about *action* – and there is no way to know beforehand if it will succeed.

My hope is that we are now at the moment in our collective history, through billions of years of natural selection and morphic resonance and the growing of consciousness and the more recent attempts at civilization, where we may be on the threshold of doing something that I don't think has ever been done – at least in this round of civilization - the creation of a very real window or doorway to a higher level of consciousness *from this direction*, making it available on a global scale for the first time . . . *just at the moment that we are on the brink of the abyss.*

So, here we are, as Stan Grof describes in *Psychology of the Future*:

> 'The task of imbuing humanity with an entirely different set of values and goals might appear too unrealistic and Utopian to offer any real hope. Considering the paramount role of violence and greed in human history, the possibility of transforming modern humanity into a species of individuals capable of peaceful coexistence with their fellow men and women regardless of race, color, and religious or political conviction, let alone with other species, certainly does not seem very plausible. We are facing the necessity to instill humanity with profound ethical values, sensitivity to the needs of others, acceptance of voluntary simplicity, and a sharp awareness of ecological imperatives.'

He goes on to say that transpersonal psychology has proven to make inroads into this seemingly unsolvable problem and that this transformation is already underway. I know that this is so. Yet, *'The question is only whether it can be sufficiently fast and extensive to reverse the current self-destructive trend of modern humanity'*[51] is a very big 'whether.'

I'm afraid that we are on our way to failure and *unnecessary* suffering as never experienced before . . . sans some miracle. However we came to be in this place, humanity – what Arthur Young calls 'Modern Man' - is stuck, and is trapped within yesterday's mind not quite capable of seeing what's actually happening, locked in the habit of hope and the assumption that the Sun will of course come up in the morning. We're in the process of evolving, and one aspect of this is the difficulty to even imagine what the next level will be, as well as the tendency to hold on to what has worked in the past. Inherent in this statement is that moving to the next level, initiating the next octave, by definition cannot be something that we will be able to prove nor be certain about, before acting – and *action* is what this is all about.

I cannot look into the future and see how the results of such an action would manifest. The best that I can do is to believe that if we could somehow *trigger* the exponential jump in evolution now required, that the dark, powerful momentum of the last age will shift to the 'Good'. To my mind, this means that all of the good works and good intentions of so many for so long will not have been in vain, and will be empowered in ways that have never occurred before. I can imagine a momentum building for the Good, while the seductive darkness loses its appeal and stranglehold. The mechanism for all this will be *frequency, vibration* and *resonance*.

These and the following pages make the case that evolving consciousness is responsible for all that is - that 'God' did not come ready-made, omnipotent and omniscient, but that he/she/it *is . . . we are . . .* in the process of growing, of learning how to be this fully awakened, fully aware, compassionate pure consciousness. We have evolved far past our own present level of collective consciousness and, in fact, out of this eons long process of awakening has emerged higher consciousness with a very hard-earned compassion, hand-in-hand with the pure, awakened awareness of the Unity of everything that is . . . the awareness that we are One.

More than this, the evidence suggests that this almost infinitely evolved consciousness relative to our own is acting in bodhisattvic fashion, outside of the normal expression of indifference we witness, and has not only intentionally created - through this meticulously and methodically *learned* Universe - the enlivened dance of matter and energy and archetype called the solar system for the *acceleration* of the evolution of consciousness, but that it has encoded and embedded the most sacred and profound information it has to share (the very heart of sacred geometry . . . and a geometry *that does not manifest*) in the most unlikely ('scientifically' speaking) of places - the proportions/dimensions of the Earth, Moon and Sun, so that it could not be ignored. This suggests that it was done in order that evolving humanity would discover it and *use* it for accessing higher consciousness – precisely when that evolving humanity needed it the most . . . now.

⊕ A Response: the Mesa Temple

The Stillpoint geometry at the heart of our solar system was embedded there as a message – not simply a message containing information, but as an invitation to *use* this information. I am calling this 'use,' or *action*, a response.

Essentially, the Stillpoint is an opening . . . the window between the Great Mystery and/or the Akashic Record, and the manifestation of the evolution of consciousness that the Universe is – it is Kether, the *Point of Creation*, the interface between the *limitless* of the Ayn and the remaining Sephirot of the *Tree of Life*, and what Buckminster Fuller calls 'the nearest approach we will ever know to eternity and God.'

In the most important experience of my life I have verified this and know it to be true. Because of this I feel that we are being asked to recreate this blueprint to gain access to that higher consciousness that created this corner of the cosmos for *us*. I also feel that it is only logical that we use it as 'they' have modeled – by expressing it within the five dimensions of our world – consciously bringing Spirit into Form.

It is also very odd, and worth noting, that this information is only *now* making itself known. Why? And why has it escaped the scientific scrutiny of hundreds of years and the notice of the best and most brilliant minds? It has been there the entire time and yet not one person has observed it as far as we know. Amazing.

John Michell, in his book *The Dimensions of Paradise*, offers an answer to this question while speaking of the restoration of the Heavenly City, the New Jerusalem on Earth, upon which the response I'm suggesting is based:

> 'One day, it is said, the Temple will be restored, the sacred world order will again be established and harmony between men and nature will once more prevail. That event, according to all prophecies, will take place at a period of extreme need and desperation.'

At no time in our thousands-years-old collective history have we ever experienced such a 'period of extreme need and desperation.'

Up to this point I have focused on the empirical or scientifically based evidence available concerning all this, as well as a sampling of some of the thinking by some of the most respected observers. To go further though, I want to access a time before our minds were constricted by the rigid mindset of materialistic science, when ancient tradition spoke in the language of magic and beauty and spirit. It is only here that any justification for what I propose will have any credence in today's scientifically obsessed world – the construction of a profound technology[52] that I call the Mesa Temple.

Before I share the details of this attempt to access this wisdom, triggering the quantum leap in consciousness that is so desperately needed now, it is important that I make it very clear that this is simply *my* best idea of how to respond. Because of the many synchronicities that conspired to guide my life, I have formed a very precise idea of what this response might be. Most importantly though, at this stage it is this new and startling information upon which it is based that is most important.

And if there is a better idea of how to respond, then that is what I want.

To begin this last section concerning the response I feel is being invited by this message in the heavens, I'd like to share a little of my personal history as regards all of this as it may be appropriate for the reader to know how I came to this place.

Just after I'd been accepted into architecture school, and because I'd become stagnated by the seemingly endless years of intellectual study and introspection and needed desperately to use my *hands* . . . I quit school and hitchhiked to Lake Tahoe from Berkeley, in California, and became an apprentice carpenter. This was a joyful time for me and I eventually became a journeyman, learning well the craft of building. Then, seemingly out of nowhere, I became fascinated and obsessed with studying sacred architecture and sacred geometry. This came from deep within me.

I taught myself how to draw and design the architecture I'd learned how to build, and the first house I designed was based entirely on the Great Pyramid . . . three stories within this shape, filled with moving water and gardens and light from the south. This led, in natural fashion, to making my living designing architecture. This journey eventually defined itself in 1986 with the design of a meditation retreat center in west Marin County, California, the center of which was a mediation hall. That project as I'd designed it never happened . . . but the hall was the genesis of what has evolved into the Mesa Temple – the response alluded to above.

At the time, in the 80's, I was doing many 10-day Vippassanna meditation retreats and it was during that time that this community was looking for land for their own retreat center. At one retreat, while talking with the head meditation instructor and the force behind the new Buddhist retreat center about the design of the meditation hall, he casually mentioned that the building should be approximately 80' across, with a surrounding covered walking meditation area that should be approximately 20' across.

This conversation happened in 1986 and, while planning a trip to Europe that year to visit the megalithic stone circles and the sacred architecture of the cathedrals, I was reading a book called *The New View Over Atlantis*, by John Michell. I turned a page and there it was . . . a mandala-like floor plan – *to scale* – of the meditation hall I was about to attempt designing. It was based upon what he considered to be the sacred proportions of the New Jerusalem . . . loosely translated as a Heaven on Earth . . . with an inner circle of 79.2', based upon the Earth's diameter of 7,920 miles, surrounded by an outer area with a dimension of 21.6', based upon the Moon's diameter of 2,160 miles – dimensions also reflected in the megalithic stones of Stonehenge and in the proportions of what remains of the St. Mary Chapel, part of the ruins of the fabled monastery, Glastonbury Abbey.

This fit perfectly, and to me magically, with the requirements I'd been given and I knew I had found my blueprint for the building I wanted to create. In many ways, I felt I was being guided and had little choice but to follow this guidance.

Still, back in 1986, the initial design seemed fixed and stagnant to me until I stopped by to visit with a friend at his home. While I was talking with him, I was transfixed by a painting behind his desk of a geometric shape that I was not familiar with . . . the torus. He also introduced me to the Stillpoint at its center – the Vector Equilibrium. He also introduced me to the work of Arthur Young. All together an incomparably important moment in my life for which I am very grateful. My experience with sacred geometry before this time had to do mostly with number, triangles, squares, circles and the Golden Mean. I felt transformed because I knew instantly that this would cure all that was blocked in my process with the design of the meditation hall and that now this design would become dynamic and alive . . . representing the movement of nature with the Stillpoint at its center – everything based upon the proportions of Stillpoint geometry and 'that which is naturally so' . . . 'the numbers, measures, shapes, proportions, and musical harmonies that are constant in nature.'

I wrote the following words 30 years ago:

> 'Because of the nature of the subject and its focus on the principles that give the meditation hall its *form*, it may appear that the form itself is the goal being sought. While the forms and shapes that engender the meditation hall are used as consciously and skillfully as possible with the intention of re-creating the precise essence of 'that which is naturally so,' each of these forms and shapes, representing nature in its swirling, constantly changing and moving manifestations, are generated from and return to one *still*, prototypical *point* at the center. The forms of the temple represent in archetypal symbolism the inherent *impermanence* of nature . . . *with its eternal, unchanging, still, heart at the center.'*

I had come to understand deeply what this Stillpoint was, but I was years away from completely *knowing*. Still, I was obsessed and as the years went by I could not turn my eyes away from all the beauty that this idea represented to me. In the 1977 film *Close Encounters of the Third Kind*, that was *me* building that Devil's Tower[53] of mud in my living room, not knowing exactly why I was drawn to do it.

In late 1986 I was in the middle of building my first model of the temple – the meditation hall. About seven years later, I built another, more evolved model. I had no choice but to follow my passion/obsession. I've now completed my third and last attempt at this.

The Mesa Temple is based upon the dimensions of the Earth and Moon, and all dimensions are proportioned to the Stillpoint geometry. I recently discovered that

that the mandala that I have redrawn below, from John Michell's 2009 book *How the World is Made: The Story of Creation According to Sacred Geometry,* was the basis of the diagram called the Heavenly City mentioned above. A remarkable feature of this geometry is the marriage of 7 (soul) with 12 (body), as well as the squaring of the circle, the marriage of Heaven and Earth – all attributes long ago woven into the design of the temple. Just think - all this exploding out of the void. Imagine that.

THE HEAVENLY CITY
(Recreated from *How the World is Made* by John Michell)

This is the underlying pattern upon which the temple is based . . . and the reader can see the beginning of the movement.

From *How the World is Made:*

'The Heavenly City is a geometer's name for the traditional diagram that represents the order of the universe and the numerical code that underlies it. There are many allusions to this diagram in ancient relics and writings . . . which provide the key to the long-lost science and philosophy that has, at various times and places, created the conditions for a golden age. The Heavenly City contains the numbers, measures, shapes, proportions, and musical harmonies that are constant in nature. It presents a universe which reconciles all the opposite and disparate elements that comprise it. It is an image of paradise, of the immanent perfection that can be found in every order of existence, from the cosmos to the individual. This perfection is by no means obvious, but many people in all ages have glimpsed it, with the effect of changing their lives. In the Heavenly

City diagram visionary experience is combined with numerical and scientific reason to produce an active symbol of divine wisdom restored to Earth.'

The Mesa Temple is the only response that *I* know of to the information placed in the heavens for all to see. It is not difficult to imagine the reader's skepticism of such an outlandish idea . . . a part of me shares this skepticism. On the other hand, this is not simply a building.

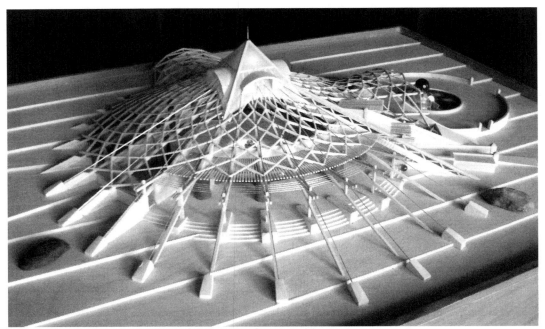

THE MESA TEMPLE MODEL

Vibration/frequency/resonance . . . Silence . . . and Love.

A temple is born.
A basket.
A womb.
Pregnant with the seed of Eternal Stillness.
A hill in the desert. A portal. Still. Silent.
A courageous act.
A communal prayer.[54]

If the Stillpoint truly is the Alpha and Omega . . . *beginning/end/birth/death /All/One* . . . then consciously recreating this Stillpoint brings Spirit into Form and completes . . . opens . . . a cosmic circuit. This is certainly the intention - to create an opening - a portal - to access this consciousness that is communicating to us in such bodhisattvic beauty. Building the temple is a brave attempt to create this opening.[55]

The model above shows the energy/structural/skeletal aspects of the temple . . . two golden-ratio born helices interpenetrating each other - the rafters and the cables - moving in opposite directions – the double torus energy . . . penetrated and balanced by the axis-mundi of the Phi, or Great Pyramid shape. This structure captures the invisible movement of energy that imbues all things. All of this is centered around one, still, *point* at the very center – the Stillpoint opening through which all of Creation emerged, dimensionless, non-existent, prefiguring all of manifestation.

It is a lightning rod. It is the acupuncture needle of the ancient standing stone. It is the release and reception of energy between Earth and Sky. It is conscious connection. It is *Invocation* based upon this incomparable information presented on the grandest of stages . . . felt always but now seen and recognized.

Before discussing the design of the temple in some detail, I'd now like to attempt to answer the question posed at the beginning of this section - why is this information appearing now and why to *me* of all people?

The timing is clear . . . there is nothing in our present paradigm that has any chance of healing what has gone so wrong with our world except some kind of shift in global consciousness. I grow daily more convinced of this. The appearance of this 4.6 billion-year-old empirical information at this very particular time - information that has *only* to do with fully awakened consciousness - only serves to make me more certain. All I can come up with in regards to why this information came to *me* is that it had to come to someone and I was prepared to understand deeply the meaning of the transcendent experience I had because of my love for sacred geometry. The experience opened my normal perception of reality into a whole new world of revelation. Different from others who may have had similar experiences, I was steeped in the vocabulary of the world of sacred geometry, and what I experienced in that other world made sense only to me, leading directly to the discovery of empirical fact.[56] Still . . . how this has escaped the notice of others remains a fascinating mystery.

The consciousness responsible for this beautiful machine we call the solar system has left a message in a bottle, orphaned to the expanse of the infinite Universe . . . until now. By using the information they've given us in one pure act we may have a chance to shift to our healthy future.

In its essence, the temple is an attempt to permit that higher dimension to be available to our density in a new way . . . its influence raising the consciousness of humanity and effecting the natural arising of compassion . . . alleviating the world's unnecessary suffering. The mechanism is *resonance*.

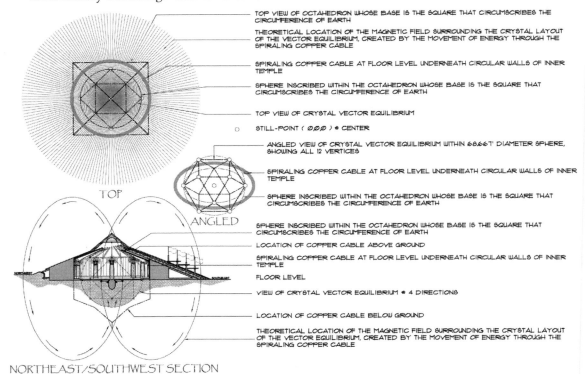

TOP VIEW OF OCTAHEDRON WHOSE BASE IS THE SQUARE THAT CIRCUMSCRIBES THE CIRCUMFERENCE OF EARTH

THEORETICAL LOCATION OF THE MAGNETIC FIELD SURROUNDING THE CRYSTAL LAYOUT OF THE VECTOR EQUILIBRIUM, CREATED BY THE MOVEMENT OF ENERGY THROUGH THE SPIRALING COPPER CABLE

SPIRALING COPPER CABLE AT FLOOR LEVEL UNDERNEATH CIRCULAR WALLS OF INNER TEMPLE

SPHERE INSCRIBED WITHIN THE OCTAHEDRON WHOSE BASE IS THE SQUARE THAT CIRCUMSCRIBES THE CIRCUMFERENCE OF EARTH

TOP VIEW OF CRYSTAL VECTOR EQUILIBRIUM

STILL-POINT (Ø,Ø) = CENTER

ANGLED VIEW OF CRYSTAL VECTOR EQUILIBRIUM WITHIN 68.66' DIAMETER SPHERE, SHOWING ALL 12 VERTICES

SPIRALING COPPER CABLE AT FLOOR LEVEL UNDERNEATH CIRCULAR WALLS OF INNER TEMPLE

SPHERE INSCRIBED WITHIN THE OCTAHEDRON WHOSE BASE IS THE SQUARE THAT CIRCUMSCRIBES THE CIRCUMFERENCE OF EARTH

SPHERE INSCRIBED WITHIN THE OCTAHEDRON WHOSE BASE IS THE SQUARE THAT CIRCUMSCRIBES THE CIRCUMFERENCE OF EARTH

LOCATION OF COPPER CABLE ABOVE GROUND

SPIRALING COPPER CABLE AT FLOOR LEVEL UNDERNEATH CIRCULAR WALLS OF INNER TEMPLE

FLOOR LEVEL

VIEW OF CRYSTAL VECTOR EQUILIBRIUM = 4 DIRECTIONS

LOCATION OF COPPER CABLE BELOW GROUND

THEORETICAL LOCATION OF THE MAGNETIC FIELD SURROUNDING THE CRYSTAL LAYOUT OF THE VECTOR EQUILIBRIUM, CREATED BY THE MOVEMENT OF ENERGY THROUGH THE SPIRALING COPPER CABLE

TOP

ANGLED

NORTHEAST/SOUTHWEST SECTION

MESA TEMPLE'S VECTOR EQUILIBRIUM – QUARTZ CRYSTAL MATRIX & ELECTROMAGNETIC FIELD

Centered on this sacred, still, point are twelve quartz crystals in a Vector Equilibrium matrix that define a 64.8' sphere above and below ground, as well as circling copper cable that is embedded deeply in Mother Earth and swirls and coils around the Stillpoint at the temple's floor level in opposing directions and travels up the four columns at the SE, SW, NW and NE directions and through the four pyramid hip rafters, connecting to the golden spire – the model of the temple grounded to Earth on Summer Solstice 2013 - creating a subtle electromagnetic vortex/field of naturally occurring Earth/Sky energy.

This subtle electromagnetic field is the enhancement of the *Schumann Resonance* . . . sometimes called the heartbeat of the Earth, and connected to mental, emotional and physical health – and to the *alpha/theta* state of consciousness associated with mediation, mystical states and the reception of extrasensory information. The importance of this enhanced, naturally occurring electromagnetic field that surrounds and protects the inner temple is discussed in more detail later.

John Michell, in *Dimensions of Paradise,* speaks about such a process:

> 'The Temple of Solomon was designed on the principle that 'like attracts like,' on the understanding, as Plotinus put it, that if one wishes to attract any aspect of the universal spirit one must create a receptacle in its image . . . . the symbolic features of the temple were intended to assist the purpose for which the building was designed, as an instrument of *invocation* . . . [and] the legend and symbolism of, for example, the temple at Jerusalem indicate that two forms of natural energy were involved, one terrestrial, the other from the atmosphere. Through the ritual fusion of these two elements at certain seasons, a spirit was generated and spread like an enchantment from the temple . . . . [functioning as] an alchemical generator and storehouse of sacred energy . . . . [performing] the sacred marriage of the electrical current from the atmosphere [with] the magnetic current of the Earth, the mercurial, serpentine spirit which corresponds in the life-system of the planet to the subtle energies of the human body.'

The masculine/matter/fixed nature of the pyramid element, establishing the *axis mundi*,[57] is perfectly balanced by the feminine/whirling/watery movement of the helical, swirling lower roof. The model shows the skeletal structure without its covering. The spiraling rafters are surfaced with 3x12 t&g decking running down the slope of the roof, covered with 12" of living roof,[58] running to and merging with the Earth everywhere except to the South, which is open to the path of the Sun. The pyramid and arched window areas are surfaced with granite. It is:

> *A rock outcropping on the top of a small hill rising out of the desert.*
> *A kiva or womb in Mother Earth.*

Within this living roof are spiraling lines of growth that mirror the spiraling rafters and cables . . . sporadic, hinting, coming into and out of existence – alluding to the subtle energy running through the temple and through *everything*. The 14 columns that surround the circular inner temple – with planters and small skylights at their tops – form a circle of trees surrounding the black granite fountain/oasis at the center (besides quartz crystal, water also stores and transmits thought-forms).[59]

Anchoring the Stillpoint is the *Omphalos stone* – a twenty-ton polished black granite megalithic stone held close by Mother Earth. There is a concavity in the center of its horizontal top - the impression of an invisible sphere in the black granite, with the non-existent Stillpoint that generates all forms at its center. The entire inner temple is filled with the subtle light coming through emerald, amber, burgundy, indigo, violet and cobalt stained glass windows. Imagine!

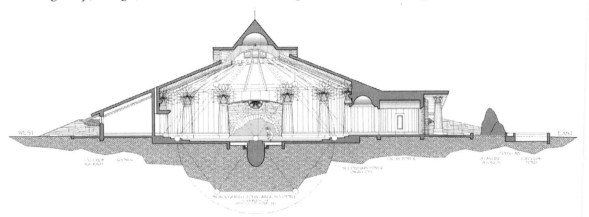

MESA TEMPLE - WEST / EAST SECTION

The temple is also a resonator . . . a technology designed to produce a certain kind of resonance, and to amplify, accelerate, and transmit that resonance, with a precision of intention mirroring the incomparable beauty and depth of the *Point of Creation*, the gateway from our origin and to our destination. The emphasis here is on the quality, precision and depth of this cosmic spark. *Resonance* is the mechanism of transformation. Everything that is vibrates. Everything has a frequency. At its essence, this is magical Invocation. From Mouni Sadhu's: *The Tarot*:

> 'Tourbillons, or vortexes, are astral creations of force which are the bases of all astro-mental realizations. Tradition ascribes the funnel like forms to them. Knowledge of the laws ruling over the tourbillons and their construction is one of the foremost principles of magic. The most guarded secrets of Hermetic magic are: finding the point of support for the tourbillon on the physical plane, and the formula of transition from the astral to the physical world.'

Some years ago I came across the work of David Hawkins through his book *Power vs Force* - revelatory to my understanding of what may be possible through the temple. The reader may be familiar with his work . . . ostensibly the objective verification of phenomena of all kinds, throughout time, through kinesiology – a simple arm muscle testing technique which he uses to objectively test the truth of any statement - a feedback mechanism to the unconscious.

This is totally dismissed, of course, by mainstream science. I find his work compelling and I include it because of this and because of my own personal experiences with kinesiology – very strange, but accurate. My body, my arm, was responding to *something* beyond my conscious knowing that turned out to be true. I offer this here, not because it's irrefutably proven to be so, but because it presents a credible model regarding the general level of consciousness of the world population. That is, there are all sorts of people in the world with all sorts of levels of awareness. Some of us are mired in anger, prejudice and hatred, while others exude compassion, caring and love. It is how we function as a *whole* that is so critical for us today. Hawkins' work presents a way that may gage what our general level of consciousness presently is . . . and has been. Apparently, it has been evolving.

The heart of what he's discovered has been the detection of powerful emotional/astral archetypes in the collective unconscious, or Akashic Record. He calls them, in the language of chaos theory, 'attractor fields' or 'strange attractors' of different levels of consciousness, around which the next octave of evolution (or devolution), whether on a personal or collective level, organizes itself.

The energy of these archetypes can be calibrated through kinesiology. He says:

> 'Calibrated levels, we suggest, represent powerful attractor fields within the domain of consciousness itself, that dominate human existence and therefore define content, meaning, and value, and serve as organizing energies for widespread patterns of human behavior.'

As the *frequency* of these attractor fields of human consciousness ascend from basic survival, at the number 1, through shame, guilt, greed, etc., at the lower end of the spectrum, up through the higher frequency/calibrations of courage, acceptance, love, joy and enlightenment, calibrating at 1000 at the upper end, the calibrations grow in power *exponentially* – meaning that the energy of very evolved, expanded thoughts, feelings, prayers, and meditations are almost infinitely more powerful than those at the bottom of the spectrum. Hawkins says that the parameters he used were limited due to its focus on *human* consciousness (from 1 to 1000), but that 'the entire field or phase space of consciousness itself is unlimited, going on to infinity.'

From *Power vs Force*:

> 'Great works of art depicting individuals who have reached the level of Enlightenment characteristically show they teach with a specific hand position, called mudra, wherein the palm of the hand radiates benediction. *This is the act of the transmission of this energy*

field to the consciousness of mankind. This level of divine grace calibrates up to 1000, the highest level attained by any persons who have lived in recorded history – to wit, the great Avatars to whom the title 'Lord' is appropriate: Lord Krishna, Lord Buddha and Lord Jesus Christ . . . these individuals set up attractor patterns of enormous force to which *the mind, with its holographic capacity to react globally to attractor fields, is subject.*'

Hawkins calibrations stop with these enlightened beings . . . but, as I've mentioned, Arthur Young's theory had yet another level, or dimension – fully evolved enlightened awareness with no need to manifest - the 'energy field' or bodhisattvic consciousness referred to above being transmitted by the Avatars . . . 'to the consciousness of mankind.'

The physicist David Bohm 'postulates a source that is beyond both the explicate and implicate realms, very much like the state of pure awareness described [and transmitted] by the sages.'[60] What the Buddha and others were transmitting was energy coming *through* them, the human form having been made clear, I assume, from lifetimes of learning 'how things work.' This is energy infused with love, compassion and wisdom and has an extremely high and refined frequency and vibration. Once again, I believe that this is the mechanism through which this transformation will occur - *resonance.* This is precisely the portal the temple would hope to open . . . permitting the transmission of infinite pure love, compassion and awareness into the density of our plane of existence.

Regarding the most transcendent art and architecture, and the energy fields associated with them, Hawkins also says:

'The great works of art, music and architecture which have come down to us through the centuries are enduring representations of the effect of high attractor patterns. In them we see a reflection of the commitment of the master artists of our civilization to perfection and grace, and thereby to the ennoblement of humanity. The fine arts have always provided the venue for man's highest spiritual strivings in the secular realm. It has been the role of the arts to realize, in physical media, ideals about what man could and should be, to set down in tangible form, accessible to all, a distilled expression of the human spirit.

But of all the arts, architecture is the most tangible and influential in the lives of men everywhere. Among all world architecture, the great cathedrals elicit a special awe. Their energy patterns have calibrated the highest among architectural forms. Our experience of cathedrals can combine a number of arts simultaneously: music, sculpture, painting, as well as spatial design. Moreover, these edifices are dedicated to the divine; that which is begotten in the name of the creator is aligned with the highest attractor patterns of all. The cathedral not only inspires, but unifies, teaches, symbolizes and serves all that is noblest in man.'

Having said that, the spiritual energy so evident coming through the cathedrals – energy that I have experienced and is obvious - is restricted and limited by a dogmatic and oppressive belief system, regardless of the perceived comfort it may give to millions. I know that the Catholic Church is also evolving . . . but it still carries a sordid history based upon a threatening dogma.

Hopefully, the temple itself has nothing to do with any belief system . . . and is based solely upon the eternal stillness of our origin, the Stillpoint, and supported by the sacred geometry of movement – the torus . . . the archetype without the dogma. As mentioned above, what I've referred to as an infinitely evolved consciousness communicating to us in such a grand bodhisattvic manner, could also be understood as an almost infinitely powerful attractor field, the transmission of which would be unleashed by the opening of the precisely designed and purely intended portal.

To repeat: the temple is an electromagnetic-field-torus-vortex centered on a consciously created, still, *point* - the ultimate portal - charging and being charged by a quartz crystal, Vector Equilibrium matrix . . . a resonator, an accelerator, an amplifier, an accumulator and transmitter of elevated human thought-forms and prayer - as well as a transmitter of the powerful 'attractor field' of the evolved bodhisattvic consciousness.

I have always understood that the 'normal' use of the temple . . . for meditation and prayer . . . as secondary and minor as compared to the opening of this portal, making pure awareness powerfully available and influential on this plane through *resonance* – the invocation and transmission. But if what Hawkins says is true: 'The influence of a very few individuals of advanced consciousness counterbalances whole populations at the lower levels' - then this secondary function is now also critical in importance. That is, if a loving or enlightened thought is almost infinitely more powerful than a hateful or envious thought, how compelling would the exponentially powerful meditations and prayers of hundreds of *focused*, spiritually expanded minds be in a technology based upon the original archetype of the Alpha and Omega, and intentionally created to expand and amplify and transmit these thought-forms?

Lynne McTaggart, in her 2007 book *The Intention Experiment*, wrote that 22 scientific studies had been done regarding the affect that focused mediation has on crime levels in various cities. In one experiment, 4000 mediators focused on violent crime in Washington D.C. From *The Intention Experiment*:

> 'In 1993, the TM's (Transcendental Meditation) National Demonstration Project focused on Washington, D.C. During a large upsurge of local violent crime in the first five months of the year. Whenever the local group reached the threshold number of 4,000, the rate of violent crime fell and continued to fall, until the end of the experiment. The study was able to demonstrate that the effect had not been due to any other factors, such as police efforts or a special anti crime campaign. *After the group disbanded, the crime rate in the capital rose again.*'

What interested me most about this phenomenon are the words I've placed in italics. It is clear that focused thought-forms affect matter, as shown so powerfully in Masaru Emoto's photographs of water crystals before and after being flooded with both loving and destructive energies *(see Water Crystal Photography, page 236)*, as well as other examples of such that are covered in more depth later. What struck me about the above experiment is that when these thought-forms are absent, the effect disperses. This is why 'permanently'[61] creating an opening to higher consciousness or to the Akashic field is so critical, and emphasizes the importance of employing an array of quartz crystals that *store* and *transmit* loving, compassionate, aware thought-forms – in concert with the moving water surrounding the Stillpoint at the very center.[62]

Most essentially, this is the creation of a new and permanent *morphic* field. Once a paradigm has evolved through its own limitations at the end of its octave, a new and unimagined octave of possibility emerges and what was once difficult or impossible, now becomes easier.

From *Power vs Force*:

> 'Another useful concept is Rupert Sheldrake's notion of morphogenetic fields, or M-fields. These invisible organizing patterns act like energy templates to establish forms on various levels of life. When Roger Bannister broke the four-minute mile, he created a new M-Field. The belief system prevailing in human consciousness had been that the four-minute mile was a limit of human possibility. Once the new M-Field was created, many runners suddenly began to run sub-four-minute miles. This occurs every time mankind breaks into a new paradigm, whether it is the

capacity to fly, an M-Field created by the Wright brothers, or the capacity to recover from alcoholism, an M-Field created by Bill W., the founder of AA. Once an M-Field is created, everyone who repeats the accomplishment reinforces the power of the M-Field. *An M-field is a standing energy field which is everywhere present. Once it is created it exists as a universally available pattern throughout the invisible universe.'*

I share this because it's helped me experience some hope regarding the future – the very real possibility of creating a new and powerful *field of compassionate, expanded consciousness* here on Earth. What is critical is the creation of the new paradigm and opening a floodgate . . . after this happens, it will have its own, growing, momentum towards the 'Good.' This is one of the principle intentions underlying the geometry of the Mesa Temple. The other is opening a portal to the other side of the veil, aligning in resonance with the higher frequency/consciousness that created this place in the vastness for the acceleration of the evolution of consciousness.

The incomparable importance of making contact with higher consciousness at this particular time in our history is beyond critical. The life or death issue of somehow dealing with the so-called elite – that .1% of humanity that are in almost total control of life as we know it - and the fact that humanity is now . . . right *now* . . . at the crossroads of its collective history . . . *screams* how critically important the Stillpoint information is to life on Earth . . . *now.*

For many of you who've read this far, your beliefs about reality may well have been challenged – this is, after all, startling, new information that contradicts thousands of years of organized religious belief and hundreds of years of materialistic scientific belief. Our beliefs, if self-created to bolster ego's[63] need to see the world in a certain way, are often limiting and harmful. Because of the complicated and difficult ways that life can so often be, we humans scramble for ways to explain away our fear . . . and these efforts become the beliefs we hold that protect us – we think.

Sometimes I hike up into the mountains to the east of where I live and look back down into this vast valley. Tiny cars and trucks and houses lose their significance, disappearing into to the vastness of Mother Earth. The tedious world of my worries and fears also disappears . . . replaced by something silent and huge and sacred. The Universe is vast . . . infinite . . . and it contains realities that are beyond our present capability to understand. These pages challenge long-held beliefs embedded deeply in our culture – both conscious and unconscious beliefs . . . beliefs that often prevent us from discovering a deeper truth.

I recently met a very young Christian couple with five very young children who I mentioned all this to – very briefly but enough that they looked like deer in the headlights. They asked if this was about the Intelligent Design movement of fundamentalist/creationist Christianity. When I said, 'No, not really,' they looked away and asked no questions. They said that God had a plan for their lives and that everything that happens is God's will. They didn't want to know.

I also recently shared this with a young man who is a new practitioner of Vippassanna, a Buddhist form of meditation. When I shared with him my understanding of the dire situation we find ourselves in and the possibility that there may be a way out of it, he told me that he'd learned to 'be in the moment, and to accept things as they are' . . . that there really was no need to do anything except to be non-attached – and he also was uninterested in hearing more.

Another long time practitioner of Tibetan mediation, a very old friend of mine, didn't want to hear about it because, regarding the world's suffering, she thought it enough to practice meditation not for one's own sake but for the sake of all beings - I didn't tell her that I considered my own prayers to be but an infinitesimally tiny spec of dust on the scales of an infinite Universe. Another person, after I'd shared some quotes from Gandhi and Tesla and Krishnamurti regarding the crisis we're in, told me not to worry . . . he was from the future . . . and it all worked out.

Another person told me that the only way to evolve is to suffer and learn from it, and that it was impossible to change human nature . . . that what I was proposing was fairy dust – or, as my dad would say . . . horsefeathers.

I told this story to another person when we were having lunch. He pointed at the salt shaker on the table and dismissed everything I'd just said by saying: 'It's just what it is . . . like that salt shaker.' Many dismiss all this as coincidence with no interest to take the time to examine the evidence, an example of the equally misguided and close-minded scientific belief. Most couldn't care less. There are a myriad of other beliefs that may have pushed some buttons. In all these examples, the reactions held by these various people – while completely different from each other - would not permit them to open to what has just been shared. And so it is.

I wish to make it clear before proceeding that my principal purpose in making this information public is to put it into the hands, hearts and minds of concerned and expanded people so that if there is a better idea than my own to address the information and perhaps to respond, then that is what I want.

Personally, I want only to build this Temple and remove myself from the endless conversation . . . and I invite anyone interested this idea to contact me. For now though, it is sharing this information with the world that is all important.

Pilla

THE PILLARS OF CREATION

A photograph of gigantic gas clouds - incubatiors of new stars - named The Pillars of Creation, 7,000 light years from Earth, taken by the Hubble Telescope in 1995. The lefmost pillar is about four light years in length. The finer-like protrusions at the top of the clouds are larger than our solar system.

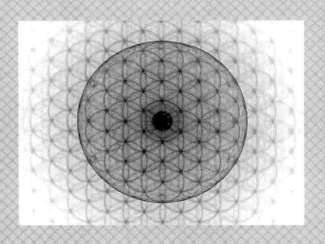

The Larger Context

The following information is the principal *theoretical* context for the Stillpoint's *practical* expression in the Mesa Temple, and is critical to understanding the importance of the temple and the possibility it holds.

The nature of scientific inquiry is that it is ever discovering new solutions, leaving old theories . . . some long held to be gospel . . . aside. It evolves. The information in this section focuses on some of the current science that defines our understanding of the world, and is critical to comprehending the depth of the information underlying the Mesa Temple . . . adding an empirical element to a subject that is ultimately metaphysical.

It is intended that some day all these words translate into a physical structure. If it were a structure to house a family, the words would be different. Their context would concentrate on different concepts . . . most essentially, concerns surrounding the daily necessities of living: rooms for gathering, dining, sleeping, and bathing, as well as a place for cars, storage areas and all the rest of the practicalities associated with daily life. *These* words describe the context for a sacred structure intended to be an access to the mysterious world on the other side of the veil.

The context for the action proposed includes cosmology, consciousness, the evolution of consciousness, physics and quantum physics, astrology, sacred geometry, ancient geodetic measurement, sacred architecture, theology, psychology, vibration, frequency and resonance, the effect of consciousness upon matter and other related subjects, as well the the impetus for caring about any of this in the first place – the seemingly endless suffering of sentient beings and the threat of the annihilation of life itself. Each of these subjects have been written about in books that fill libraries and it would be impossible here to do any of them the justice they deserve. The intention is to provide a general review of the related science to give the reader a sense of the intellectual context for what is essentially an *experience*.

Material has been taken from a variety of sources and organized in a way that leads inevitably, I believe, to a logical conclusion that supports the audacious claim made in the previous chapter – that our solar system was created intentionally, outside the 'normal' way of Universal creation, by an *evolved* higher consciousness. The majority of the content of this section comes from an investigation into the world of cosmology and physics. Generally, those of us not a part of that scientific club whose vocabulary is filled with incomprehensible terms and enigmatic mathematical equations are considered Philistines - yet we must not be shy.

⊕ IMPOSSIBILITY:

The best place to begin this discussion is to make it as clear as possible that, scientifically speaking, it is almost 'impossible' that we are here at all. Science tells us that our existence is a 'happy coincidence' – a random accident. That is, other than some kind of chance occurrence, it appears that it is 'impossible' that we *are*. 'Impossible' refers to the lack of any credible scientific explanation, and is in quotes because we *are* here . . . me, you, Beethoven, Shakespeare and Einstein. Ironically, through the brilliant discovery of the precise mathematical constants included in the equations that form the physical basis upon which the Universe is based *(see Sir Martin Rees's Just Six Numbers, page 86)*, it is science itself that provides us with the astounding body of evidence leading directly to this unlikely reality. Pages are dedicated to the wonder of the scientific mind and its remarkable accomplishment of unraveling much of the superstition of past centuries because I feel that this perspective needs to be understood, at least on the layman's level. But it will be made clear that a much more expanded and spiritual understanding of why we're here is desperately needed . . . now.

The *chance* that we exist at all, according to the scientific worldview, is a *pinpoint in the midst of infinity.*

Think about that image . . . the vastness of it, the improbability of it.

That we're here simply by coincidence is absurd. Yet . . . here we are - and common sense screams that this is not an accident. While the Universe within which we exist is almost impossible according to any law of materialistic scientific inquiry sans chance, Ilya Prigogine, the chemist-physicist and recipient of two Noble prizes and whose work has already been mentioned, put it this way:

> 'The statistical probability that organic structures and the most
> precisely harmonized reactions that typify living organisms would be
> generated by accident, is *zero.*'

Within *this* reality, *this* Universe, there are very specific principles - *laws* - that govern or organize all that we see, all that we are, all that is. Many of these laws are based upon the incomprehensibly precise constants covered in Sir Martin Rees's book *Just Six Numbers,* mentioned earlier and discussed further below. Within these laws, within *this* almost impossible reality, there are also phenomena *within our own solar system* that are '*impossible*' themselves . . . that is, absurdly outside of the laws of chance . . . as well as physics, geometry, mathematics or *any* kind of science.

So . . . why? If not any of the above . . . what? Evidence will now be presented that suggests an answer to this ancient question. The scientific improbability of it all will be made evident, as will a deeper reason for why we *are* at all.

⊕ SCIENCE, BELIEF . . . AND A MIDDLE WAY:

Science, the often brilliant light of the rational mind, continues to penetrate deeper and deeper into this oldest of Mysteries – why are we here? Why is any of this here? Yet the deeper the discovery, the more obvious it is that this is not the territory of the reasoning mind limited by the need to capture and interpret all information in the old, accepted, reductionist way. From Massimo Citro's *The Basic Code of the Universe*:

> 'In the illusory world, we live according to points of view, none of which can describe bodies in their entirety, only parts of them. The truth cannot be divided, dissected, or categorized because it lacked material forms; *we can only classify illusions, and science is limited to describing patterns of fictions.* An atom is a probabilistic aspect of the vast sea that continuously exchanges atoms between stage and backstage, removing and replacing them with such speed that we cannot notice it. The universe is a single fabric whose parts are only appearance. A particle 'is' only when we observe it and translate it into an image: this is the 'curse' by which the senses present us with a distorted reality, but unfortunately this is the program. The truth beyond the senses is unknowable. *Deus absconditus,* 'hidden God' (as Thomas Aquinas put it) is behind all forms hidden by the illusions of Maya.'

I would only say that while everything that we experience as 'truth' is experienced through our senses, it is possible to journey to the other side of this materialistic, sense dominated world, into the implicate world, to experience a deeper understanding of reality, or truth, through meditation, breathwork, hallucinogens, etc. Still, science's discoveries, and the words used to describe them, continue to come closer and closer to the indescribable.

The scientific mind has learned to steer clear of any answer having anything to do with that which it cannot quantify . . . and understandably so given how humanity's fear and greed, in the name of God, has created so much delusion and suffering. Until only very recently, the answer to this inescapable question – why are we here? - and the power it wields in the hands of those who claim to have the answer, has been clouded in the superstitions of the world's organized religions. These religions didn't start out this way. Most all of us, at some point in our lives, feel a presence of something far far greater than ourselves. Sometimes these experiences take the form of revelation. The problem comes when we dogmatize the experience . . . when we co-opt it for more ego-oriented purposes. Rupert Sheldrake makes this clear:

'Spiritual traditions in general, and religions in particular, were not founded on irrational propositions, or on blind faith, or on dogma, or on fear. They arose from states of consciousness that go beyond normal everyday experience. Shamans, Indian rishis, the Buddha, the Jewish prophets, Jesus and Muhammad spoke from their direct experiences of connection to a greater consciousness.'[64]

From these pure experiences, organized religion has devolved into what we see today in the more fundamental expressions of Christianity, Islam, and Judaism. Buddhism is in a realm of its own.

The rational mind, finally freed from this oppression and able to seek objective verification for phenomena, only came into prominence in the last few hundred years - a blink of an eye in human existence – an infinitesimal slice of a moment of Universal experience. But, rather than being a cure to the superstition problem, it has succumbed to its own inherent flaws – human consciousness as we now experience it - and is, essentially, the other side of that coin.

Science, the intellectual and practical discipline studying physicality through observation and experiment, has proved itself to be extraordinarily adept at accurately describing our physical world. These descriptions are based upon what is called the 'standard model,' which is 'a mathematical description of the elementary particles of matter and the electromagnetic, weak, and strong forces by which they interact,'[65] as well as Einstein's theory of General Relativity, which defines all motion to be *relative* to a frame of reference and that 'space and time are *relative*, rather than absolute concepts.'[66] All of this is now defined in one quantum field theory: what physicist and Nobel laureate Frank Wilczek terms the *Core Theory*: 'The quantum field theory of the quarks, electrons, neutrinos, all the families of fermions, electromagnetism, gravity, the nuclear forces, and the Higgs [Field]' . . . and accounts for 'every experiment ever performed in a laboratory here on Earth.'[67]

From this ultimate description of the way our world works, flow the brilliant discoveries that fill our world: automobiles, smart phones, rockets to outer space, the Kepler telescope, television, fiber optic cable, toasters, garden hoses, the Large Hadron Collider . . . as well as the stuff that makes up our daily experience: people, trees, air, water, giraffes, stars, fire, grapefruit . . . *everything* in the physical Universe. The claim is that 'the laws of physics underlying everyday life are completely known.'[68] Very impressive.

Another way of saying this is that science *reduces* all of Universal phenomena and human experience to an 'explanation of complex life-science processes and phenomena in terms of the laws of physics and chemistry'[69] . . . this is called reductionist science. And herein lies the problem: *the assumption that physicality is all there is.* The materialistic/reductionist science so prevalent in our daily lives exists entirely within the world . . . or box . . . that its vocabulary defines. The most primal vocabulary upon which science is able to make its profound discoveries is mathematics, the essential core of which is the differential calculus worked out by Newton and Leibniz. But the vast territory defined by empirical science, vocalized most elegantly by calculus, is limited by the very 'real' Planck Constant[70] . . . beyond which it cannot go: past this barrier is the world of the Stillpoint . . . the world of zero and infinity - the world on the other side of the veil.

The theoretical physicist Sean Carroll, an interpreter of science for the rest of us, calls this world of reductionist science the 'big picture' . . . and named his 2016 book after it. It's been very useful to have read this book, as it is an excellent explanation of the materialistic side of the scientific view . . . the dominate view in science today. He calls it 'poetic naturalism.' The 'poetic' adjective suggesting that while there's only one world – that defined by physics – there are 'many ways of talking about it.' The implication in the title – that this scientific view is the big picture - is backed up by over 400 pages of extensive scientific explanation. Carroll says that 'the universe, and the laws of physics, aren't embedded in any bigger context, *as far as we know.*' Hence, the title of this chapter, which attempts to make clear that the other side of the veil is critical to understanding the world in its wholeness. Like Carl Sagan before him, he helps the masses to understand the work of the brilliant discoveries of those bastions of materialistic science Steven Hawking (*A Brief History of Time, The Grand Design* and *The Universe in a Nutshell*) and Richard Dawkins (*The Selfish Gene* and *The God Delusion*), and so many others who make their studies within the ideological view that the physical world is all there is – seemingly totally unaware that they are thinking within the box called 'science.'

The Big Picture is a book filled with descriptions of ways of thinking or, as 'poetic naturalism' might say, ways of talking about our world: property dualism, substance dualism, panpsychism, eliminitavism, moral constructivism, relativism, humanism, naturalism, dysteleology, determinism, deduction, induction, abduction, Bayesian thinking, entanglement, strong emergence, coherentism, methodological naturalism, methodological empiricism, theism and atheism, Darwinian, Newtonian, realist and antirealist, quantum field theory, standard-model of particle physics, general relativity, crossing symmetry, entropy and negative entropy, epigenetic phenomena, quantum indeterminacy, anthropic principle, dual process theory, downward causation, compatibilism, consequentialism and deontology, moral constructivism, synchronic and diachronic meaning. Whew! Totally overwhelming . . . but you have to hand it to the human mind.

I couldn't help thinking, when I came upon all these isms and ology's, of when my son was in his skateboard phase and came back one evening after zooming around town and asked: 'Dad, why are there *two* First Baptist churches?' He came upon the announcement along the highway at the entrance to Bishop that listed the many churches in town, just in case someone passing through needed religious sustenance: Seventh Day Adventist Church, Calvary Chapel, St. Timothy's Rectory, River Church, Our Lady of Perpetual Help, First Southern Baptist, First Baptist, Father's Heart International, Church on the Mountain, Church of Religious Science, Kingdom Hall of Jehovah's Witnesses, The Neighborhood Church, Christian Fellowship, Calvary Baptist Church, Sierra Baptist Church. Apparently the Presbyterians, Muslims, Episcopalians, Mormons, Jews, and Lutherans have not made it here yet. No doubt, human beings are certainly *trying* to figure all this out – whether scientifically *or* religiously.

Carroll acknowledges that the Big Bang 'marks the end of our theoretical understanding,' and that we are only aware of what happened 'soon after the Bang,' and that the 'Bang itself is a mystery' . . . and acknowledges that 'it's a label for a moment in time that we currently don't understand.' Yet this exposes the myopia of the scientific view. While acknowledging that the *moment* of the Bang demarcates the moment *in time* when 'we can't see any further into the past' - and is simply the 'beginning of our observable part of the cosmos' - correctly defining the territory that science is so expert at discovering - he casually dismisses the timeless world on the other side of the veil marked by the Planck Constant, where he acknowledges science cannot venture. In fact, 'nothing' is the word science uses to describe this implicate world - the world known by sages throughout history . . . the world of the eternal present moment, outside of space and time – that true Mystery from which all of manifestation emerged.

Evolution, in the scientific view, occurs by 'marching from one moment to the next in a way that depends solely on [the Universe's] present state. It neither aims toward future goals nor relies on its previous history,'[71] and is fully determined by patterns described by the laws of physics. In this view, everything that *is*, happens in a deterministic way, completely defined by the established laws of physics . . . yet at the quantum level, the process 'seems to happen randomly . . . the laws of physics describe how everything was created in a random quantum fluctuation out of nothing'.[72] That is, when this initial quantum randomness grabs hold, its future is determined by the established laws of physics, eventually expressing themselves in the organic world through the process of natural selection - the *unrandom* process whereby plants and animals better adapted to their environment tend to survive - the idea introduced by Darwin and made into dogma by Dawkins' in *The Selfish Gene*.

Here, the scientific and existentialist views merge: there is no meaning, nor purpose to any of this. It is the atheistic worldview, with no need for any 'God' to get it all going or manage it. The mistake made here, and understandably so, is that when the debate rages between 'God' and 'No-God,' it always refers to the God of the major religions . . . the old Creator-God of myth. The Stillpoint phenomenon merges the established scientific certainty of evolution in the Universe, with the now scientific certainty of an evolved higher consciousness that most would consider God – the critical difference being that this 'God' *evolved* to be so.

In the materialistic scientific view, consciousness itself is not exempt: particles randomly bump along for who knows how many billions of years and, what do you know, consciousness just happens to *emerge* from this randomness . . . *from* the *brain*, that just happened to emerge itself. 'Emerge,' as meant here, means that it is 'not a part of a detailed *'fundamental'* description of the system . . . a naturalist believes that human behavior [and consciousness itself] emerges from the complex interplay of the atoms and forces that make up individual human beings.'[73] The model presented in these pages is that consciousness is *the* fundamental context . . . and from that *everything* in manifestation emerges.

Given the empirical proof that our own solar system . . . unlike anything else we have witnessed in the observed Universe . . . was created *intentionally*, outside of the way creation normally happens, and that a message was embedded at its heart having *specifically* to do with expanded awareness, expanded consciousness, it seems pretty clear that the evolution that science has championed has everything to do with *consciousness* – that in fact, the *evolution of consciousness* is the true big picture, the true context within which science's place is fully understood . . . a worldview that includes not only the brilliant scientific discoveries of the *physical* Universe on this side of the Planck Constant, but the vast and *more real, implicate world* on the other side of that veil. The hubris of the scientific view loves to take pot-shots at this essentially unverifiable world – that is, unverifiable within the vocabulary of science.

But now there *is* empirical evidence that verifies this world. It is a new day.

I am reminded of the words with which Jane Goodall – a truly extraordinary human being who also happens to be a Cambridge-trained scientist – used to begin her book *Reason for Hope*:

> 'Many years ago, in the spring of 1974, I visited the cathedral of Notre Dame in Paris. There were not many people around, and it was quiet and still inside. I gazed in silent awe at the great Rose Window, glowing in the morning sun. All at once the cathedral was filled with a huge volume of sound: an organ playing

magnificently for a wedding taking place in a distant corner. Bach's Toccata and Fugue in D Minor. I had always loved the opening theme; but in the cathedral, filling the entire vastness, it seemed to enter and posses my whole self. It was as though the music itself was alive.

That moment, a suddenly captured moment of eternity, was perhaps the closest I have ever come to experiencing ecstasy, the ecstasy of the mystic. How could I believe it was the chance gyrations of bits of primeval dust that had led up to that moment in time – the cathedral soaring to the sky; the collective inspiration and faith of those who caused it to be built; the advent of Bach himself; the brain, his brain, that translated truth into music; and the mind that could, as mine did then, comprehend the whole inexorable progression of evolution? Since I cannot believe that this was the result of chance, I have to admit anti-chance. And so I must believe in a guiding power in the universe – in other words, I must believe in God.'

Yes. How in the world can the scientific view be so limited as to see this – the vision and manifestation of the cathedral of Notre Dame, put to music by the genius of Bach – to unexpectedly emerge from accidental, random, phenomena? The brain does not create consciousness – the evolution of consciousness created the brain, the *instrument* capable of merging with the evolved *field* of *learned awareness* - the Akashic Record, if you will, of everything learned throughout the eons of distant past.

It is the common scientific understanding that Geordano Bruno established in theory . . . and Copernicus, Kepler and Galileo soon afterwards with empirical verification - that our geocentric worldview, placing us at the center of the Universe, was not true – implying, according to *The Big Picture*: 'Cosmically speaking, there's no indication that we matter at all.'[74] The first statement is certainly true . . . but to imply that we don't matter is a knee-jerk kind of intellectual jump. We certainly *do* matter. We are a highly evolved aspect of creation, proving the incomparable significance that the larger context *is* the evolution of consciousness in the Universe. The Stillpoint phenomenon proves that we are a point on a line extending from the unconscious *whim* to awaken from the vast oblivion – the Great Mystery - to the fully awakened awareness evidenced by the intentional creation of our corner of the cosmos, our solar system . . . for the acceleration of the evolution of consciousness. Emerging consciousness exploded out of the Stillpoint – the Bang – and moved through various stages of evolution: Light, particles, atoms, molecules, stars, planets, plants and animals and us . . . to the higher consciousness witnessed in the crop circle/UFO phenomena . . . and far beyond. The Universe is waking up.

Science considers the bottom line of Universal existence to be a 'quantum wave function' . . . after that it is simply random emergent systems that we establish convenient ways of talking about. It is too often the study of separate parts. It is this idea of a dualistic, separatist world that is so dangerous. It is this concept – the separation into parts, rather than the truer reality of unbroken wholeness underlying this illusion of separateness,[75] that is behind the insanity of building a wall to keep people in – the Berlin Wall, and building a wall to keep people out – the Trump Wall, endless war and all the rest. In *The Big Picture*, Carroll addresses this issue:

> 'Another example – controversially – might be human consciousness. People are made of particles, and we have a successful picture of how particles behave, the Core Theory. You might think that we could fully describe a person if only we knew the complete state of all of their particles. We have every reason to believe that the domain of applicability of particle physics includes the particles that make up human beings. But it's possible, *however unlikely,* that there is one set of rules obeyed by particles when there are only a handful of them interacting with one another, as studied by particle physicists, and a slightly different set of rules that they obey when they come together to make a person. There is no evidence that this is true for human beings . . .'

Really? This seems a classic denial that the whole is more than the sum of its parts. But there it is. To support this claim, Carroll dismisses the entire realm of psychic ability and phenomena. In a classic example of what Carroll refers to as confirmation bias - 'our tendency to latch on to and highlight any information that confirms beliefs we already have, while disregarding evidence that may throw our beliefs into question'[76] - Carroll many times refers to all such experience with the pejorative term 'spoon bending' . . . clearly using the exposure of some of Uri Geller's[77] more deceptive showmanship in decades past to smear all of the many valid verifications of such phenomena. From *The Big Picture*:

> 'Many people still believe in psychic phenomena, but they are for the most part dismissed in *respectable circles of thought*. The same basic story holds for other tendencies we sometimes have to appeal to extraphysical aspects of what it means to be human. The position of Venus in the sky on the day you were born does not affect your future romantic prospects. Consciousness emerges from the collective behavior of particles and forces, rather than being an intrinsic feature of the world. And there is no immaterial soul that could possibly survive the body. When we die, that's the end of us.'

I am grateful to Mr. Carroll for laying it all out so clearly. For what it's worth, when I was born, Venus was in mid-heaven (at 12 o'clock noon, straight up, in the sky above), aligned with the constellation Pisces. I was told that this meant that I had an inherent and deep appreciation for beauty. Nothing could be more true. I somehow was attracted to making my living designing architecture that sought to be beautiful whenever possible, and was mysteriously pulled toward, and ultimately obsessed, with sacred geometry – the purest . . . and most beautiful . . . archetype for all that is. Further, I was blessed by being taken into the very heart of sacred geometry . . . the Stillpoint . . . which has to be the absolute definition of Beauty. Maybe just a coincidence. Astrology is *not* that found in terse newspaper back pages, as Carroll would like us to believe. I refer the reader to the work of Arthur Young in *Science and Astrology* and Richard Tarnas in *Cosmos and Psyche*.

When I first read this dismissal of all psychic experience and phenomena in such a casual and transparent way . . . in the midst of a tour de force of reductionist science . . . I was incensed. Such a cheap shot. '*Respectable circles of thought*' indeed. Did he know nothing of the mountain of evidence supporting and validating such experience? Or was he so completely blinded by his own 'confirmation bias' that it simply didn't exist for him. I think the latter is true – and that this is dangerously true for materialism in general - he unconsciously ignored it because it didn't exist within his own belief system. Had he not heard of Hal Putoff, Russell Targ, Pat Price, Helen Hammid, Edgar Cayce, or Ingo Swann or the work current work of Dean Radin, etc.? If these are unfamiliar names to you, and if you're interested in countless examples of more going on in the human experience than the Core Theory can possibly account for, read any of Russell Targ's many books, and look for Targ's first film regarding all of this – *Third Eye Spies* – hopefully 'coming soon', or visit his website at www.espresearch.com. Hal Putoff and Russell Targ were physicists employed by the C.I.A. to investigate extra-sensory perception at the Stanford Research Institute in Menlo Park, California during the Cold War, in an attempt to co-opt what the Russians were doing. Pat Price, Helen Hammid and Ingo Swann were psychics who assisted them. Edgar Cayce should be a familiar name to most. Dean Radin is the chief scientist at the Institute for Noetic Science. The many examples of the work they and others have done puts to rest any doubt regarding the non-local, non-physical aspects of reality . . . something materialistic oriented science ignores – *because the Core Theory simply has no room for them.*

Soon after I'd read Carroll's off-handed denial of all such phenomena, lumping it all into the 'spoon bending' category, I found myself talking to a friend I'd known for almost thirty years at a gathering. Somehow the conversation developed such that I shared my upset about all this to her . . . and she responded by casually saying: 'I've bent a spoon.' 'Have you now?!', I blurted out. The timing of running into someone who'd actually done this shocked me into the present. She proceeded to explain, in glorious and uninhibited detail, how this came about. An evening was offered to those interested. About 30 people came and were asked to sit on the floor in a circle . . . and a large bag of spoons was spilled out in the center. The group leader asked each person to pick out 12 spoons that they felt some kind of connection . . . and from these 12, to pick 4 that especially 'talked' to them. Once this was done, the group was instructed as to how to proceed . . . essentially rubbing the spoon on the handle (much as I'd seen Geller do), and focusing the mind on bending it. As she told the story, everyone focused on the purpose at hand . . . until, sure enough, someone yelled out 'It's happening!' . . . and then, one after another, spoons began to bend all around the circle. I don't know if everyone was successful . . . but I know for sure that I was being told the truth. As though reading my thoughts . . . ;-) . . . she said: 'And, yes, they were all *real* spoons.' Now . . . this is not a 'scientific' experiment. . . and it's hearsay to boot – but it was told to me so spontaneously and from a person of such impeccable character that I have no reason at all to doubt it. And the *timing* was synchronistic.

The same dismissal by orthodox science of astrology and telekinesis also goes for life after death. 'There are no particles or fields that could store [such information] and take it away.'[78] Since there's no such particle, force, or field of particles within the Core Theory, such realities simple cannot exist – thank you very much.

There is, of course, a plethora of evidence . . . all discounted by the scientific world . . . that lends support for each of the subjects mentioned above – but I will include only one story regarding an extra-sensory experience having to do with death from another impeccable source – Jane Goodall once again. When Jane's second husband died in Germany, two children – one in England and the other in Tasmania – dreamed about it the night he died. The first, her son Grub, had the same dream three times that night . . . in the dream, Jane's sister Olly came to his school to tell him that Derek had died. Each time, the dream woke him, the last time upsetting him so much that he couldn't go back to sleep.

> 'In the morning Olly arrived at the school. Vanne [Jane's mother] was in Germany with me, having arrived the day before after an urgent feeling that she needed to see Derek. Olly took Grub outside into the garden and told him she had some sad news. 'I know,' he said. 'Derek is dead, isn't he.' Olly was stunned – until he told his dream.'[79]

The other child who dreamed of Derek that night was a young girl with down syndrome named Lulu. She was the child of close friends, and Lulu loved Derek and he her. The night Derek died, she woke up and ran to her nanny, who was sleeping:

> 'Mary,' she said, urgently. 'Please wake up. That man has come, and he likes me. He is smiling.' Mary, half roused, told Lulu she had been dreaming, and to go back to bed. But Lulu persisted. 'Please come, Mary. I want to show you. He is smiling.' In the end Mary sat up, resigned.
>
> 'Lulu, tell me who you mean. Who is this man who is smiling at you?'
>
> 'I don't remember his name,' said Lulu. 'but he comes with Jane, and he walks with a stick [Derek used a cane]. And he likes me. He really likes me.'
>
> Two children, in two parts of the world. Two children whom Derek had loved. It is so easy in a skeptical, reductionist scientific world to explain away these sorts of things as coincidental dreams, hallucinations, or psychological reactions triggered by the onset of pain, stress, or loss. But I have never been able to discount such experiences so easily – there have been too many events in my life, and in the lives of my friends, which have defied any kind of scientific explanation. Science does not have appropriate tools for the dissection of the spirit.'

How true. Core Theory says all this is not possible, and consequently all this evidence and experience is a lot of hooey to those who've turned the Core Theory into *belief* - but there's so much more to life than the best of science can offer . . . and consciousness is at the *core* of this issue. In *The Big Picture*, Carroll at least tips his hat to the idea:

> 'But the universe, and the laws of physics, aren't embedded in any bigger context, *as far as we know*. They might be – we should be open-minded about the possibility of something outside of our physical universe, whether it's a nonphysical reality or something more mundane, like an ensemble of universes that make up a multiverse [discussed later]' . . . and 'finally, there is the manifest loophole that describing the world in terms of physics alone might not be good enough. There might be more to reality than the physical world.'

Yes, Sean, you *should* be open-minded towards something outside of our physical universe. Perhaps if this information enters your world, you'll reconsider.

Quantum reality is based upon the idea that, at the most basic particle/wave reality, the world exists as as *superposition*[80] . . . 'in all possible states simultaneously' - that is, random and unpredictable. But once 'observation' occurs, *however that may be*, 'position' clicks into place and the physical world is then determined by the established laws of physics, more and more so as it evolves into complexity. If consciousness is fundamental . . . is, in fact, the context within which all of manifestation occurs . . . *and is evolving* . . . than whatever primordial form consciousness may have taken in the distant past, relative to what we consider it to be today, must have determined its direction, its future.

This eons long evolutionary process has figured out . . . during a vastness of 'time' that is incomprehensible to us . . . how to create the electrons and protons and neutrons and all the rest that make up matter, as well as the fields within which matter and energy operate – the entire physical world that science is so adept at discovering. Science believes that they're fixed . . . and always have been.

Once this process reaches the organic world, it is called 'natural selection': 'The mechanisms of evolution - like natural selection and genetic drift - work with the *random* variation generated by mutation,'[81] and: 'The genetic variation on which natural selection acts may occur randomly, but natural selection itself is not random at all.'[82] That is, while the core process generated from the quantum level is random, this evolves into a process that is decidedly *not* random.

> 'What [natural] selection operates on is more or less random: new genetic traits arise though numerous mechanisms (recombination, mutation, transposition, fusion, etc., as well as truly epochal changes like the incorporation of mitochondria into the first eukaryotic cells). These are 'random' in the sense that they're hard to predict.
>
> The combination of the two acts a bit like a ratchet: new genetic combinations are presented in a way which is essentially random, but only the fit combinations survive. The net result is that organisms tend to get more fit for their environment, until that environment changes. That also tends to produce increasing complexity over time, since everything started very simple and there was no place to go but up.'[83]

What could possibly pull direction – evolution - from randomness? Identical to the process that establishes *position* from the *random superposition quantum core* of the physical world, the evolutionary process is, I believe, pulled into this reality by overriding *purpose* to continue the process, governed now by the laws of physics – which also have been *learned*. This 'purpose' is, in fact, the innate desire to awaken, purpose-driven by the evolution of consciousness. The orthodox scientific view is that consciousness *emerges* unexpectedly (that is, could not be predicted by the present state of the Universe at the time . . . randomly) from the evolutionary process . . . and is *not* fundamental.

> 'Emergent evolution is the hypothesis that, in the course of evolution, some entirely new properties, such as mind and consciousness, appear at certain critical points, usually because of an unpredictable rearrangement of the already existing entities.'[84]

Primordial, relatively *unconscious* consciousness – seen in the worlds of particles, atoms, and the inert molecular expressions - has evolved to become the awareness of surroundings, exhibited by plants and animals, to self-awareness - the state of being aware that we are aware . . . the state of consciousness witnessed in humans and beyond. From the philosopher David Chalmers:

> 'Even a photon has some degree of consciousness. The idea is not that photons are intelligent or thinking . . . but the thought is maybe photons might have some element of raw, subjective feeling, some primitive precursor to consciousness.'

If consciousness is fundamental . . . in fact, is the context for all of manifestation . . . and is evolving, it makes total sense that the 'consciousness' we witness in our human reality would have 'primitive precursors.' Evolution plods along as these precursors *learn* how to better provide a platform for consciousness to grow and proceed. There is ultimate purpose in all this. Instead of the scientific maxim based upon the arrow of time that 'our progress through time is pushed from behind, not pulled from ahead,' it is, in fact, this inherent purpose that pulls us into our future.

Through billions of years (and who knows what kind and how many other universal expressions in the infinite past) of primitive evolution with no *care* as to how it's done – just that it moves forward – we see the beginning of care in the phenomenon of *life*. As life evolves from the most basic organisms, through more and more complex expressions, it is still solely survival-oriented. Then, in the higher life forms, this survival mechanism begins to express itself as *care* . . . or is it all simply selfish genes? I think not. Finally, as far as the world we know is concerned, this evolution of consciousness blossoms into the ethics and morality of human existence . . . and ultimately into the compassion of the Buddha. This certainly has not happened by accident . . . but through purposeful action.

Later in these pages (*see Who Are We?, page 136*), an example of this Buddha-level compassion is offered. Here is a more current example:

'Acts of self-sacrifice in the hell of the death camps were frequent. There was a moving incident that took place at Auschwitz when a Pole, facing a death sentence, sobbed and begged that his life might be spared so that he could stay with his two children. At this moment, the great priest Saint Maximilian Kolbe stepped forward and offered his life instead. After surviving two weeks in the starvation bunker, Kolbe was then murdered by the Nazis but the story lived on, serving as an inspiration to surviving prisoners: a beacon of hope and love had been lit in the dark confines of the concentration camp.'[85]

The story of the Buddha, told later, is very similar, and focuses on the idea of *bodhisattva* . . . 'a being that compassionately refrains from entering nirvana - a transcendent state in which there is neither suffering, desire, nor sense of self, and the subject is released from the effects of karma and the cycle of death and rebirth'[86] - in order to lift others out from suffering. The term represents the ultimate level of compassion, where the individual consciousness - free from the need to incarnate and free from the realm of suffering – chooses to return to the wheel of evolution to help those still caught in the web. This appears to be exactly what we are witnessing: the Stillpoint information embedded in the dimensions of the Earth, Moon and Sun is a compassionate message from an evolved, enlightened, bodhisattvic higher consciousness - a message intended to help humanity bridge the abyss before us.

This section has been an attempt to explain materialistic science's role in inhibiting us from claiming our true role as stewards of this miraculous planet that has been so carefully designed and provided for us. Sometimes poetry can cut to the heart of an issue when many words can't, and I offer the following poem by one of the most compassionate human beings I've ever run across, once again, Jane Goodall. I was stunned when I first read this. She sums up everything I've been trying to say in a few short words that speak from and to the heart. In these short verses she accurately represents science's conflict with Love, introduces the concept of bodhisattva in the most beautiful of ways, and alludes to the 'meaning of the moon's existence' with a prescience I don't believe even she was aware. Incredibly, for me, she even expresses compassion for those that stand on the Bridge – the bodhisattvic awareness responsible for our existence that she calls Saints . . . the Buddha with a tear running down his cheek:

ONLY THEY CAN WHISPER SONGS OF HOPE

The world has need of them, those who stand upon the Bridge,
Who know the pain in the singing of a bird

And the beauty beyond a flower dying:
Who have heard the crystal harmony
Within the silence of a snow-peaked mountain -
For who but they can bring life's meaning
To the living dead?

Oh, the world needs those standing on the Bridge,
For they know how Eternity reaches to earth
In the wind that brings music to the leaves
Of the forest: in the drops of rain that caress
The sleeping life of the desert: In the sunbeams
Of the first spring day in an alpine meadow.
Only they can blow the dust from the seeing eyes
Of those who are blind.

Yet pity them! Those who stand on the Bridge,
For they, having known utter Peace,
Are moved by an ancient compassion
To reach back to those who cry out
From a world which has lost its meaning:
A world where the atom – the clay of the Sculptor -
Is torn apart, in the name of science,
For the destruction of Love.
And so they stand there on the Bridge
Torn by the anguish of free will:
Yearning with unshed tears
To go back – to return
To the starlight of their beginnings
To the utter peace
Of the unfleshed spirit.
Yet only they can whisper songs of hope
To those who struggle, helpless, towards light.

Oh, let them not desert us, those on the Bridge,
Those who have known Love in the freedom
Of the night sky and know the meaning
Of the moon's existence beyond
Man's fumbling footsteps into space.
For they know the Eternal Power
That encompasses life's beginnings
And gathers up its endings,

And lays them, like Joseph's coat,
On the never changing, always moving canvas
That stretches beyond the Universe's
And is contained in the eye
Of a little frog.

Between the dogmas of religion and science there is a middle way. Threads of pure truth have always been a part of the esoteric heart of the paths of all who genuinely seek a true answer to the question of why we are here. And now the preponderance of scientific evidence suggests, as the following will make clear, that it is a scientific metaphysical stretch that our reality could have happened by coincidence or accident or, as orthodox science would put it, simply by random acts determined by chance and necessity, or natural selection. Historically, the other alternative is that it happened by design and this, of course, has always implied a Designer – always with a capital 'D.'

The old Creator-God paradigm, at least the one commonly accepted by so many billions of people over the years, is *not* what is being suggested here.

Rather . . . and this of course is where words begin to fail . . . what *evidence* suggests is an Awareness, a Consciousness, that is so unimaginable to the current human mind that it is invisible in its wholeness. For thousands and thousands of years our ancestors . . . like the old story of the blind man describing an elephant while only touching one of its legs, tail or trunk . . . have explained existence in terms limited by half truths and a mind tempered by fear. The Creator-God of that paradigm simply is and has always been, fully formed, omniscient and omnipotent – the Architect of the Universe. The critically significant difference, as so much new evidence suggests, is that this Consciousness, while it has *always existed*, has not always been fixed or held the same awareness, and is itself in a continual process of evolution. *As above, so below* . . . just like us – because that is who we are. In the quote below by Max Planck he acknowledges *consciousness* as the force behind all form and movement in the Universe – the matrix of all that is:

> 'As a man who has devoted his whole life to the most clear headed science, to the study of matter, I can tell you as a result of my research about atoms this much: There is no matter as such. All matter originates and exists only by virtue of a force which brings the particle of an atom to vibration and holds this most minute solar system of the atom together. We must assume behind this force the existence of a conscious and intelligent mind. This mind is the matrix of all matter.'

83

What he does not address is that all this has been and is a work in progress . . . and that as this 'mind' evolves, it learns how to communicate and share the wisdom earned for further evolution. I believe that he is correct when he says that this 'mind is the matrix of all matter.' It is, in fact the matrix of everything . . . and this background matrix could well be the Zero Point Field already mentioned - what Ervin Laszlo refers to as a *universal information field* – or Akashic Record. It is the purpose of the following to demonstrate - through the discoveries of the rational, scientific mind - that there not only seems to be a Primal Intention, but that also within this Intention – at this stage of its progression - are designed *precise* macro and micro *cosmic markers along the way to guide the process towards its goal.*

The innate, essentially unimaginable quality of the infinite is present everywhere. While the word 'goal' implies a point in time, an ending, the scientific community, as well as the human mind in general, has long struggled with this idea, as well as the idea of a beginning. The deeper the inquiry into the heart of the Mystery, the closer one comes to an understanding that it is in fact beginningless . . . and endless - and ultimately this understanding is beyond the ken of the rational mind.

The following recounting of science's attempt to explain how we got here will attempt to make clear the exact horizon where science stops, and peers into the Mystery . . . and in doing so, describes what it sees on the other side of this threshold in terms inherent to the deepest spiritual wisdom. More importantly, it will make clear that what is glimpsed beyond the limits of science is the same image found purposefully encoded throughout the physical world as *communications*, that it has been in front of us forever without our recognizing it . . . *and* that the method of the expression of this embedded code (and the inconceivable statistical improbability of its expression over and over pointing to its significance) leads us back to the source . . . to the state of Oneness that is both the beginning and the end.

Also, to my mind, the empirical evidence presented affirms that there *is* a purpose – that a Primal Intention has initiated and drives all of this. Certainly we are not here through accident. But, in truth, whether or not it is purposeful or by chance is less important than the existence of a third reality - a possibility beyond its happening by chance *or* by design will become all too evident – just as Planck alluded above . . . simply that there IS a presence that has always existed, eternal, outside of time; that this presence is growing in awareness and that there now exists empirical proof that this awareness has become fully enlightened, totally aware, able to manifest and create space and time at will. It is this reality that is central to what we will discover as this story unfolds.

. . . that there *is* an Infinite Presence. That it simply IS and that it *grows* in awareness. That this is essentially verifiable on the most profound yet rational level, mirroring perfectly that which we already know in the deepest parts of our healthy being . . . and that now is the time in our particular history when we are finally able and ready to assemble the pieces of this timeless puzzle and make this journey, this opening, this return, in a dynamically new way.

⊕ THE BEGINNING . . . THE ORIGINAL INTENTION:

As already mentioned, over fifty years ago I read the following words by Meher Baba,[87] and through their poetry they come as close to our mind's understanding of infinity and beginning/ending as I think is now possible with words. I repeat them here as they serve as a very non-scientific beginning to a factual discussion . . . leading back to a conclusion that is essentially non-scientific, and outside the rational mind:

> 'Before the beginning of all beginnings, the infinite ocean of God was completely self-forgetful. The utter and unrelieved oblivion of the self-forgetful, Infinite Ocean of God in the beyond-beyond state was broken in order that God should consciously know his own fullness of divinity. It was for this sole purpose that consciousness proceeded to evolve. Consciousness itself was latent in the beyond-beyond state of God. Also latent in this same beyond-beyond state of God, was the original whim (lahar) to become conscious. It was this original whim that brought latent consciousness into manifestation (form) for the first time. Slowly and tediously consciousness approaches its apex in the human form,[88] which is the goal of the evolutionary process, and thereby an individual mind gradually differentiates itself from the sea of oblivion . . .'

The essence of this Intention . . . this *whim* . . . was to become aware, awake, fully conscious . . . and this Intention is *our* intention because *this – all* of this - is who we are and are becoming. I believe that this cosmic spark that got it all going, this whim, exploded though the Stillpoint – the interface between the unrealized and realized God. Poetically speaking, this cosmic spark initiated the eons long evolution of consciousness. For whatever reason, it is a process experienced in time, a process evolving us into the awakened state . . . and this Universe as we know it is the perfect expression of that growing consciousness, mirroring in its every precision and in its most grand and tiniest dimensions its . . . *our* . . . original

Intention. Also . . . as already stated . . . embedded within it are clues *from* us *to* us, signposts leading back to the heart of the truth – the eternal present infinite moment living in *the stillness at the center* – always there for us when we are ready to open to it, when we have grown or evolved to a being of readiness.

This *still center* is in fact an opening into the great Mystery, what has now evolved from Meher Baba's 'beyond-beyond state' to fully awakened, aware consciousness. It is a doorway accessible by fully enlightened awareness to this eternal present moment, outside of and finally free of the rational mind bound to physical survival. The 'physical' properties of this portal have finally been made clear, and we are now ready to open to it - and there very well may be a means that we can. This is the intention.

⊕ WHAT ARE THE CHANCES THAT THIS UNIVERSE COULD EXIST AS IT IS, PERMITTING LIFE AND CONSCIOUSNESS AS WE KNOW IT?

Mentioned earlier, Sir Martin Rees, Britain's Astronomer Royal, published a book in 1999 called *Just Six Numbers*. In the book Rees argues that six numbers underlie the fundamental physical properties of the Universe, and that each is the *precise* value needed to permit life . . . and that, in fact, if any *one* of the numbers were different 'even to the tiniest degree, there would be no stars, no complex elements, no life.' Ervin Laszlo suggests that there may be as many as 36 of these constants. From physicist Paul Davies:

> 'The really amazing thing is not that life on Earth is balanced on a knife-edge, but that the entire universe is balanced on a knife-edge, and would be total chaos if any of the natural 'constants' were off even slightly. You see, even if you dismiss man as a chance happening, the fact remains that the universe seems unreasonably suited to the existence of life - almost contrived - you might say a 'put-up job.'[89]

Rees has stated that 'The physical laws were themselves 'laid down' in the Big Bang,' but admitted that 'The mechanisms that might 'imprint' the basic laws and constants in a new universe are obviously far beyond anything we understand.'[90] Indeed. Rees' six numbers are found in the Universe's smallest and largest structures. Two relate to basic forces, two determine the size and large-scale texture of the Universe, and two fix the properties of space itself. In order that this section is more easily readable, a more complete explanation of each of these phenomena is contained in Appendix C. They are:

1. **E** (Epsilon) - The strength of the force that binds atomic nuclei together and determines how all atoms on Earth are made.
2. **N** - The number that measures the strength of the forces that hold atoms together   divided   by   the   force   of   gravity between them.
3. **Ω** (Omega) - The number that measures the density of material in the Universe - including galaxies, diffuse gas, and dark matter.
4. **Λ** (Lambda) – The   number that describes the strength of a previously unsuspected force, a kind of cosmic anti-gravity that controls the expansion of the Universe. It is sometimes refereed to and the cosmological constant.
5. **Q** - The number representing the amplitude of complex irregularities or ripples in the expanding Universe that seed the growth of such structures as planets and galaxies.
6. **D** - The number of spatial dimensions in our Universe – that is, three.

Each of these critical numbers signify a precision upon which, if any *one* of them were not *precisely* as it is, the Universe as we know it would not exist . . . *we* do not exist. The *precision* of each one of these numbers also means that the *chance* of our being here to witness all of this is infinitesimally small.

More so, if each of the six numbers Rees has identified were dependent upon the others . . . that is, if the existence of any one of the numbers was inherently related to any of the others . . . the chances of this Universe being just as it is would *still* be infinitesimally small. *But this is not the case.* 'At the moment, however,' says Rees, 'we cannot predict any of them from the value of the others' . . . that is, each number *compounds* the unlikelihood of each of the other numbers. If a one in ten chance is given to each of these constants happening by chance (an absurdly low number), than the chance of all of them happening by chance is obtained by multiplying $1/10^{th}$ x $1/10^{th}$ x $1/10^{th}$ x $1/10^{th}$ x $1/10^{th}$ x $1/10^{th}$ = *one chance in 10 billion* – and there may be thirty six of them.

Another way of saying this, well within the scientific mind, is that if a from-nothing, briefly existing molecule is absurdly unlikely, a from-nothing, nearly 14-billion-year-old observable Universe based on very precise, particular limits (the six numbers, etc.) is *vastly* less likely.

This really means that this 'unlikelihood' is compounded *exponentially*, meaning that the chances of this Universe happening just like this *accidentally* is comparable to the relationship of a pinpoint in the midst of the infinity that, scientifically speaking, the Universe appears to be. A *pinpoint* of likelihood.

Again . . . please think about that image for a moment.

The book/film *The Privileged Planet*, mentioned earlier, addresses Rees's six numbers (as well as other such numbers) and much other scientific evidence pointing towards the empirical 'impossibility' of not only our Universe, but our own solar system within it. It's a scientific approach to justify a Designer . . . the old Creator-God paradigm. This is not at all what the information presented here is about . . . but the empirical facts themselves, discussed in the book/film, are powerful and irrefutable.

As mentioned earlier, there are some twenty factors that have been discovered to be essential for the existence of complex, carbon based life forms – *us*. Apparently, all of these conditions have to exist *simultaneously* for complex life to exist . . . and, like Rees's six numbers, *they are not directly related to each other*. See Appendix C for the complete list, but here is a simplified one. They include:

- The existence of liquid water.
- The distance of the host planet from the star it orbits (the Goldilocks zone).
- The kind of star necessary – a so-called main sequence G2 dwarf star.
- Protection from asteroids and other projectiles from outer space by giant gas planets (Jupiter and Saturn).
- The planet's location within a circumstellar habitable zone.
- That the planet *has a nearly circular orbit*.
- That the planet has an oxygen/nitrogen rich atmosphere.
- That the planet has the correct mass.
- That the planet is orbited by a large moon; that it has a magnetic field.
- That there are plate tectonics (for biodiversity).
- That the planet has the correct ratio of water and continents also for the biodiversity necessary.
- That it is a terrestrial planet (has rocky land masses).
- That it has a moderate rate of rotation (in our case, due entirely to the existence of our large Moon).
- And finally, that is has the right location within the galaxy (various factors).[91]

Dr. William Lane Craig, a philosopher of science, from the film:

'If you deny the process of cosmic design you're basically left
with two alternatives: either this fine tuning is a result of physical

necessity, that is to say, there is some unknown theory to explain why these constants and quantifiers have to have the values they do, or else you just have to say this just occurred by chance alone. That is, the result of sheer accident. Well, that first theory doesn't seem too plausible because *there just isn't any theory that would explain why all these constants and quantities have the values they do.* They appear to be *just arbitrarily put in* at the creation as initial conditions.

With respect to the second alternative . . . chance . . . most theorists recognized that the odds against the Universe being life permitting are just so fantastic that chance simply cannot be faced unless you say that our Universe does not represent the only role of the dice. And so what many theorists have been driven to is multiplying our probabilistic resources by saying maybe our Universe isn't the only role of the dice. Maybe there are out there parallel, unseen, undetectable universes, and that our Universe is just one in this cosmic crapshoot in which there is an infinite number of other worlds in which the constants and quantities vary randomly and so by chance alone somewhere in this infinite ensemble of universes, our universe would have appeared by chance alone, and here we are, the lucky beneficiaries and recipients of this chance hypotheses. So in order to rescue the chance hypothesis, physicists have been driven beyond physics to metaphysics, to this extraordinary hypothesis of a world's ensemble of an infinite number of randomly ordered worlds in order to explain away this appearance of design.'

While Craig says that these constants 'appear to be just arbitrarily *put in* at the creation as initial conditions,' no mention is made as to 'who or what' may have put them there . . . yet the implication is clearly that this was through the 'process of cosmic design' . . . that is, a cosmic Designer – the old Creator-God. This is essentially the best that religion or science can do . . . it's one of three choices; 1. The Creator-God; 2. An unknown and unlikely scientific theory explaining away the precision of the constants; or 3. An 'infinite number of randomly ordered worlds,' ours being the one we happen to witness. So: 1. The familiar 'Creator-God' has always been a handy way to explain away this miraculous grandeur we live within – but it is based upon thousands of years of superstition and belief and comes nowhere near explaining away the indifference witnessed everywhere in the observed Universe. 2. It seems evident that there is no 'law' that could possibly explain all of the 'coincidences' required for higher life forms to exist. 3. Infinite possibilities will be discussed below.

But there is a forth possibility regarding Martin Rees's six numbers, upon whose 'impossible' precision our Universe depends. Logic suggests that if in fact our own Universe is driven by the purpose to awaken, originating through a singular point of creation, *all* universes – and it seems by extension that there must have been others - evolve by the same principle – *the six numbers a reflection of what evolving consciousness learned as it developed.*

As mentioned in the pages that follow, the theoretical biologist Rupert Sheldrake hypothesizes that the 'laws' of nature are in fact *habits* . . . that is, the laws as we experience them were *evolved* through an interaction of awareness and experience. In the spirit of 'as above, so below,' it is only logical that evolving consciousness learned how to create a universe that would allow complex, self-aware, life forms to exist – that is, allowing the evolution to continue.

It is also worth noting here that Stephen Hawking, representing orthodox science, stated in the 2011 film *Did God Create the Universe?*, along with Martin Rees above, that he believed that the laws of nature are fixed and always have been. Yet: *'The mechanisms that might 'imprint' the basic laws and constants in a new universe are obviously far beyond anything we understand.'* Mechanisms. This is the mindset that science remains muddled in. And, because science's *only* stab at plausibility for this Universe happening by chance is infinite universes, it would be useful to investigate this idea of infinite possibilities a bit.

⊕ Infinite Universes . . . Infinite Possibility:

Faced with such overwhelming improbability, cosmologists have scrambled to offer an explanation. The simplest is the so-called brute fact argument mentioned above - the weight of an endless supply of possibilities finally offering up the world as we know it. 'A person can just say: 'That's the way the numbers are. If they were not that way, we would not be here to wonder about it. Many scientists are satisfied with that,'[92] says Rees. Typical of this kind of thinking is Theodore Drange, a professor of philosophy at the University of West Virginia, who claims it is nonsensical to get worked up about the idea that our life-friendly Universe is 'one of a kind.' As Drange puts it:

> 'Whatever combination of physical constants may exist, it would be one of a kind.'[93]

Yah . . . well . . .

Rees also objects, drawing from an analogy given by philosopher John Leslie. 'Suppose you are in front of a firing squad, and they all miss. You could say, 'Well, if they hadn't all missed, I wouldn't be here to worry about it.' But it is still something surprising, something that can't be easily explained. I think there is something there that needs explaining.'[94]

Well . . . yah . . .

The most compelling argument by far that chance is the explanation for the almost impossible likelihood of our Universe existing just so, comes from the latest theories of the origin of the Universe from physicists such as Andrei Linde, originally of the pre-Glasnost Soviet Union and presently doing research at Stanford University.

Linde's work leads to the possibility of an infinitely self-reproducing Universe . . . infinite universes endlessly producing infinite possibilities. One can see from this that the possibility exists, out of chance alone, for our own set of realities to have 'just happened.' It would be useful to summarize how Linde got there, starting with the Big Bang theory – the long accepted theory that the Universe as we know it exploded from an infinitesimally small *point of singularity* (morphing into physicality at the radius described by the Planck Constant) 14 or so billion years ago and, according to today's most accepted theory,[95] is presently still expanding.[96]

In 1929 Edwin Hubble discovered that distant galaxies were emitting infra-red light, meaning that these galaxies were speeding off away from us into the distance, and recently science has discovered that the farther away they were, the faster they were moving[97] - inferring they all exploded from a point of *gravitational* singularity - defined as: *a point of infinite density, when all distances between objects are zero* – and are still in the process of exponential expansion before gravity eventually catches up, slows it down, and pulls it all back in. Again, this phenomenon is understood through the lens of the Big Bang Theory and may very well be understood in the future as an aspect of the torus model – an infinitely huge movement expanding and contracting simultaneously and continuously, eternally spiraling outwards from the Stillpoint at its center, learning and growing, and then making the return to and through its origin, to 'begin,' yet again, another spiral of evolution.

There were two main weaknesses in the original theory. The first was the horizon problem - the puzzle that the Universe looks the same on opposite sides of the sky (opposite horizons) even though there hasn't been time since the Big Bang for light to travel across the Universe and back.

The second had to do with the fact that anything so infinitely dense as a point one-billionth the size of a proton containing all the mass in the known universe and the gravity that such a mass would generate, would by definition collapse upon itself immediately after the explosion called the Big Bang. But it didn't collapse . . . and here we are.

This problem all had to do with Ω (Omega), the third of Rees' numbers, which measures the density of material in the universe - including galaxies, diffuse gas, and dark matter.[98] The number reveals the relative importance of gravity in an expanding universe. If gravity were too strong, the universe would have collapsed long before life could have evolved . . . too weak, no galaxies, no stars, no planets.

This is also called the 'flatness' problem. The puzzle was that the observed space-time of the Universe appears to be very nearly 'flat,' which means that the Universe sits just on the dividing line between eternal expansion and eventual re-collapse . . . right between a 'closed' and an 'open' geometry, with matter, velocity, and gravity all in balance. In a closed universe, space-time curves back on itself, such that light beams that start out parallel will actually meet. In an open universe the beams will diverge. The value of Ω describes the ratio between the average density of matter in space and what that density would need to be to make the universe perfectly 'flat.' If Omega equals 1, the universe is flat. Observations of Omega are within .02% of one . . . in physic's terms – **One.**

According to classic Big Bang theory, if the Universe starts out with the omega parameter less than one, Ω gets smaller as the Universe ages, while if it starts out bigger than one, Ω gets bigger as the Universe ages. The fact that Ω is between 0.1 and 1 today means that *in the first second of the Big Bang it was precisely 1 to within 1 part in 10^{60}*. This makes the value of the density parameter in the Beginning one of the most precisely determined numbers in all of science, and the natural inference is that the value is, and always has been . . . **One.** Random . . . or *learned*?

This means that matter, velocity, and gravity were all in *balance* at the moment of the explosion. I've gone into some detail here to draw our attention to the question, *how did this precision happen?* How did *any* of the critical constants of physics come to be what they are – permitting evolution to continue such that I am now typing these words? Random accident? I am as convinced as I possibly could be that evolving purpose *learned* what was required and what was not, what worked and what did not, creating the Laws as we know them as it went along - the eons-long push of the evolutionary process, involving seemingly endless attempts at universal success. *This* Universe, this attempt, is extremely fine tuned because it has *evolved* to be this way.

Orthodox science believes that the laws upon which the Universe is based have always been just as they are . . . fixed. But this is an assumption. Why couldn't something so apparently fixed have come into being through morphogenetic causation . . . except in this case it took unimaginable eons to do it? Human beings are a blink of a blink of an eye relative to the history of *this* Universe. Could not electrons and other supposedly 'fixed' building blocks of nature also be a blink of an eye in the long process of evolution? For me, this is the most logical answer to why matter, velocity, and gravity are all in balance so close to the 'explosion' - expanding consciousness *learned* how to accomplish this *before* our Universe came into being.

But there is a problem with the *density* of this balance in the singularity just before it exploded . . . the density of the singularity at the Planck density is the highest density describable with current physics - over 10^{93} g/cm^3. How could anything that dense ever expand? It would have an enormously strong gravitational field, turning it into a black hole and snuffing it out of existence (back into the singularity) as soon as it was born. How this number could start out so finely tuned is the question that intrigued a physicist working at the Stanford Linear Accelerator, Alan Guth, in 1979.

In another example of how expanding consciousness may have figured out how to solve this initial density problem, Guth realized that Omega did not have to be preposterously fine-tuned from the start. All of these problems would be resolved if *something* gave the Universe a violent outward push *when it was still about a Planck length (10^{-35}m) in size* – about one-billionth the size of a proton. An *exponentially expanding* early universe, which he would come to call the inflationary universe, *would drive Omega toward one*, not away from it, making a flat universe . . . unity . . . *inevitable*. Here we have science saying that *something* gave the Universe a violent outward push. What could this 'something' be? While quantum physics allows the entire Universe to appear, in this *almost infinitely* compact form . . . *out of nothing* . . . inflation can prevent the immediate collapse from happening. 'Inflation' is another word for 'violent push' . . . but how did it 'know' to do this? Random accident?

Forgive me here, but until science defines this incomprehensible 'push' in its own terms (which the world 'inflation' simply labels), it seems to me the *perfect* analogy to the slumbering God of mythology and religion, expressed now as our infinite, unimaginably powerful Universe, beginning to wake up. Look out!

During this explosive period just after birth, what is now the observable Universe exploded from a region one-billionth the size of a proton to about the 'size of a grapefruit' in an infinitesimal fraction of a second (between 10^{-37} second and 10^{-34} second). The universe expanded at this exponential rate before beginning to settle down to the more sedate expansion originally described by the Big Bang theory.[99]

Such a small region of space would be too tiny, initially, to contain irregularities, so it would start off homogeneous and isotropic. There would have been plenty of *time* for signals traveling at the speed of light to have crisscrossed the ridiculously tiny volume, so there is no horizon problem - both sides of the embryonic universe are '*aware*' of each other. Space-time is smoothed out by the expansion . . . making the Universe 'flat' . . . resolving the horizon problem (most clearly indicated by the uniformity of the background radiation) by taking regions of space that were once close enough to have gotten to know each other well and spreading them far apart, on opposite sides of the visible Universe.

The beauty of this theory is that it achieved this success not only because it resolves many puzzles about the nature of the Universe, but because it did so using the understanding of quantum theory developed by particle physicists (the study of the very small) completely independently of any cosmological studies (the study of the very large). These theories of the particle world had been developed with no thought that they might be applied in cosmology (they were in no sense 'designed' to tackle all the problems they turned out to solve), and their success in this area suggested to many people that they must be telling us something of fundamental importance about the Universe - *as above, so below.*

This 'as above, so below' process may also define a *fractal* Universe. A fractal being a geometric shape that is self-similar and has fractional (fractal) dimension . . . having a *whole number proportion* – precisely representative of the original *quantum*, the *wholeness* of *action*, described by the Planck Constant, discussed below.

The reason why the quantum theories created such a sensation when applied to cosmology is that they predict the existence of exactly the right kind of mechanisms needed. These mechanisms are called *scalar* fields, and they are associated with the splitting apart of the original grand unified force[100] into the fundamental forces we know today, as the Universe began to expand and cool. Gravity itself would have split off at the Planck time, 10^{-43} of a second, and the strong nuclear force by about 10^{-35} of a second. Within about 10^{-32} of a second, the scalar fields would have doubled the size of the Universe at least once every 10^{-34} of a second.

At that point, the scalar field has done its work of kick-starting the Universe, and is settling down, giving up its energy and leaving a hot fireball expanding so rapidly that even though gravity can now begin to do its work of pulling everything back into its beginning, it will take hundreds of billions of years to first halt the expansion and then reverse it. The torus model provides the same mechanism . . . expansion as the spiraling energy moves away from the Stillpoint and compression as it reaches its zenith and begins its spiraling return.

I suppose that if there were an in-breath and out-breath of God, this would be it.

At the very center of this cosmic spiraling exploding dance of evolution is the non-existent, dimensionless, Stillpoint at the center of the singularity as science defines it – that which was latent in the mind of God and preceded the singularity.

⊕ SCIENCE'S LAST GASP TO EXPLAIN AWAY ANY INITIATING CAUSE:

But . . . back to the scientific obsession to explain God away by no-cause multiple universes. The next step forward came with the realization that there need not be anything unique about the Planck-sized region of space-time that expanded to become our Universe. If that was part of some larger region of space-time in which all kinds of scalar fields were at work, than only the regions in which those fields produced inflation could lead to the emergence of a large universe like our own.

Linde called this 'chaotic inflation,' because the scalar fields can have any value at different places in the early super-universe. The idea of chaotic inflation led to what is (so far) the ultimate development of the inflationary scenario.

Chaotic inflation suggests that our Universe grew out of a quantum fluctuation in some *pre-existing region of space-time*, and that exactly equivalent processes can create regions of inflation within our own Universe. In effect, new universes bud off from our Universe, and our Universe may itself have developed from another universe, in an infinite process that had *no beginning and will have no end*. A variation on this theme suggests that the 'budding' process takes place through black holes, and that every time a black hole collapses into a singularity it 'bounces' out into another set of space-time dimensions, creating a new inflationary universe. This seems to accurately represent *some* of what I believe to be true – the beginningless/endless aspect of what we witness - but it's a convenient way for science to dismiss the need for a Creator, or an original *cause* – that is, the system has always been just as it is, recreating itself out of itself, one random mechanistic event after another.

In the quantum world, there are particles called 'virtual' particles . . . phenomena that are here in this moment and then gone and then here again, all randomly – *fluctuations* that *emerge* and *disappear* instantaneously. Where do they come from? Where do they 'go'? Rather than assuming that they are part of a 'quantum fluctuation in a pre-existing region of space-time,' the reason science suggests for universes in general, is it not more logical – if outside of scientific dogma – that this quantum reality is none other than the constant dialog between the implicate and

95

explicate worlds, the one instantaneously informing the other and visa versa? The physicist David Bohm put it this way:

> 'Rather than suggesting a continuous entity that moves 'thought' time and space, the image of ordered enfoldment-unfoldment allows for a view of the electron as a perpetually emerging explicate structure, temporarily unfolding from an ordered implicate background, and then rapidly enfolding back into this background, in an ongoing cycle. By extension, the whole of experience can be understood as a flow of appearances resulting from such a cycle of enfoldment and unfoldment.'[101]

So . . . we have arrived at the end of a long yet very simplified explanation of how science came to include the possibility of infinite universes . . . and infinite possibility. Rees:

> 'The idea is that a possibly infinite array of separate big bangs erupted from a primordial dense-matter state. If there are many universes (which are being continuously created) each governed by a differing set of numbers, there will be one where there is a particular set of numbers suitable to life. We are in that one.'[102]

The concept of infinite universes is the best that science can do to explain away the infinitesimally tiny chance that the precision of the six numbers sited by Rees, upon which the existence of *this* Universe depends, could have happened by chance. It is interesting to note that this is not the kind of falsifiable hypotheses that science is built upon. In fact, it approaches metaphysics in its philosophical rebuttal to the God question.

While concocting infinite universes to justify the precision of the constants that make the equations that this Universe is based upon work, science will have to offer more than a metaphysical hunch to explain other 'impossible' phenomena. That is, regardless of whether the multi-universe theory is true or not, it simply cannot account for the 'coincidental' anomalies that occur *within* this Universe. It turns out that there are equally 'impossible' phenomena within our Universe that cannot be explained away in this manner. Thus far, science's only answer to what follows is to ignore it . . . or call it 'coincidental.'

⊕

⊕ POINT OF SINGULARITY:

All of science's theories of the nature of the Universe are based upon a limit beyond which science cannot yet go – a threshold to the doorway of the Mystery itself. This is the Planck Constant, the eminently prototypical constant that describes in infinitesimally tiny physical *and* temporal terms, the time/size/density constant that quantum physics as we know it can grab onto and make sense of this Universe – and beyond which science knows *nothing*. Science, stopped at this threshold, peered through that doorway and what it saw – the 'singularity' – is as close as it can come to the Stillpoint . . . which is precisely what the Mesa Temple seeks to reach – and the doorway it seeks to open.

The work of Andrei Linde and others essentially made credible the idea that multi-universes could exist and that from this array of possibility even our own Universe, based as it is on a set of incredibly precise numbers, could exist within the realm of chance. But, as mentioned, this 'solution' approaches metaphysics and is not based upon experimental science.

But it has just been suggested that there was a limit upon which all of cosmological scientific theory is based – the Planck Constant - and beyond which it cannot go. I'd like now to address that limit, this threshold at which science stops and peers across . . . beyond which lies the Mystery.

Here's a very scientific, sometimes incomprehensible and rather beautiful summary of this primal constant, from Encyclopedia Britannica and Wikepedia (italics mine) Full description found in the endnotes:[103]

> 'The Planck Constant is a physical constant that is the *quantum of action* in quantum mechanics. It . . . [is] behavior associated with an *independent* unit – or *particle* - as opposed to an electromagnetic wave and was eventually given the term *photon*. Its relevance is now integral to the field of quantum mechanics, describing the relationship between energy and frequency, commonly known as the Planck relation. In physics, *action* is an attribute of the dynamics of a physical system . . . . *the action must be some multiple of a very small quantity* (later to be named the 'quantum of action' and now called Planck Constant). This inherent granularity is counter-intuitive in the everyday world . . . this is because the quanta of action are very, very small in comparison to everyday macroscopic human experience. Hence, the granularity of nature appears smooth to us. Thus, on the macroscopic scale, quantum mechanics and classical physics converge at the *classical limit*. The classical limit is the ability

of a physical theory to approximate classical mechanics. The *dimension* of The Planck Constant is the product of *energy* multiplied by *time*, a quantity called action. *The Planck Constant is often defined, therefore, as the elementary quantum of action.'*

The 'inherent granularity . . . appearing smooth to us' is reminiscent of Buckminster Fuller's earlier quote regarding the illusion of time:

> 'The whole of physical Universe experience is a consequence of our not seeing instantly, which introduces time. As a result of the gamut of relative recall time-lags, the physical is always the imperfect experience, but tantalizingly always ratio-equated with the innate eternal sense of perfection – thus the mind induces human consciousness of evolutionary participation to seek cosmic zero.'

What all this means to me is that the laws of physics only begin to work at this ultra-tiny threshold or original quantum – when o transforms into 1, where *nothing* becomes *something* – where *stillness* becomes *action*. This is the interface between the physical and spiritual worlds – and the limit that science approaches but does not have the imagination or tools to go beyond. This critical constant represents length, time and mass – the length aspect equaling the Schwarzschild Radius . . . defining the size of the tiny *sphere* – or *tetrahedron* that defines that sphere - beyond which physics cannot enter and the center of which is the world of the Stillpoint.[104] Science calls this tiny sphere the 'point of singularity.'

⊕ Light . . . and the Torus:

Regarding materialistic science's limitations, I've tried to make clear that all of science is based upon information often derived from mathematics which can only work this side of the limit of the Planck Constant (*see above, and Science, Belief . . . and a Middle Way, page 69*) – this *spacial* mathematics cannot go beyond this veil, into the world of the Stillpoint. This inherent weakness - that the mathematics that science is based upon is a *spacial* mathematics - was recognized by Aurthur Young.

> 'My purpose is to give a formal description of Life, including the universe of matter, the universe of thought (idea), and an analytical method that would enable all known truths to be stated, and other truths to be predicted. To do this, it is necessary to make a study of everything the human race does, thinks, wants, etc., both from the scientific, or outside, point of view and from the point of

view of one who participates, one who is a human being. It is also necessary to invent a *nonspatial* mathematics which will be capable of solving all problems of any kind . . . Its important feature is that it has nothing to do with space, or rather the sense of space, from which the old Euclidean, Newtonian logic evolves.'

Arthur Young, age 22, in 1927

Further, he realized that the 'descriptions' that his research would focus upon would include – in fact be based upon - the all-important process of time . . . hence, *process* theory. His theory begins with the first quanta . . . the photon of light . . . whose qualities include no mass, no dimension and timelessness – *outside* of space and time . . . essentially the 'manifested' aspect, the 1 aspect, of the Stillpoint/Light phenomenon . . . 0/1 . . . Stillpoint/Light. Light - the original spark or burst of energy from the infinity of the Stillpoint - descends into matter, losing energy and freedom (superposition) all the while as particles, then atoms, then molecules are created as the Universe cooled. This is the deterministic aspect of the world. Young felt that we needed to have a determined Universe to supply the platform for the upward ascent of growing free will and the ability to create. That is, one must first understand the laws upon which the Universe is built, and then, from that understanding, use those laws – in a marriage with imagination – to create. This is how he discovered how to make the helicopter fly.

This transition – from the deterministic world to that of free-will and creation - occurs during what Young termed 'the turn' – when inert matter figured out how to store energy, eventually creating the possibility of life . . . and ultimately of self-consciousness and enlightenment *(see Arthur Young's quote in Critical Mass, page 215)*. This is the moment in the long process of evolution where inert matter transforms into organic life, and entropy shifts to negative entropy, determinism gradually giving way to creative free will. Then, through the development of plants, then animals, and then humans, there is an ascent via the development of consciousness, returning to the freedom of Light . . . Enlightenment.

In this section, I would like to re-focus on the beginning, in particular on what was *fundamental* to this beginning. On the other side of the veil separating manifestation from the Great Mystery is the Zero aspect of the Stillpoint. Just on this side – still outside of space and time, with no mass, no dimension, and timeless, is the One aspect, the first quanta, or First Action, or . . . the photon – Light.

The Bible begins with Genesis 1:1:

'In the beginning God created the heavens and the earth. 2: The earth was formless and void, and darkness was over the surface of the deep, and the Spirit of God was moving over the surface of the waters. 3: Then God said, 'Let there be light'; and there was light.'

After that, everything was open to thousands of years of interpretation. Here's yet another. Whether the author intended it or not, the word 'God' – in the Western religious traditions – has often been interpreted to mean an entity-like, person-like, *being* . . . that has always existed in perfection, just so, dispensing the job of creation in regal fashion. I interpret the word 'God' in these passages to mean the *process of becoming*. I refer the reader once again to the Meher Baba quote *(see The Beginning . . . the Original Intention above, page 85)* and a quote by Adyashanti *(see The Temple in the Context of the Last 11,500 years:, page 128)*. In this context, I'm not sure I'm aware of a more profound statement from ancient wisdom than the opening lines of Genesis.

The purpose-driven 'whim' to awaken becomes 'aware' that in order to 'know the fullness of His own divinity,' the oblivious 'God' (the process of becoming) of the 'beyond-beyond' void needed to create 'other,' to become aware of what It was *not* . . . hence, the dualistic worlds of heaven and earth, or – the Limitless Mind of God (the Ayn of the Kabbalah), and the manifested world of the 10 Sephirot, 'the 10 attributes or emanations through which God reveals himself' . . . all of this manifesting through the portal of the Stillpoint - Kether, the Point of Creation.

Light.

The photon of Light . . . the first quanta, the first action, the original *monad* . . . holds a very special place in Arthur Young's *process theory* of the evolution of consciousness in the Universe. While Young has been an unequaled influence on my growing understanding of what truly is the 'big picture,' my focus had been elsewhere and I'd forgotten what I'd learned about his understanding of Light so many years ago in *The Reflexive Universe*. I don't know how that happened . . . but I give thanks to a talk by Dr's. Bob Whitehouse and Mike Buchele for re-introducing me not only to Young's profound perception regarding this subject . . . and also to their own work.

Any conversation regarding Light must include the torus . . . the way that energy (ultimately Light) *moves* in the Universe. As it turns out, there is something very special regarding both the uniqueness of the photon as well as the shape of the torus. Keep in mind that the Stillpoint geometry does not manifest . . . while it is the genesis of all that does. The *first* manifestation, as the Bible correctly tells us, is Light: The 'Spirit' transformed itself into 'Light.' Or, as stated later, regarding the I Ching symbol of light and dark merging into one another *(see Number, page 195)*: 'If the Stillpoint can be understood as the point where the Great Mystery enters

Creation . . . or more precisely, where the Great Mystery (the Kabbalistic *Ayn*, the Egyptian *Nun*, Meher Baba's '*beyond-beyond*') becomes manifest as light . . . the Zero/One phenomenon can be visualized as the I-Ching symbol, having a Zero aspect (darkness) and a One aspect (light)' – the two forming wholeness.

Since this is also the shift from *Stillness* to *Movement*, it seems profoundly logical that the geometric shift would be from the vector-equilibrium, or Flower of Life/Stillpoint geometry, to the primal geometry of movement: the torus. The following quote is found in Arthur Young's last writings . . . in a beautiful book published after his death entitled *Nested Time:*

'The neatest trick of modern physics was achieved by Eddington in his unification of relativity and quantum physics. According to Eddington, ordinary space is described as a sphere, $4/3\pi R^3$. When we include uncertainty we multiply this by 2π, a circle or an added circularity. This is equally true at the quantum or at the cosmic level, and is not dependent on size (size is given by the value of R). However, we must also include what Eddington called limitation of scale, a factor of ¾. I interpret this as meaning that we limit what could be called outgoingness and begin to deal with ingoingness. We then have $¾ \times 2\pi \times 4/3\pi R^3 = 2\pi^2 R^3$, which is the formula for the Einstein-Eddington hypersphere [the shape of the Universe]. It is also the formula for the torus with an infinitely small hole [the Stillpoint], which I have dealt with in *The Reflexive Universe.*

The torus is a vortex which has two directions of rotation, horizontal and vertical, *and is a likely candidate for the 'form' of a photon.*

It does so by equating the curvature of space-time with the curvature of uncertainty. Both the universe and the photon are going through a cycle – very short for the quanta of action exchanged by atoms and molecules, and very long for the universe. *Our human time span is about midway between.* The curvature of the universe is its closing in on itself, its expansion and contraction, its breathing, its inspiring and expiring. But everything goes through its cycle, the shorter encompassed by the larger, as the day is encompassed by the year. Einstein was unable to unify relativity and quantum theory because he was looking for a 'field.' A field is two or more dimensional, whereas the unification contributed by Eddington is *a more fundamental level – that of rotation, or zero dimensionality.*'

The Stillpoint is the infinite portal at the center of the torus. It is a whole composed of two aspects: Zero (the Vector Equilibrium/Stillpoint itself), and One (the toroidal nature of moving Light). This is the shift from Spirit/Stillness (the Stillpoint) to Movement/Manifestation (the torus) at the most primal level . . . 'Let there be Light.' So, Light is the first 'form,' and it's likely expression . . . as well as the shape of the Universe itself . . . is toroidal. Both Light itself (the photon) and its likely shape (the torus) are unique among particles and forms.[105]

Science considers the photon – the original quanta – to be simply just another particle. But it has properties that no other particle comes close to displaying. In the beginning, or, in the Beginning, if you like . . . it was all Light. Total freedom, superposition, everywhere. Unlike the other sub-atomic particles that began to appear as the Universe cooled down - eventually leading to atoms and molecules and gas clouds and stars and galaxies and planets and plants and animals and humans and evolved higher consciousness – photons exhibit the following unique properties:

- The photon has the mysterious quality of being both a particle and a wave . . . depending upon *observation* . . . that is: some form of *consciousness*.
- The photon is the unit of *undivided wholeness*, aka packet of light, aka Quantum of Action whose formula is ML^2/T (Mass, Length, Time). Science tries to break things apart to reach the smallest particle . . . division upon division upon division. The photon *is* the original particle.
- Unlike all other particles, 'the passage of light through space is accomplished without any loss of energy whatever.'[106] Regardless of how far a photon travels, it loses no energy – '*Light is pure action*, unattached to any object, like the smile without the cat.'[107]
- Light has complete freedom of when and where to go (uncertainty) and by the path of Least Action, or the *quickest possible way* – primordial, or primal, intelligence. In this sense, it seems to know where it is going . . . no mistakes, no wrong turns, no backtracking or trying to figure it out *(see the quote from Max Planck, in* The Process Taking Us from Here to There, *page 205).*
- Unlike other particles, Light is 'outside of space and time and has no rest mass' - no mass, no dimension, timeless.
- Unlike the other particles, with light, the *smaller* the photon, the more energy it carries: it's frequency times its wavelength always equals the same constant speed of 186,000 mps, so for light there is no Space or Time – hence, the smallest photon (smallest wavelength) would have the highest frequency, and contain ALL

the energy in the Universe (smallest photon = Stillpoint = one/zero). That is, the infinitely small photon has infinitely large energy, the energy to create the Big Bang, to start the Universe.

• Light has rotation, and the formula for this Angular Momentum is the same formula as the formula for the Quantum of Action above . . . ML^2/T. This rotation is also *fundamental*.

• Light is pure action: 'This light energy is everywhere, it fills the room, it fills all space, it fills the Universe, connecting everything with everything else. It includes much more than the light we *see* by, for *all exchange of energy* between atoms and molecules is some form of what used to be called electromagnetic energy, which extends over a vast spectrum and would be better named interaction. Visible light covers just one octave in that spectrum.'[108]

• 'Objects [particles] can be at rest or move at a variety of speeds. Light, on the other hand, has but one speed (in any given medium) and cannot be at rest.'[109]

• Light has the unique ability to 'provide us knowing about other things,'[110] that is, the ability to *see* other things.

• The photon is the only particle that we actually *see* . . . the example of radioactive material shooting of a photon of light when decaying. The *seeing* process is an inter-active process that interacts w/ the body – becomes absorbed, part *of* you.

• Other particles can be seen, and photographed and studied by more than one person . . . but Light 'can be seen only once: its detection is its annihilation. Light is not seen; it is seeing.'

• The photon is active at every Stage of the 7 Stages of Arthur Young's model of the process theory of all evolution in the Universe – it interacts with particles, with the atom, with molecules, with cells (triggering chemical reactions), with animals and with humans. Light is interacting throughout the entire process of evolution.

• Unlike any other particle, the Higg's Field does not impart mass to the photon.

The appearance of Light has no *cause* . . . it simply *arises*. Another way of saying this is that Purpose pulled this reality from the future. This seems identical to me to Meher Baba's *whim* that was latent in the Beyond-beyond state of God (*see page 23*) . . . at the moment that the awakening process began. This 'whim' arose in the oblivion . . . with no cause pushing it – rather, purpose pulling it.

Let's now address the 'form' and movement Light expresses. First, I have referred to the photon as a particle, but a more accurate way of talking about this is to say that 'photons are not separate; they are all part of Light and so shift their energy up or down.'[111] That is, particles aren't 'things,' they don't have 'identity,' they have movement, they are dancing points of energy, their frequency times wavelength equaling the speed of light.'[112] The photon is best described as a flowing wholeness. The following quote by the physicist David Bohm, who has been mentioned regarding the implicate world on the other side of the Stillpoint, expresses this perfectly:

> 'The new form of insight can perhaps best be called Undivided Wholeness in Flowing Movement. This view implies that flow is, in some sense, prior to that of the 'things' that can be seen to form and dissolve in this flow.'

Light would be the first of these 'things' to manifest this flow, and the torus models this idea best . . . inward flow being the enfolding, into the implicate, into the Stillpoint, the outward flow being the unfolding, emerging into the explicate, the manifested world. There is a continuing reciprocity between the enfoldment/unfoldment world of the torus and the implicate world of the Stillpoint, the one in-forming the other, and visa versa.[113]

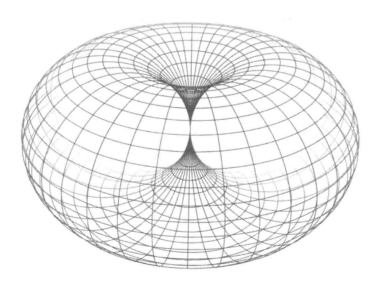

THE UMBILIC, OR HORNED TORUS . . . WITH THE STILLPOINT AT ITS CENTER

As Light's form is toroidal in the vibrating, pulsing original Quantum of Action at its smallest scale, the photon, and toroidal at its largest scale – the shape of the hyper-sphere, or Universe, it would follow that all forms of energy and movement in between these realms would also be toroidal . . . and so it is. As Light descends into matter, according to Bohm's view, the whole is in continuous flux - vortex structures in a flowing stream - and hence is referred to as the holomovement:

> 'The processes of matter, organic life, and consciousness are all seen as flowing from the reciprocal ordering principles of enfoldment and unfoldment. It is through these ordering principles that the holomovement – the 'ground of all that is' – expresses itself in particular forms and experiences.'

It is the *fundamental* patterning for energy flow – that is, fundamental to all manifest entities in a dynamic state'"[4] and displays unique qualities on par with those of the photon. Much of the following is from a remarkable website called *Cosmometry*, created by Marshall Lefferts, as well as from the work of Arthur Young. For the serious reader interested in all matters relating to many of the subjects briefly covered in these pages, I highly recommend the site . . . http://cosmometry.net/ . . . and *The Reflexive Universe* for further study.

DUAL-TORUS . . . THE TREE WITH ITS ROOT SYSTEM

Like the uniqueness of the photon, here are some of the unique qualities of toroidal systems that are, like Light, *fundamental* to universal expression:

- The torus is the only self-sustaining system of energy flow and is found at all scales, micro and macro. At present, the actual shape, or form, of the quantum reality is only theory . . . but by 'extending the observation of the consistent presence of this flow form into the quantum realm, we can postulate that atomic structures and systems are also made of the same dynamic form.'[115] At the atomic level, the torus manifests with electrons moving in and out of the implicate/explicate relationship/experience, creating toroidal dynamic vortex flow patterns (proposed as the fundamental model of atomic structures - these are 'energy dynamics,' not 'things' flying around). Up in scale, the toroidal shape of red blood cells. Then, into the macro-world of the dual-torus expression in trees, with leaves pulling in light, roots pulling in water/nutrients, oxygen exuded to atmosphere.

Expanding into the cosmological realm, the Earth's magnetic field is designed to receive energy from the Sun, bringing it into the Earth at the poles. The Sun itself is contained in a larger field called the helio-sphere and the helio-sheath which receives enormous energy from the *center* and the field of our galaxy – absorbed into the Sun by it's magnetic field lines. We receive all the energy from the Sun and all the galactic energy, into our bodies into our cells into our atoms. Then the galaxy itself, galaxies being dual toroidal structures themselves . . . 'nested, mutual influence . . . the pathways through which the qualitative infinity of nature may manifest.' This is the holomovement, seamlessly embedded at all scales – universal, atomic and quantum. Totally interconnected torus flow process transferring energy from the Universe, from the unified field, from the zero point, the infinite potential field. . . . atoms, blood cells, tornadoes, hurricanes, whirlpools, solar systems, magnetic fields around planets and stars, and whole galaxies themselves are all toroidal energy systems . . . the classic, fundamental, way energy moves and is formed in the manifested Universe.

• The torus is made from the medium from which it exists – water, air, smoke, stars, etc.

• The torus is centered by a Stillpoint singularity – called a black hole by science - and a central axis of rotation.

• The torus is self-generating and self-creating.

• The torus is self-reflective – it is reflective upon itself and learning from itself (the Universal expression of the evolution of consciousness).

• The torus is ever-evolving and transforming. Not static, but dynamic. 'Alive.'

    • Information is shared throughout the whole – the key to any living system.

    • There is reciprocal and balanced exchange from self to environment and back again.

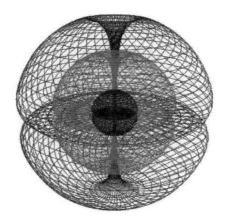

ENERGY PATTERNS OF THE DUAL-TORUS

I would like to end this section with an amazing insight by Dr. Bob Whitehouse, given towards the end of his and Dr. Mike Buchele's talk on Arthur Young's process theory, in Berkeley in 2013. This is just a sampling of Dr. Whitehouse's thinking – for a more complete experience of his ideas, please refer to his forthcoming book *UNFOLDING NOW: One Pattern, One Source, One Universe*:

> 'Einstein was looking for a field theory to resolve relativity with quantum theory. What Arthur Young is showing us is that the quantum of action or the nature of light is not a field. It is something more primary - and the field then emerges. Young had a very profound sense of what the Universe is made of – *Light*. He also observes that Light attracts and radiates . . . and I am noting that these are the two motions that make up the torus. Young was showing, based partly on Eddington's work, that the Universe, the hyper-sphere, has the same formula as a particular torus [the umbilic or horn torus with a point at its center], though no one else had realized that. He says that Eddington showed, in a way of reconciling quantum and relativity theory, that whether you're looking at the quantum or Universal level, they're all the same; it's

just a difference in scale. In other words, the nature of Light and the nature of the Universe are just a difference in scale.

Going further, the smallest possible Universe, if you will, is the quantum of action, or Light, and that Light's smallest photon contains the energy of the whole Universe and starts it all off. What I'm suggesting is that Light itself is toroidal, as Arthur was wondering about [see Arthur Young's quote above, from *Nested Time*], and that it both radiates (goes out) and attracts (brings in). If Light is toroidal then it must move within itself, because we know that anything that moves, moves toroidally in its own medium, like air within air and water within water.

What Einstein and others were looking for was the ether that was the wavelike carrier for Light. *But it would turn out to be Light itself.* Light would be the real source, by whatever name you want to know that or have for that - from outside of space and time. For the nature of Light – for the particle-wave photon - there is no space or time. If we put that together with the notion that the smallest possible photon of Light would be invisible, could appear anywhere, anytime, have all power, and start the whole thing, create All. *As it creates, it spirals out in this divine proportion [see comments on the torus, in Implications, page 45[16]], which is built right into every aspect of nature.*

The inanimate side comes out with the left hand spiral. The DNA and everything in life starts with the right-hand spiral. These are the two spirals that come out of the torus and to me represent the going out and the coming back. The journey out into the world I consider to be the masculine half of the journey, and the bringing it home part of the journey, is the feminine side. I'm suggesting that Light not only starts it all, it spirals out into creation of time and space, filling the entire Universe with its nature. We know that Light is everywhere, and that, as it's creating the Universe, the Universe is toroidal[117] – so, this is the Light within.

I'm saying that the purpose of this toroidal journey of Light into Matter and then back throughout all of Creation, is to know or to enlighten Itself. To the scientist this would mean that Light is the long sought Unification, and that it does have a medium through which it travels - its own nature, Light itself, and that it has Purpose - otherwise it wouldn't be going out into form, and returning. This Light of Spirit both creates and then exists within all that is, aiming to bring Light 'home' [Enlightenment] as ultimate Purpose. This could be stated in all these different ways: 'coming home,' 'knowing oneself for the first time,' 'knowing who we really are,' recognizing and realizing the 'Light within.'

What I'm suggesting here is that the question of immanence or transcendence can be resolved with this idea, because we can say that it starts with Light, outside of space and time, creating All . . . filling itself, filling everything with Light itself, and that the purpose is to fill the Universe with Light and come back home, to complete the journey home. It is the unseen Ground of Being that precedes all and that through its nature creates and comes into everything everywhere.'

How extraordinary! Still theory and not yet scientifically proven . . . but an inspired and beautiful idea. Light and Movement . . . photon and torus: primordial . . . *fundamental*. The only aspect of the the Absolute more fundamental is consciousness itself.

Let us return now to the factual phenomena that suggest that this primordial consciousness has not only advanced to the fully enlightened state suggested by Arthur Young and considered 'ineffable' - a level of consciousness no longer needing to re-enter the long wheel of evolution, the world of suffering . . . finally free from all constraint - now with the ability to manifest space and time at will – a consciousness that has left us a message written in the stars.

⊕ INFINITE UNIVERSES ASIDE: BEYOND THE UNIVERSAL CONSTANTS:

While considering the possibility that our Universe is but one of an infinite array of possible universes, Martin Rees suggested that the fact that our Universe displays a certain 'ugliness and complexity' is also an argument *against* there being a 'Designer.' An example of this complexity, he suggests, is the fact that the Earth's orbit is elliptical . . . not a circle. If its orbit were a circle - which would permit life but is not required by life - this would raise suspicions that either *God* or *chance* had fixed its course; we would have to accept that such fine-tuning was due to either brute fact or providence. But an elliptical orbit, and similar less-than-elegant aspects of the Universe as we find it, suggest that, as Rees puts it, 'our Universe may be just one of an ensemble of all possible universes that allow our emergence.'[118]

What is missing from this misleading viewpoint is that the eccentricity (the percentage that a particular orbit deviates from a perfect circle) of the planets (excluding Pluto which in some 'circles' is no longer considered a planet, and Mercury, which is within a couple of hundredths) is within *thousandths* of being a perfect circle . . . the Earth's eccentricity being 0.0167 ten-thousandths from perfection. All of the remaining six planets orbits are very close to the perfect circle . . . hardly 'ugly.' One can almost see the original design - with the existing phenomena being the imperfect expression of the ideal – and one can forgive Mercury for having to dance so close to the massive Sun every moment during its orbit for the last 4 ½ billion years.

That aside, imagine again for a moment the sheer unlikelihood of our Universe existing just as it does . . . suggested by the exponential improbability of Rees's six numbers . . . letting the mind move past the pat explanation of infinite universes making its existence possible. A literal pinpoint of possibility in the midst of infinity. It appears that within this almost impossibility of realities - *this* Universe - there are a series of phenomena that not only stand far outside the explanation of 'brute fact,' but suggest that they are signposts directing us towards the purpose of the initiating Intention – to awaken into total awareness.

Beyond the attempt by science to explain away the uncanny precision of the Universal constants through the infinite possibility of infinite universes, there exist phenomena that are well outside of the laws of geometry, physics or any type of scientific inquiry, coincidence or chance, some of them seemingly more so than others, *within our own solar system.* That is, these phenomena exist *here*, but are not universal. All of them will be considered by many or most to be 'happy coincidences' . . . but all of them reflect a kind of order that doesn't just happen by chance. All of them together, unrelated as they are to each other yet all ultimately related to the geometry of the Stillpoint - *the stillness at the center* - seem to suggest a Designer – the conscious intention in these instances deviating from the general layout of Rees's 'ugliness and complexity' (common to the rest of the Universe as we observe it) in order to lead us back to the center, the stillness within which the Eternal Present lives. Much will be said about this *stillness* . . . but it is sufficient to say for now that this is essentially and uniquely different than the expression of *movement* that all of creation is. It is that important.

Each of these phenomena, in its own way suggests the circle rather than the ellipse . . . and intention rather than chance.

⊕ WITHIN OUR OWN UNIVERSE: THE MILKY WAY AND THE SOLAR SYSTEM:

We will now move from the hypothetical sea of infinite universes into our own world, and quite a bit closer to home – our galaxy and the solar system within it - where astounding and 'impossible' phenomena will lead us to this same Stillpoint.

There is another possibility that makes even this remarkable concept of infinite universes moot. This is the core idea presented in these pages that initial intention, within the primal state of oblivion, initiated the evolution of consciousness such that 'God should consciously know his own fullness of divinity.' It is becoming more and more clear to the more expanded within the scientific community that consciousness precedes manifestation. This initial intention, the original spark of emerging consciousness, enters our world at the Planck Constant – the quantum of action denoted by the number 1. It enters through that tiny sphere just beyond the radius the Planck threshold where science stops - the dimensionless *Stillpoint* at the center of the Schwarzschild sphere, when all distances between objects are zero, the geometry of creation and of consciousness . . . the same Stillpoint geometry *that does not manifest, symbolically* embedded in the heart of our own solar system, the 'impossible' proportions of the Earth, Moon and Sun.

The Universe that we live within consists almost *entirely* of the *extreme heat* of stars, and the *extreme cold* of space itself. The tiny window where liquid water can exist, and which leads to the kind of life that exists here and where the complex organisms that it consists of are supported, is infinitesimally small.

> 'Earth sits squarely in the planetary comfort zone – the narrow margin in space and time where the right kind of star can give rise to the right kind of planet with the right conditions for life. Most scientists agree that this criterion applies to higher life-forms.'[119]

Critical to life appearing here on Earth is the existence of *liquid* water. While water in its many forms is somewhat plentiful in the Universe, liquid water is extremely rare. When one thinks about it, it is clear that the vast majority of the Universe exists as either extremely hot – stars, or extremely cold – space. There is a distance-window from the Sun, or any star, where liquid water is possible – in our solar system it is just beyond the orbit of Venus, extending outwards to very near the orbit of Mars.[120]

Time is also a critical element that our lives are dependent upon. Six billion years after the Big Bang, the heavy elements needed for planets to support life began to be created.[121] In several billion years from now, these heavy elements will begin to disappear. We exist within the window where life is possible, time-wise.

Our solar system also exists in the correct distance-window from the center of the galaxy. Too close to the center of the galaxy or within one of the spiraling arms, stars are so densely grouped that they may give off too much radiation and create too many collisions to support life, while, if we were too far from the center, our Sun may contain too few metals to create massive enough planets to hold on to an atmosphere.[122] We are situated for the perfect conditions permitting life to evolve.

All this, fortunate for us as it is, is within the scope of possibility – yet still *required* for us to be here. Add to these 'fortunate' phenomenon the list of requirements for life to exist listed above on page 88, and we are back to relying on the infinite possibility argument that science uses to justify the existence of our Universe. That being said, the following phenomena are outside any of the laws of physics, geometry, or mathematics common to this Universe . . . or chance. That is, they truly are 'impossible' to today's scientifically influenced common sense. The brute fact, or infinite possibility excuse cannot enter here.

What is even more remarkable, is that each of these phenomena, in one way or another, has one ultimately profound aspect in common . . . a similarity that implies a purpose – a purpose that will make itself evident as this story unfolds. Each of them includes either the geometry of the Stillpoint *(see Number, page 195)* or a reference to unity . . . from tiny to huge. They suggest an intelligence far beyond our own capability which seems to be guiding us somewhere, teaching us something immeasurably important. The validity of all I'm proposing is based upon these phenomena, so please forgive the repetition.

Some of these phenomena are arguably more 'impossible' and uncanny in empirical terms than others, but each are included because, to a greater or lesser degree, there is no known reason for their having to be this way *(again, to be more easily readable, a more complete explanation of each of these phenomena is found in Appendix C, page 325).*

1a. The precise dimensions of the Earth and Moon fit into some of the most sacred geometry known . . . squaring the circle, phi, the Great Pyramid, *(see Earth/Moon Relationship, page 7 & 179).*

1b. The precise *proportions* of the Earth and Moon and Sun are encoded within the Flower of Life, or Stillpoint, geometry – a geometry that does not manifest.

2. The Sun and the Moon are *precisely* the same size in the sky at total eclipse. This phenomenon exists because the Sun's diameter is 400 times larger than the Moon's while being 400 times farther from the Earth than the Moon during only 2% of its existence – the window containing human existence . . . the total eclipse of the Sun by the Moon being evidence of the precision. This 1 to 400 ratio is an aspect of the Stillpoint geometry.

3. Amongst an infinity of possibilities, the Ecliptic, the more or less flat plane described by the planets' orbits around the Sun, is almost *precisely* at a 60-degree angle from the plane of the Milky Way

Galaxy . . . the angle most integral to the Stillpoint geometry. This could be *any* angle of an infinity of possibilities.

4. The Ecliptic aligns *precisely* with the center of the galaxy – making possible the Solstice alignments with the center. Again, the possibilities of this relationship are infinite. The transit of the center of the galaxy by our Winter Solstice began in 1987 and ends in 2018. It is happening *now*.

5. The fact that it takes the wobble in the 23-½ degree slant in the Earth's axis 25,920 years to make one rotation (called the Great Year) . . . *one* cycle . . . and it takes light *almost exactly* 26,000 years to travel from the center of the galaxy to the Earth.

6. The fact that 1x2x3x4x5x6x7 = 7x8x9x10 = 5040 = the sum of the radii of the Earth (3960 miles) and the Moon (1080 miles), accenting the 'unborn' significance of the number seven *(see Number:, page 195, and Appendix D, page 335)*, as well as the seven-ness aspect of the geometry of the Great Pyramid, an aspect of the Earth-Moon relationship and the Stillpoint geometry. In fact, this uncanny relationship between number and measurement in the solar system continues – please see the empirical information regarding this *(see Earth Moon Sun, page 19)* from John Martineau's 2001 book *A Little Book of Coincidence in the Solar System* .

7. The fact that in *seven* days, a quarter of the cycle of the Moon's orbit around Earth, the original *single* cell that eventually becomes who *you* are divides and re-divides until it once again achieves a sphere consisting of 64 spherical cells *(see When We Were Small . . . page 188)* . . . the same number as the 64-tetrahedron/sphere configuration which is generated by the Vector Equilibrium and is geometrically connected to the Chinese 'I Ching,' the Egyptian 'Flower of Life,' and the Kabbalah's 'Tree of Life.'

8. Kepler's discovery regarding the mean orbits of the planets and their relationships to the Platonic solids.

9. The entire panoply of geometric relationships by the orbits of the planets in our solar system displayed in John Martineau's *A Little Book of Coincidence in Our Solar System (see Earth Moon Sun, page 19)*.

Added to all this is the remarkable 'coincidences' recorded in Christopher Knight and Alan Butler's book *Who Built the Moon?*, and included here in Appendix B (please make sure you see this) – yet another list of 'impossible' phenomena in our solar system. Newton, speaking of the almost ineffable elegance of the solar system – while within the limitations of his own belief in the old Creator-God – said:

'This most elegant system of the sun, planets, and comets could
not have arisen without the design and dominion of an intelligent
and powerful being.'

Science's official explanation for all the above, when it isn't ignoring it, is
random, accidental, coincidental occurrence. This is absurd. Infinite universes
cannot explain all this phenomena. While it is possible (even if this possibility is
infinitesimal), that if infinite universes produce infinite possibilities, that *this*
possibility, *this* Universe, is simply the one that permits us to witness it . . . *these*
phenomena, produced as they are *within* the laws of *this* Universe, are well *outside* of
any explanation due to physics, mathematics, geometry – in fact all of science - or
chance . . . and are not required for life. Something else is happening here. All of
this evidence points towards *intention*.

Just as Martin Rees's six numbers, when considered one-at-a-time, are already
unimaginably unlikely - and almost *infinitely* unlikely as they are unrelated to each
other (the improbability of the precision of each number exponentially increasing
the improbability of the others) - so does the unlikelihood of each of these
phenomena geometrically increases the unlikelihood of each of the others.

Taken as a whole, given that each of the above examples leads directly or
indirectly to the geometry of the Vector Equilibrium (*see Number:, page 195*), the
Stillpoint or Unity itself . . . we are even deeper into the chance of all of this
happening in any 'normal' way is . . . *a pinpoint in the midst of infinity*.

As I've made clear, there is another explanation . . . the evolution of
consciousness itself. And if this is the case, one has to begin to see this incredible
array of empirical evidence as a communication, a message of some sort, from
awakened consciousness to our own evolving consciousness.

Before exploring this more expansive explanation for all that is, we now turn to
yet another fascinating aspect to these remarkable phenomena witnessed in our solar
system . . . the aspect of *time*. Most of what has been shared above has to do with
uncanny *spacial* phenomena that form this *message* from higher consciousness. But
the message is also found within empirical data concerning *time*.

⊕ Time:

There is something very interesting happening not only with all of the
spacial/dimensional/proportional aspects of the geometric phenomena observed in
our solar system . . . but also with *time*. According to Einstein's widely accepted

Theory of Relativity, time and space are part of the same medium . . . $E=MC^2$ . . . and are intimately and inherently connected and inseparable: space-time.

But orthodox science sees this as a kind of fixed, 'block universe' . . . 'In it, there is no *motion* of time. All is given, past, present, future, in one giant block.'[123] The common image of this 'block' is a loaf of bread with infinite, infinitesimally thin slices, each a slice of three-dimensional space in an *instant* of time without motion, with one instant existing *relative* to the next, depending upon the velocity of the observer, etc. In other words, once again reducing experience to focusing on its parts rather than the whole. In fact, it is now well known that matter is experienced not only as a *particle*, but also as a *wave* – and waves *vibrate and move*, creating a *continuum* of space-time, where the reality is an eternally and always connected *whole*.

The dual particle/wave reality, according to quantum mechanics, permits a particle to be described as a wave that spreads out over a great distance . . . but remains one particle and through *observation* the wave is collapsed and its location is known. This is consciousness engaging with energy/matter. The interaction is primal. It is *fundamental*. This has recently been validated by new research[124] showing that this collapse actually happens, backing up years of research into quantum entanglement, where 'particles are connected in a mysterious way even when separated, so that observing or affecting one instantly affects the other'[125] . . . *simultaneously*. This is what Einstein famously referred to as 'spooky action at a distance,' apparently breaking the 'law' that nothing travels faster than the speed of light. This is all to say that the bread slice model is inaccurate, and that space-time is a continuum, a whole of interconnected energy-matter-time, with the *motion* of time fluidly connecting the past to the present . . . and to the future. Even in Einstein's model of time, all frames of time, or instants of time, exist all at once. We assume that the past influences the future . . . but in this model, why can't the future influence the past . . . in an endless feedback loop? That is, all time affecting all time all of the time?[126] In my own experience, in the pure awakened moment of awakened *consciouness* in the Stillpoint reality, outside of time, this fluidity expresses itself as infinity in all directions . . . in space *and* time . . . *everything* all at once, now.

Regarding the *future:* A reality that forcefully makes itself known within the myriad of 'coincidences' in our solar system has to do with an uncanny anticipation of the future 4.6 billion years before the future 'arrived.' Evidence has been presented that our solar system was created intentionally by an essentially infinitely evolved consciousness. This would have happened around 4.6 billion years ago. It's clear also that this intentional creation was done according to the 'rules' . . . that is, by using the established laws of physics and morphic resonance with inert matter, then continuing through many billions-year-long process of natural selection/morphogenetic resonance throughout biology/life, etc. Empirical evidence suggests that the higher consciousness responsible for our being here was also aware of at least the *probable* general progression of the evolution of life and of consciousness up to the time that Man, with the newfound use of free-will and his capacity to create, appeared on the scene. It is here that the future is malleable and *uncertain*, but always filled with both *possible* futures based upon the direction of the evolution, guided by more and less powerful evolutionary *chreodes*[127] - or evolutionary ruts - and *probable* futures depending upon deeper and more ingrained chreodes.

It appears that this higher creative consciousness was very aware of how this was likely to progress until the arrival of consciousness capable of creation and free will . . . *us* . . . and, how much *time* it would take. The Moon is gradually moving away from the Earth. The total eclipse of the Sun by the Moon - because the Sun's diameter happens to be 400 times larger than the Moon's - only occurs in the window when the Sun is exactly 400 times farther away from the Earth than the Moon. The Moon slowly and steadily has been moving away from the Earth, entering the window where perfect total eclipse is possible and when the likely evolution of life on Earth was well along, some 50 million years ago, to last another 50 million years – a window that is a mere 2% of the entire elapsed time of our 4.6 billion-year solar system - making sure that its message could be read when humans had reached a level of readiness to understand it.[128] The timing is . . . 'fortunate.'

Another, still unexplained, aspect of all this is the *timing* of the *discovery* of the Stillpoint phenomenon message – a discovery ultimately about consciousness, expressed at the heart of the solar system. This discovery, apparently, is restricted to the tiny window where the *global* situation has become critical . . . *now*. It appears that this possible or probable global crisis was anticipated as a very real possibility. It is likely that this was foreseen because the crisis we now find ourselves in is *archetypal*: consciousness, evolving through billions of years of dualistic, survival-based, separatist reality, will inevitably reach this crossroads if given the *time*. It is the crossroads of sustainability on a finite planet, forcing the evolution from survival-based, me-or-you consciousness towards a holistic awareness that we are all one organism dependent upon each other and our healthy environment of endless connectedness to permit future evolution to occur – we are all either in this together and evolve, or each for himself and disappear.

Clearly, the Stillpoint phenomenon could not have been detected until humanity had reached a stage where it had advanced enough to invent and use the telescope and make precise distinctions regarding what was observed. Brahe, Kepler and Galileo produced extraordinarily precise data. Their own capabilities were exponentially magnified when modern telescopes and the computer entered the scene. But it was not until the consciousness problem that has always been an isolated, provincial affair for the last 11,500 years or so, became the global crisis that it is today, that suddenly this information (that has been in front of us for all this time) somehow makes itself known. This new information includes the brilliant work of John Martineau, Alexander Thom, Christopher Knight and Alan Butler, John Michel and others, as well as the discovery of the Stillpoint phenomenon . . . almost all of which has come to light very recently, between 2000 and 2009, where any number of man-made crises could destroy all life on Earth – as well as within the tiny time span when our Winter Solstice transits the center of the galaxy.

But even more uncanny and amazing is the phenomena surrounding the term 'mile.' As discussed in 'Scale:' (see page 350), the common understanding of the fact that 5,280 *modern* feet equal a 'mile' is based upon is purely arbitrary. As the story goes, King Henry I decreed that the yard should be the distance from the tip of his *nose* to the end of his *thumb*, the foot then being one-third of this royal length. This led Queen Elizabeth I to declare, in the 16th century, that henceforth the traditional Roman mile of 5,000 Roman feet would be replaced by one of 5,280 of Henry's feet. Who knew? But it turns out that the modern mile is also made up of whole number segments of the geodetic Megalithic Yard (accurate too on the Moon and Sun!) . . . 1,920 MY equal one *modern* mile.[129] Hmmmm. So . . . the modern mile is related to the Megalithic Yard and somehow replaced the traditional Roman mile. This is convenient because of the uncanny relationship between this unit of measurement and the dimensions of the Earth, Moon and Sun. It becomes yet another message. Here are some inexplainable and incredible *facts* relating to the mile, from John Martineau's *A Little Book of Coincidence in the Solar System:*

Miles of Moon and Earth

Radius of Moon = 1080 Miles = 3 x 360 Miles
Radius of Earth = 3960 Miles = 11 x 360 Miles
Diameter of Moon = 2160 Miles = 3x1x2x3x4x5x6 Miles
Radius of Earth + Radius of Moon = 5040 Miles
= 1x2x3x4x5x6x7 = 7x8x9x10x11 Miles
Diameter of Earth = 7920 Miles = 8x9x10x11 Miles
There are 5280 Feet in a Mile
= (10x11x12x13) – (9x10x11x12)

Random accident? One can easily see the precision in the design of the Earth and Moon . . . and these same 'coincidences' are also found in the dimensions of the Sun - see 'The Message in Detail,' Appendix B.[130] There can be no doubt regarding the intentional creation of our solar system by a consciousness infinitely greater than our own. All this could only be known once accurate measurements, not only of the Earth, but of the Moon and Sun, were available . . . in the last few hundred years . . . and, equally importantly, when it would be clear that this intentionality is only witnessed *here – now*.

But how could this higher consciousness, responsible for creating our solar system 4.6 billion years ago, possibly know that Man, with free will, would possibly assemble 5,280 'feet' (related to the geodetic Megalithic Yard) into something he would call a 'm*ile*' . . . the same mile that fits so perfectly within the information presented above and in Appendix B? The only answer that I can imagine is that this fully enlightened consciousness lives in the eternally present *moment* where the fluidity of time, intimately connected to the entanglement of matter through the wave function, makes the past, present and probable future transparent . . . making not only the past and present knowable, but also the future – at the level of probability. Perhaps somehow they *knew* that, given the possibilities and probabilities evolved through morphogenetic resonance and the paths (chreodes) carved within time that would drive future events, and knowing the general timing, there was some chance that Man would somehow decide that a certain number of geodetic *feet* would be grouped into what is now called a *mile*. But this seems a huge stretch and beyond probability. It *seems* that 'they' were able to look into the future and anticipate the possibility. But how could they be certain? Perhaps the Megalithic Yard that appeared 5,000 years ago – after civilization had been destroyed by the Flood - can be seen as guidance. That is, this unit of measurement that divides not only the Earth, but also the Sun and Moon, into perfect whole numbers could not have been 'guessed' by evolving humanity 5,000 years ago. Perhaps it was deduced by creating the length of a pendulum relating to the 366 day orbit of the Earth around the Sun, or perhaps *given* to our ancestors by those more knowledgeable – affirming the myths suggesting extra-terrestrial influence 5,000 years ago. Either way, today's 'mile' was considered at the inception . . . 4.6 billion years before it came into official use as a measurement of distance.

But why this explosion of consciousness during this particular *time* so many thousands of years ago? Given that the evidence is overwhelming that our solar system was created intentionally and then left to evolve within existing law, it's tempting to wonder just what this law involves. Certainly the laws we experience now upon which the universe is based – Martin Rees's six numbers and others, the laws of chemistry and physics and the evolution of life based upon natural selection and morphic resonance - and certainly suffering has always been an integral part in this for all sentient beings, and suggested as law by the famous words 'the poor shall always be with us.' It has always been, after all, tooth and claw . . . earthquakes, volcanoes, disease, tigers, tidal waves, hurricanes, locusts, war, tornadoes, etc. But evolve and survive we did, and it appears that everything went *more or less* according to plan until around eleven and a half thousand years ago. There is a vast amount of evidence compiled by Graham Hancock in his books *Fingerprints of the Gods*, *Supernatural* and *Magicians of the Gods* that make it clear that civilization had made huge advances up to this time[131] . . . but then catastrophe struck – and civilization was destroyed . . . globally.

Was the Flood, etc., a part of the 'plan'? I don't think so – but it may at some point have become anticipated *(see also the discussion on the Great Pyramid, page 181, in the section '. . . and Moon', page 177)*. I think this was all part of the natural law that was not interfered with – the popular myth that Atlantis was being punished for its wayward ways aside. Certainly we lost our footing . . . but we lived on in the caves and high ground, gathering ourselves as best we could. And perhaps just as certainly, once this global annihilation had occurred, destroying all our advances, 'they' knew that the opening to the inflow of spiritual wisdom that would accompany the winter solstice alignment with the center of the galaxy was fast approaching – the end of the Great Year, the return of the light – and that a boost of some kind was now required (again . . . compassion) in an effort to prepare us for this time. Perhaps they found it necessary to provide the tools required to rebuild by seeding Earth's struggling humankind with the knowledge needed . . . agriculture, monumental architecture, mathematics, geometry – and perhaps most important of all, the *message* encoded in the geodetic unit called the Megalithic Yard found in ancient megalithic stone circles . . . a unit of measurement based upon the circumference of the Earth – yet found also when measuring the Moon and Sun. An encoded message letting us know 'they' existed and were communicating with us.

This higher consciousness knew that in a relatively precise window of 'future' time, consciousness would evolve on this planet in the normal way such that the above information would be discovered, making clear the existence of this higher consciousness to the inhabitants of this planet. Also, in some way that I do not understand, the heart of this communication – that having only to do with consciousness and the blueprint for a possible global shift in consciousness . . . the Stillpoint geometry found in the proportions of the Earth, Moon and Sun – has also remained hidden until now . . . or more precisely, at the exact time that our Winter Solstice's alignment with the Sun points directly to the center of the galaxy, the source of the Mayan's spiritual wisdom and, we assume, the calender that covers many many thousands of years and ending during that transit, on December 21st, 2012. It is only *now* that this information is critically needed . . . and, while it's been here all along, we are only aware of it *now*.

This discussion of time and of the future is closely related to the creation of morphogenetic fields and their projection into the future. Morphic fields form within and around the system they organize and are vibratory patterns of activity that interact with electromagnetic and quantum fields of the system.[132] These fields 'are shaped by morphic resonance from all similar past systems, and thus contain a cumulative collective memory . . . [and] are local, within and around the systems they organize, but morphic resonance is *non-local* . . . [they] are fields of *probability*, and they work by imposing patterns on otherwise random events in the systems under their influence . . . they contain attractors [goals], and chreodes [habitual pathways toward those goals] that guide a system toward its end state, and maintain its integrity, stabilizing it against disruptions.'[133] They work through a resonance with the past and contain attractors that pull them into possible and probable futures.

Henri Bergson, one of the most famous and influential French philosophers of the late 19th century-early 20th century and a major critic of the model of time as presented by Einstein and his Theory of Relativity - the idea that 'matter lived in an eternal instant, and had no time within it.'[134] He saw time intimately connected to the vibratory, fluid, wave aspect of matter/energy . . . inherent to all matter in the sense of process – no dead 'instants,' but an alive flow of energy from past to present to future, with *consciousness providing the connection.* Bergson:

> 'Duration is existentially a continuation of what no longer exists into what does exist. This is real time, perceived and lived . . . Duration therefore implies consciousness; and we place consciousness at the heart of things for this very reason that we credit them with a time that endures.'[135]

Imagine a fully awakened, fully enlightened consciousness no longer needing to manifest in the physical world as we know it, with full access to the complete process of the evolution of consciousness available in the Zero Point Field or the Akashic Record – or perhaps *being* the embodiment of that wisdom - able to read accurately (or simply *know*) all of the major and minor chreodes of development, or evolution, as well as the goals, or attractors, pulling the field into its possible or probable future. Such a consciousness would be able to see into 'our' possible and probable future and place signposts to guide us along the way and, ultimately, through this critical interval, or crisis, in our evolution. Significantly, physics allows for this reality, as 'both directions of time are created equal.'[136]

The arrow of time is created because the Universe began with a very low entropy and extremely high equilibrium (the Stillpoint – the geometry of which is the Vector Equilibrium, all vectors being equal, in perfect balance, or equilibrium) and progresses steadily to higher entropy. In Sean Carroll's *The Bigger Picture*, there is this interesting quote:

> 'Nobody knows exactly why the early universe had such a low entropy. It's one of those features of our world that may have a deeper explanation we haven't yet found . . .'

I would suggest that this is a sure sign that the larger context, the *real* 'big picture,' is that the arrow of time was required for the *process* of the evolution of consciousness to occur – beginning with the Vector *Equilibrium*. It would seem too – as mentioned above - that the interval we're now experiencing is archetypal and thus eminently predictable. In Dr. Steven Greer's talk in London in September of 2015,[137] he mentions this interval. When addressing the status of our present level of consciousness, he said: 'As [physicist] Michio Kaku[138] points out, we're not even a Level 1 civilization. We're at level zero. Level 1 is: You're living peacefully and you're not cannibalizing your biosphere and have sustainable energy systems that don't endanger the environment.' *This* is the transition we now face . . . the transition from all belief systems based upon separateness to a knowing that everything is connected – and our role shifts from the ignorant assassin of all life on Earth to the conscientious and aware steward of an intimately connected ecosystem.

We will now return to the attempt by science to understand all of this - and to the threshold that science has reached . . . and to what is seen on the other side of that threshold.

⊕ What Came Before the Big Bang?

We live in a world molded by materialistic scientific understanding, and it is precisely this flawed worldview that prevents us from experiencing the depth of who we truly are. I am not speaking now of the wonders that science has brought us, but of its inherent reductionist limitations – it tends to reduce the world to a study of its parts, disconnected from the whole . . . and it is just here that the truth is missed. Just as Newton's view of the world was valid as far as it went on the macro scale, the newly discovered quantum world had its own set of laws that probed deeper into the mystery. This is a mysterious world where Newtonian physics is no longer valid. I believe that we are at the threshold of a yet deeper understanding of our world . . . and the Stillpoint reality is where this understanding will be found. Before discussing that world of wholeness and possibility, I would like to return briefly to this threshold as science sees it . . . the Planck Constant that defines the point of singularity from which the Universe exploded.

It may be that the reader is most interested in what all this information implies about where we're *going*, but I feel that it is important to understand the Planck reality, as it is where we came *from*. Science describes this pre-reality as 'nothing.' This is the worldview we live within. I'm suggesting that it is much more than that.

Science is the study of the quantifiable . . . and the concepts of infinity or zero, while relegated as placeholders, are still a part of the mystery outside of consideration. But, on the one hand science has been forced to use the idea of the infinitely huge to explain away the precision of the Universal constants. On the other end of the spectrum, zero – or the infinitely small – is protected from consideration by the quantum threshold of density, length and time that permits the quantum equations to work – the Planck Constant.

The great, unanswered question in standard Big Bang cosmology is *what came 'before'* the supposed gravitational singularity – this Planck-sized density one-billionth the size of a proton, containing the *entire* Universe . . . an ultra-tiny sphere whose *radius* is the Planck length – with the Stillpoint at its center. Compressing an object of mass to a size smaller than this radius results in the formation of a black hole, where time and space disappear. This is called the Schwarzschild radius and is roughly the 'distance scale at which general relativity becomes crucial for understanding the behavior of an object of a given mass"[139] . . . anything smaller than this radius and matter collapses and disappears into a black hole.

Taken at face value, the *observed* expansion of the Universe implies that it was born out of this singularity, this point of *infinite* density – but quantum physics tells us that it is meaningless to talk in quite such extreme terms, and that this 'makes no sense,' as its best physical theories, including general relativity and quantum mechanics, *stop working* when they try to describe matter that is almost infinitely dense - and that instead we should consider the expansion as having started from a region no bigger across than the so-called Planck length (10^{-35}m), *when the density was not infinite* but 'only' some 10^{94} grams per cubic centimeter. The Planck Constant represents the absolute limits on size, density and time allowed by quantum physics.

The theories break down when the concepts of the infinite or zero are introduced. At the Planck density, the highest density describable with current physics - over 10^{94} g/cm^3 . . . which corresponds to roughly, perhaps *500 billion* galaxies (this number was reached through a recent German computer simulation – the 'observed' number is up towards 200 billion) squeezed into a space a billionth the size of an atomic nucleus - physics begins to 'work' . . . but before that moment, as zero/infinity is approached, physics can tell us nothing. Past this horizon is the world of the dimensionless Stillpoint itself – the Alpha and Omega and the opening to the infinite world of the eternal present moment, including all that ever was, is or will be. *It is the entrance to the other world.* *This* is what the Universe exploded out of – the plenum of the meta-universe, the Great Mystery of the Lakota, the Ayn of the Kabbalah, emerging through the Stillpoint, Kether, the Kabbalah's *Point of Creation.*

The current model of the universe begins when the density was somewhere below the Planck density . . . *and virtually nothing can be said about what the Universe was like before that.* We therefore take as our initial condition a Universe at or just below the Planck density, *and any questions about the instant of the Big Bang itself are eliminated from consideration.*

This is the horizon mentioned earlier . . . the threshold through which reductionist science cannot go. But what does science see on the other side? What is infinitely small? . . . and why is that important?

All we can know at the present time comes from information received from the *observable* universe, which is a subset of the total Universe. What we observe is a Universe that appears to be expanding at an accelerating rate. The *age* of the Universe is determined from what we *observe* . . . that is, either by projecting back into time from estimations of the ages of old stars or star clusters, or by extrapolating backwards factoring in the rate of acceleration, the so-called Hubble Constant. According to data from these sources, the visible Universe, from which this data comes, extends fourteen or so billion light years out into the vastness.

Imagine a sphere, with you at the center, that extends to the fourteen billion light year mark . . . the distance that light has traveled at 186,000 miles per second for just this side of 14 billion years. This is the distance at which we are presently able to perceive the light from distant stars.

What is beyond that? What is infinitely large? What is beyond the singularity? What is infinitely small? But small and large lose their meaning when talking about this point that includes *everything.* Simply put, it was infinite in both 'directions.' In both scenarios the space was completely filled with matter that began to expand . . . a koan.

The perhaps not so obvious correlation to this scenario is that there is no *one* center of the expansion; the universe is simply expanding at all points. The only answer to the question 'Where did the Big Bang happen?' is that it occurred *everywhere.* This uniform/everywhere expansion means not only that we should see every other galaxy moving away from us, but that observers in another galaxy should see exactly the same thing. Again, a concept beyond our ken.

In a uniformly expanding universe, every observer is at the center of the expansion, with everything else moving outwards . . . *the center . . . IS . . . everywhere.* *Much like we experience consciousness.*

My brief glance of what lies beyond our world through the lens of the Stillpoint does me little good in understanding what that world looks like in scientific terms. I believe that the dimensionless Stillpoint, the interface between our world and the world we came from, expanded exponentially, possibly through the mechanism of what science calls inflation, in time, ultimately to infinity . . . becoming the very matrix within which the manifested Universe exists – the matrix possibly expressing itself through the so-called Zero Point Field containing all memory of past experience.

I intuitively feel that the Zero Point Field that science considers the primal matrix underlying our Universe is the information repository learned from all universal expression, and so must be related intimately to that other world beyond the veil. It is, I feel, the Mother of all fields - this field that connects all 'consciousness and entanglement' through, I believe, the Stillpoint/Flower of Life matrix - and provides our Universe with the blueprints necessary for the evolution of consciousness to continue its expansion.

It appears, if what we're being told is true, that one of the first blueprints to manifest is the newly discovered *Higgs Field.* Science has now discovered a basis for this field, founded upon the discovery of the Higgs Boson by CERN, the huge circular particle collider in France/Switzerland. *Just* after the Planck moment, at 10^{-32} of a second, this field emerges – with the Higgs boson being science's present understanding of the initial elementary particle, the smallest possible *ripple* in the Higgs Field. It is *here* within this primal *field,* that particles apparently get their mass and, I assume, the laws imparted to them, in-formed from past universal experience – it is *here* that our Universe comes into being.

All matter, energy, space and time are presumed to have exploded outward from this original singularity – meaning that all of this matter, energy, space and time *began* at this *horizon,* and that past this horizon is the world of the unknown. That is, the 'singularity' is, in fact, the density of the Universe at the event horizon – and beyond that is a no-man's land leading to the infinite world of the Stillpoint.

But science likes to hang on to its 'stuff' – in this case everything that came *after* the Planck moment – and casually dismisses whatever came *before* this as a *question that is meaningless, since time itself began at the singularity* . . . and sometimes logic can make a fool of itself.

In the 2011 film already mentioned, *Did God Create the Universe?,* Stephen

Hawking, one of today's preeminent cosmologists, presents his opinions that there is no 'God' that created this Universe. *Why?* – He couldn't have – there was no *time!* The narrator, also propounding Hawking's views, scoffs at the idea of a pre-existing 'God' because . . . drum roll . . . since *time* was only created at the moment of the Big Bang, there would be no time preceding that moment – so 'God,' who had no time within which to work, couldn't possibly have created the Universe – even if She hurried. *This* is science's inane argument discounting the entire implicate world discussed by David Bohm, as well as discounting the reality behind the veil known so intimately by advanced spiritual beings throughout history and the thousands of experiences so well documented and explored by Stanislav Grof and many others – as well as countless other examples of a dimension or reality beyond our own. Many have experienced this level of consciousness – the eternal present moment – and time does not exist *as we experience it, but past-present- future instantaneously – just as unreasonably as does everything else exist, in one present, timeless moment, infinite in all directions.* As well as anything, this astounding lack of imagination and myopia represent the heart of today's orthodox science.[140]

The primordial 'stuff' of inflation, Guth and other cosmologists contend, is very likely a spontaneous creation that boiled out of absolutely nowhere by means of an utterly random but nonetheless scientifically possible process. Still, from Alan Guth:

> 'It is rather fantastic to realize that the laws of physics can describe how everything was created in a random quantum fluctuation out of nothing, and how over the course of 15 billion years, matter could organize in such complex ways that we have human beings sitting here, talking, doing things *intentionally.*'

How very similar this statement is to Meher Baba's 'Before the beginning . . . this original whim brought latent consciousness into manifestation (form) for the first time . . . as slowly . . . consciousness approaches its apex in the human form.'

⊕ THE STILL POINT:

We are finally coming to the reason for this long description of where science ends and the heart of this information begins. I know that this last bit, filled with almost incomprehensible, koan-like scientific description may have been difficult to stay with. I persevered because I wanted to establish, from the scientific worldview, the incomparable importance of the Stillpoint. What science 'saw', but could not accommodate, when it reached the Planck threshold was this Stillpoint, the non-existent point containing nothing and everything that permitted the Universe as we know it, by the *learned* laws of physics, to come into being - what science must quantify by calling it the point of gravitational singularity, one-billionth the size of a proton: a point of *infinite* density, when all *distances* between objects are *zero* – and all inquiry into what lies beyond that threshold is considered meaningless and not worth consideration.

Imagine a dimensionless point. No objects, no distance between objects, no dimension . . . non-existent. When there are no objects, there is no movement. A point of no movement . . . still. The Stillpoint. When there is no thing moving through no space, there is only stillness . . . but a dynamic stillness that is timeless and dimensionless yet includes and is pregnant with all existence . . . the Absolute *(see Rodney Collin's definition of the Absolute, page 198).*

Present data suggest that the size of the total Universe is infinite, and its 'stuff' is uniformly expanding everywhere from every point of observation, meaning that it 'originated' at each of these points . . . the energy/matter/density in each point is *infinite* – within an *infinitely small, dimensionless point.* This is the definition of a 'point' in our world . . . no dimension. Literally *everywhere* . . . yet *nowhere.* So, by our three-dimensional definition of reality it doesn't even exist. It is not 'born,' similar to the number seven *(see Number:, page 195)* . . . the number inherent to the Phi-pyramid geometry connecting the dimensions of the Earth and Moon and Sun (exhibiting the proportions of the geometry of the Stillpoint) with Ø (Phi), the golden proportion of life *(see page 178),* and π (Pi).

This dimension of no dimension - the point - is also the only dimension common to *all* universes, no matter how *infinite* and no matter how many dimensions they may have . . . or, rather, common to the *genesis* of *each* of them.

When science finds the Grand Unified Theory, as the engineer, author, designer, inventor, and futurist Buckminster Fuller originally suggested, it will likely be found within the geometry of the Vector Equilibrium (Stillpoint) that is nowhere yet everywhere.

⊕ The Holographic Universe:

The following is an excerpt from a review of Michael Talbot's book, *The Universe as a Hologram* by Michael Kisor:

> 'In 1982 a remarkable event took place. At the University of
> Paris a research team led by physicist Alain Aspect performed what
> may turn out to be one of the most important experiments of the

20th century. Aspect and his team discovered that under certain circumstances subatomic particles such as electrons are able to instantaneously communicate with each other regardless of the distance separating them. It doesn't matter whether they are 10 feet or 10 billion miles apart. Somehow each particle always seems to know what the other is doing.'

The problem with this of course is that according to Einstein's theories, nothing can travel faster than the speed of light. So what is going on? While reductionist science studies *parts,* this suggests a holographic unity where everything is instantaneously connected, a collection of *wholes* . . . suggesting a 'more complex dimension beyond our own.'

'Such particles are not separate 'parts,' *but facets of a deeper and more underlying unity* that is ultimately *holographic* and *indivisible.* If the apparent separateness of subatomic particles is illusory, it means that at a deeper level of reality *all things* in the universe are *infinitely interconnected* . . . . and all of nature is ultimately a seamless web. At its *deeper level* reality is a sort of *superhologram* in which the *past, present, and future all exist simultaneously.* This suggests that given the *proper tools* it *might be possible to someday reach into the superholographic level of reality.'*

And what fits the definition of this 'superhologram' more than the Stillpoint? I've used the term 'eternal present moment' to describe this reality, and called it 'a moment of complete, pure awakening to the eternal timeless moment where past, present and future exist at once, outside of/including all time, infinite in all directions - the essential Oneness of all that is and isn't.'

'Allowing, for the sake of argument, that the superhologram is the matrix that has given birth to everything in our universe, at the very least it contains every subatomic particle that has been or will be - every configuration of matter and energy that is possible, from snowflakes to quasars, from whales to gamma rays. *It must be seen as a sort of cosmic storehouse of 'All That Is.'*[141]

Others, including physicist John Bell, have performed experiments to verify this phenomenon. Those experiments had weaknesses, or loopholes, but recent experiments – University of Delft researchers, led by Ronald Hanson, US National Institute of Standards and Technology researchers, led by Krister Shalm, and the University of Vienna researchers, led by Anton Zeilinger - claim to have closed these loopholes.[142]

'Things get really interesting when two electrons become entangled,' said Ronald Hanson from the University of Delft. 'They are perfectly correlated, when you observe one, the other one will always be opposite. That effect is instantaneous, even if the other electron is in a rocket at the other end of the galaxy.'[143]

This is the world of quantum entanglement . . . everything is connected, all the time, instantaneously.

Talbot goes on to point out that Karl Pribram, a neurophysiologist from Stanford, believes that memories are not composed of *particles* or *bits,* but are 'patterns of nerve impulses' and in fact believes that the *brain* is a hologram – that is, the individuated hologram of our brain is instantaneously interconnected with the infinite hologram of the Universe. This general model is known as the holographic paradigm . . . everything is connected, all the time, and we interface with the larger whole through the holographic nature of our own brain. Further, this holographic model is a storehouse of *everything* that has ever been . . . the Akashic Record.

⊕ THE TEMPLE IN THE CONTEXT OF THE LAST 11,500 YEARS:

This chapter, entitled *The Larger Context,* has been a description of some of the more important scientific ideas that provide the context within which the Mesa Temple exists. In the excerpt from *The Universe as a Hologram* were a couple of sentences that read:

> 'At its deeper level reality is a sort of superhologram in which the past, present, and future all exist simultaneously. This suggests that given the proper tools it might be possible to someday reach into the superholographic level of reality.'

It is this scientific reality in particular that provides the context for the Mesa Temple being one of these tools designed 'to reach into the superholographic level of reality.' It certainly is intended to be. Before going into more detail about the temple itself, I'd like to review the process that got us here.

I think it has been made clear that there is a great deal of intentionality in our solar system, outside of the normal way creation happens in the Universe. Since consciousness precedes creation, it must also be clear that at some point there must have been an initial, or primal, intention . . . Meher Baba's 'whim' expressed through the singularity as science understands it. That all of what has been shared above could possibly have happened by chance or, as science would have us 'believe,' by the existence of infinite universes providing infinite possibility, borders on the laughable. But what is most important, is that it IS happening . . . or rather, that it *is*.

From Adyashanti:

' . . . this innocence waking up, realizing that everything is itself, even all the confusion and all the ignorance . . . everything . . . *this* is the dissolving of confusion, of ignorance, of karma . . . awareness yielding to itself, to its inherent creativity, to its expression in form, to experience itself' . . . and finally . . . 'only awareness remaining.'

The purpose of the Mesa Temple is to accelerate this innocence waking up.

It seems that science will inevitably unlock the mystery to time travel, anti-gravity,[144] and interplanetary travel if given time . . . that is, if its darker shadow doesn't annihilate us first – by the prostitution to those who would use it to control by putting its discoveries into the hands of those who are committed to using this knowledge for their own ends . . . Hiroshima and Nagasaki being the best examples of this, despite Einstein's naïve yet understandable intentions. But the Mesa Temple is about none of that – although it is predicated upon the same next-paradigm Stillpoint/torus geometry that may lead to breakthroughs in those more pragmatic areas – including that of 'free' energy.

The temple is literally about making an attempt to open a doorway to the Infinite that the Stillpoint surely is and accessing the Akashic record of all experience – and to respond to the intelligence that embedded this geometry of creation in the dimensions of our Earth, Moon and Sun. I use the phrase 'surely is' because this is Kether, the Point of Creation at the top of the Tree of Life, the very interface between our world and the 'Limitless mind of God'; because Buckminster Fuller called the Stillpoint geometry 'the nearest approach we will ever know to eternity and God'; and because I experienced this directly myself and know it to be true. If totally awakened self-awareness has been the Primal Intention all along ('consciousness slowly and tediously approaching this apex'), then it now truly appears that evolving consciousness has reached the stage where *tools* can intentionally be used to assist greater humanity in experiencing the awakened state that has been its goal for all time – or, more accurately said . . . always.

This by itself is enough justification for making this 'temple experiment.' But there is another equally important reason for doing so . . . one that speaks more directly to us personally and to our fate.

The *experience* of the awakened state inherently brings with it the natural arising of compassion, the bodhisattvic compassionate understanding which transcends the seeming separateness of 'other' - complete in the realization of One Being. If there were a way to accelerate this process that has crept along for so long, and at a time in our long history when we need it the most - this is that time.

Yet we are immersed in a world that *believes* in the purely scientific view, as earlier times were immersed in religious superstitions. And please keep in mind . . . orthodox science, *especially* physics, deals with the inert, material world and is the flag bearer of materialism. Life and consciousness are hardly a part of that discussion – except perhaps as consciousness intrudes into the quantum world – something that has inspired many physicists, and scientists in general, to expand their worldviews. As has been mentioned, the incredible and 'impossible' phenomena that Kepler discovered relating the orbits of the planets to the Platonic solids was and is ignored - science moved on because it could make no falsifiable predictions based upon the information. Like someone holding up a cross to ward off a vampire, science completely ignores the information because it can't explain what it implied within its own terms. Today, the study of biology has become as reactionary as has physics been in the recent past.

But Arthur Young made the case in his process theory that the answer to 'why are we here?' was exactly the same answer, in eminently scientific terms, as the poetic definition offered by Meher Baba *(see quote, page 23)* . . . because primal purpose initiated the process of the evolution of consciousness – Young's monad driven by purpose manifesting first as the photon, the first quantum, and evolving through all the seven stages of evolution to our own 'Dominion' stage: 'Thus Creation comes at last to recognize itself' or, in Meher Baba's words, 'in order that God should consciously know his own fullness of divinity'.

Our Universe is the ultimate expression of the evolution of consciousness.

> 'It has occurred to me lately - I must confess with some shock at first to my scientific sensibilities - that both questions [the origin of consciousness in humans and of life from non-living matter] might be brought into some degree of congruence. This is with the assumption that mind, rather than emerging as a late outgrowth in the evolution of life, has existed always as the matrix, the source and condition of physical reality - that stuff of which physical reality is composed is mind-stuff. It is mind that has composed a physical universe that breeds life and so eventually evolves creatures that know and create: science, art, and technology-making animals. In them the universe begins to know itself.'　　George Wald (Noble laureate, biology)

Or, in Sri Aurobindo's words:

'But what after all, behind appearance, is the seeming mystery? We can see that it is the Consciousness which had lost itself, returning to itself, emerging out of it giant self-forgetfulness, slowly, painfully, as a life that is world-sentient, to be more than sentient, to be again divinely self-conscious, free, infinite, immortal.'

As mentioned earlier, in Arthur Young's process theory, he couldn't even theorize about its grand finale, its zenith - fully realized enlightened awareness transcending the need for physical manifestation, outside of time - it was, simply, 'ineffable.' Yet the beauty and accuracy of his theory predicted it.

The facts presented in this section, as well as the general summary that preceded it, are tangible, objective evidence proving the existence of such a fully awakened consciousness. There is no other way to explain it - or, if there is, I haven't heard it and I've been listening for it in earnest for some time. Our solar system was created intentionally for the evolution not simply of life . . . but of consciousness - by the highly *evolved* consciousness predicted by Young's process theory. This is an important distinction. While what is spoken about here is the same *presence*, all references to such consciousness that I know of refer to the 'Creator-God' - an always-existing, fully-formed presence, omniscient and omnipotent, ready-made and complete. This *presence* has always existed . . . but it, as everything, is in the process of evolving. The Stillpoint evidence suggests a 'god' in the *process* of growing and learning - innocent and unaware at the beginning of the journey; purpose gradually fueling awakening, driving evolution through the fixed laws of inert matter to the completely self-deterministic, free-will zone we humans find ourselves within.

Arthur Young's detailed process theory, entailing the seven stages of the evolution of consciousness from the photon of light to fully awakened awareness with no need to manifest and with the ability to create at will, included only two question marks in all of the forty-nine sub-stages - at the beginning of the last stage, the Dominion Kingdom of which Man is a part, and at its very end, the ineffable mentioned above.

It would be useful now to discuss his first question mark - how did Man, representing the kingdom just after the Animal Kingdom, appear here . . . and how did he make such advances in the last 5,000 or so years? While the 'missing link' myth still persists due to gaps in the geologic/fossil record, there really is no reason to believe that the 3 ½ billions-year-long progression of the evolution of life on this planet stopped just before Man showed up - unless, of course, one's beliefs support the idea of a 6,000 year old Earth. Perhaps as long ago as 200,000[145] years, Man in our present form, appeared. There is little in the archaeological record that demonstrates much progress until advanced symbolic cave paintings turned up at Maros in Sulawesi and in the caves of France and Spain around 35,000 years ago. Consciousness was slowly evolving – and it appears that a quantum leap in this process had occurred.[146]

In ways we don't yet know, civilization advanced in the next 20,000 years to have the capability of assembling huge, shaped, astronomically aligned megaliths . . . one weighing 1200 tons. From all around the world, similar evidence led to stories of lost civilizations . . . of Atlantis, mentioned first by Plato, from Solon 200 years previously, as a continent existing some 9,000 years before his time that vanished beneath the sea.[147] Extensive geologic evidence exists from all around the world that 11,500 years ago a great catastrophe occurred to life on Earth - most likely, a careening celestial body ricocheted into our solar system, coming too close to Earth - or perhaps a disintegrating comet hit the ice-cap.[148] Earthquakes tore apart huge areas, accompanied by vast volcanic eruption and finally by global flooding . . . what Western history calls the Great Flood, with similar stories found in prehistoric myths from peoples all around the world. Ancient stories exist, Sumerian in particular, about a rogue planet of some kind crashing though the solar system and coming so close to Earth that unimaginable fire and flooding and catastrophe engulfed the planet, essentially annihilating any advances made in the distant past.

The ruins of Gobekli Tepe in Turkey are thought to be at least 11,000 years old - before the time science tells us agriculture was invented; the discovery of man made megalithic stones called the Bimini Road off the coast of the Bahamas are thought to be over 12,000 years old; and the discovery of megalithic architecture inundated by fathoms of sea water off the coast of Japan, India, Egypt and other underwater areas. Regardless, after the Flood, civilization began to re-emerge, and rebuild itself, slowly emerging from the caves. Agriculture appeared early on, as well as the domestication of animals. Life plodded forward for the next 6,000 or so years . . . and then something happened. Myths exist from ancient cultures around the world speaking of 'gods from the sky' who provided critical initial information upon which to build a civilization, implying that these 'gods' came from the stars.

What we do know for sure is that a leap in the advancement of civilization occurred 5,000 to 6,000 years ago, along with stories of gods who came out of the sky to impart the information necessary for this advance. I believe this is related to the idea of non-intervention or non-violation mentioned earlier – this was guidance - and this transfer of information could have happened in a revelatory way. Lynn Picknett and Clive Prince mention in their book *The Forbidden Universe* that: 'A major component of the magical worldview hardwired into humanity is that specially trained individuals can enter into a state of communion with the gods in which they are given intensely *practical* information.' During this period writing appeared. The wheel was invented. Humans learned to combine tin and copper: the Bronze Age emerged. The agricultural revolution took off.

Discussed earlier in the section on *Time* were suggestions regarding *why* this jump in human consciousness occurred 5,000 years ago . . . and this is all woven into the mysterious phenomenon of the fluidity of the space-time reality. That is, probable futures can be foreseen within the awakened reality of the eternal present moment given the tendency of chreodes, or ruts in the evolutionary process, to determine these futures. And perhaps too, random events that are *not* a part of the evolutionary process as such . . . like the careening celestial body that came so close to Earth 11,500 years ago that the Great Flood of myth and history created total catastrophe on this planet . . . can be seen and anticipated: the Great Pyramid? (*See the discussion regarding the Great Pyramid, page 180*).

It is possible that *because* this random event happened, throwing civilization back into its past, we were given all this advanced information to prepare us for this uniquely particular time in history. That is:

> ' . . . 'they' knew that the opening to the inflow of spiritual wisdom that would accompany the winter solstice alignment with the center of the galaxy was fast approaching – the end of the Great Year, the return of the light – and that a boost of some kind was now required (again . . . compassion) in an effort to prepare us for this time. Perhaps they found it necessary to provide the tools required to rebuild by seeding Earth's struggling humankind with the knowledge needed . . . monumental architecture, mathematics, geometry – and perhaps most important of all, the *message* encoded in the geodetic unit called the Megalithic Yard found in ancient megalithic stone circles . . . a unit of measurement based upon the circumference of the Earth – yet found also when measuring the Moon and Sun. An encoded message letting us know 'they' existed and were communicating with us."[49]

But why did this come about so suddenly . . . and, especially, why all the monumental building? I think this is the key.

Encoded within all the building from all around the world at this time are the messages letting us know that all this was done intentionally by a higher consciousness and was done for a reason. But what reason? As has so often said now, the answer to all of this lies at the very heart of it – the blueprint for consciousness and the portal to higher awareness embedded in the heart of our solar system. A whole lot of work done on our behalf. It would be an incomprehensible shame to miss the opportunity. Many researchers suggest – outside of the orthodox scientific view - that our present civilization could never have reached such heights so quickly were it not for this kind of divine/extra-terrestrial, intervention.

As mentioned above, ancient myths from around the world speak of 'gods' dispensing essential information for survival and evolution. The 300 famous elongated *Paracas skulls* found in Peru in 1928 seem to validate the existence of extra-terrestrial beings, with cranial volumes up to 25 percent larger and 60 percent heavier than ordinary human skulls, and having one parietal plate instead of two. DNA tests have shown that they have mitochondrial DNA with mutations non-existent in any human, primate, or animal known.[150] Related to this, and also to Arthur Young's argument that the knowledge/wisdom we've come by could be due to visitations from other worlds, is Noble Laureate Francis Crick's *panspermia* theory (the word borrowed from it's original creator, the famous 17th century Jesuit monk and Hermeticist Aathanasiu Kircher[151]), where the seeds necessary for *life itself* to emerge on this planet were carried here in some kind of spaceship or asteroid: i.e. *we* wouldn't even *be* here if it weren't for higher consciousness from somewhere *else* – or by chance. The physicist/chemist Crick calculated that the chances of life simply happening by chance on Earth were one in in 200,000,000.[152] Other estimates add so many zeros that it could not *possibly* have happened by chance. But, it turns out that all the building blocks necessary for life are contained in the tails of comets . . . essentially seeding the Universe with the possibility of life – much like plants and trees who spread their seeds on the wind.

Accident? I think not. None of these estimates, incorporate the *incalculable* importance of *consciousness* and its purposeful evolution.

My point is only this: while it's important, it doesn't matter relative to this larger discussion. Evidence certainly demonstrates that extra-terrestrials have visited this planet in the distant past and most probably shared technology not available to us at the time, expressed in technological architectural feats that are not duplicable even now . . . the huge, 1200 ton Stone of the Pregnant Woman at Baalbek, Lebanon and the hundreds of tons, precision fit stones of Peru . . . and are visiting now, also sharing profound information with us through the geometric precision of the crop circles – some of them also impossible, by us, to duplicate. I feel it is undoubtedly true that the Universe is filled with a *Star Wars*, or *Star Trek* myriad of expression of varied intention. But *all* of this is an expression of the evolution of Universal consciousness manifesting at all levels of existence, including our own particular place in the process. But whether or not our present level of consciousness was aided by 'gods' from somewhere else, or that life on this planet was put here rather than evolved here, is unimportant relative to this larger context. The reality is: it *evolved somewhere*. It is the *evolution* of consciousness in the Universe that is the big picture. What *is* interesting to me about the possibility that higher, more evolved life-forms from somewhere else are responsible for where we are today, is the idea that a higher form of consciousness would help a lower form of consciousness evolve through guidance of various expressions. It is the bodhisattvic creation of our solar system all over again, although at a lower level of manifestation.

Even more importantly for our world today, is that a very highly evolved consciousness, in a purely bodhisattvic gesture (no such overt gesture is witnessed anywhere or anytime in what we observe of our Universe unless one believes the story of the parting of the Red Sea) has woven into the heavens the most important information that it has to share - the key to the doorway of our origins and of our destiny, the geometry of the stillness at the cosmic center – the eternal, present, fully enlightened *moment*. The Universe does not exhibit compassion . . . at best a benign indifference . . . but an *evolved* consciousness does.

The embedding of the Stillpoint geometry in the Earth/Moon/Sun was not done for information's sake. It was done for the most practical of reasons . . . with the intention that, when most needed, it would be discovered . . . and *used*. A response is being asked for. Unlike the previous stages of our billions of years of evolution, initially driven by unconscious purpose . . . the *whim* to awaken . . . *our* evolutionary future must be self-determined. It is in our hands.

The dimensions of the Earth, Moon and Sun expressing the non-manifesting Stillpoint geometry, as well as the Megalithic Yard which is not simply a *geodetic* unit, but a unit of measurement common also to the Sun and Moon, as well as other examples mentioned earlier, are phenomena making clear that this higher consciousness not only exists but is *communicating* to us. 'They' are telling us that they are here and want to convey extremely important information. The heart of the message is the symbolic expression of the Stillpoint geometry – the geometry of *consciousness*. In the Kabbalic wisdom, this is Kether, the *Point of Creation* and is the opening, or connection, to the world of Spirit, the implicate world on the other side of the veil, and to the Akashic Record of all of evolution's wisdom. I have verified that this is so . . . the Stillpoint *is* this opening that the Kabbalah names Kether, what Buckminster Fuller called *'the nearest approach we will ever know to eternity and God.'*

Our world is experiencing a crisis of consciousness – and it is only *now* that this information is appearing. 'They' are sharing with us the most critical information they have, and it is information critical to our future. This clearly is a bodhisattvic gesture . . . and only an *evolved* consciousness is capable of this. I believe they are asking us to *use* this information to create an opening from our direction to this enlightened awareness and to the Akashic record of all that is . . . to *create a resonance* with the wisdom and compassion inherent to such awareness – uplifting humanity by creating a shift in global consciousness.

⊕ WHO ARE WE?

Higher consciousness, an aspect of the general field of consciousness, operates at a very high frequency relative to our own. The basic aim in opening this portal is to create *resonance* with this higher frequency, uplifting our own. But who are we . . . what is our 'frequency' presently?

As a species, we cover the whole spectrum of contraction and expansion. The result of our general level of consciousness is the world we now live in. There is *so* much beauty here for which human beings are responsible. Here is a *very* short list of some of the people *we know of* that are responsible for such beauty (and please forgive my Western bias, the world I grew up in): the painting of Michaelangelo, Raphael, Cassatt, Van Gough, Kahlo, Vermeer and Rembrandt; the brilliance of the minds of Kepler, Newton, Da Vinci, Maxwell, Planck, Currie and Einstein; the writing of Shakespeare, Rumi, Steinbeck, Twain, Tolstoy, Baldwin, Wilde, Melville, Christie, Austen and Joyce; the poetry of Yeats, Blake, Naruda, Keats, Ginsberg, Plath, Thomas and Rilke; the music of Beethoven, Hildegard von Bingen, Bach, Mahler, Paganini, Tchaikovsky, and more recently the genius of Ray Charles, the blues of Muddy Waters, Lighting Hopkins, Big Mama Thornton and Jimmy Reed, and the poetry/songs of Dylan, the Beatles, Mose Allison, Joni Mitchell and Van Morrison, as well as the other-worldly riffs of Coltrane, Parker and Coleman; the architecture of Da Vinci, Barragán, Wright, Morgan, Piano and Bernini, as well as the nameless architects of the Mayan and Egyptian pyramids; the medical genius of Hippocrates, Pasteur, Harvey, Jung, Paracelsus, and Barnard; the sculpture of Michelangelo, Da Vinci, Degas, Rodin, Bartholdi and Andy Goldsworthy; the plays of Shakespeare, Williams, Euripides, Beckett, Miller, Aristophanes, Goethe and O'Neill; the courage of Lincoln, Gandhi, King, Malcolm X, Steven Biko, Douglass, Mandela, Rosa Parks and Spartacus . . . as well as countless unknown contributers.

You get the idea. Our potential for brilliance and beauty and courage and discovery and goodness seems infinite. So, what happened? Certainly those on the other end of the spectrum – the psychopaths who own, control and run everything - have stolen our future and seduce the weaker and less expanded to do their bidding. But it is the vast multitude of population somewhere in between with whom our hope lies. These are the billions of basically good people who just want to get on with their lives, wish no harm to others, care for their families and friends, and want simply to live a happy and harm-free life . . . they are the Papaganos of Mozart's *Magic Flute. These* are the people whose general level of consciousness, focused as it must be in today's world on survival, needs to be uplifted to an awareness that we are all in this together and depend upon each other – now more than ever – towards a new paradigm, a new world, a healthy future. But who are we . . . *now?*

Each of us is an individual, yet each is also a member of a larger group to which we consciously or unconsciously conform. Generally, if we wish to express our individuality, we tend to do this within accepted limits. I remember watching my son grow up and being influenced by his peers – even when he thought he was expressing his individuality. He thought that he was being unique with the clothes he wore . . . not realizing that he was conforming generally to the clothes worn by his friends. At one point, he wanted a tattoo. I would have been all for it had it been his own idea . . . but this happened when that entire generation decided that this was how they'd express their 'individuality.' Instead of individual expression, tattoos have become a way of fitting in. We *all* do this to some extent. Susceptibility to 'group-think' – as well as our ability to think for ourselves - is an aspect of our general level of consciousness. Following are three well known experiments and other examples that address different aspects of this phenomenon.

In 1951, Solomon Asch did experiments at Swarthmore College in Pennsylvania to determine just how much we are tempted to conform to the group consensus. The experiment consisted of eight people, seven of which were in on what was happening, one not. Three line segments were shown to all eight participants . . . one short, one medium and one long. Another line was presented to all of the participants that matched one of the original three lines. Each participant was asked, one by one, to say which of the three original lines it matched. The seven who were in on what was happening answered first . . . always the participant who was naive – the subject - answered last, after hearing each of the previous answers.

Those who were consciously a part of the experiment were all told which line to pick – regardless of whether or not it was the correct match. The subject, however, had no idea that the other students were not real participants. Imagine yourself in a similar situation: you've signed up to participate in a psychology experiment in which you are asked to complete a vision test. Of course *you* would not be affected by those around you to make such a simple and obvious decision! But, as it turns out, 75% of the naive subjects went along with the group at least once, with an average of 33% overall. Why? In most cases, the students stated that while they knew the rest of the group was wrong, *they did not want to risk facing ridicule*. A few of the participants suggested that they actually believed the other members of the group were correct in their answers. That is, not only is there a need to fit in, but a tendency to surrender to someone else's reality if it's perceived that they are smarter or more well informed. It could also be that individuals may have actually been motivated to *avoid conflict*, rather than an actual desire to conform to the rest of the group. Conformity can be even stronger in real-life situations – especially when money or threat to life is involved - where stimuli are more ambiguous or more threatening.[153]

Asch went on to conduct further experiments in order to determine which factor influenced how and when people conform. He found that:

- Conformity tends to increase when more people are present, but there is little change once the group size goes beyond four or five people.[154]
- Conformity also increases when the task becomes more difficult. In the face of uncertainty, people turn to other for information about how to respond.[155]
- Conformity increases when other members of the group are of a higher social status. *When people view the others in the group as more powerful, influential, or knowledgeable than themselves, they are more likely to go along with the group.*[156]
- Conformity tends to decrease, however, when people are able to respond privately or if they have support from at least one other individual in a group.[157]

I draw your attention to the words in italics above: 'When people view the others in the group as more powerful, influential, or knowledgeable than themselves, they are more likely to go along with the group.' Who are these 'more powerful, influential, or knowledgeable' people in general? They are those who've won the neo-Darwinist race to the top. *These* are often the worst and most spiritually ignorant of us . . . the psychopathic small percentage who find themselves CEOs of large corporations, presidents and dictators, the Kings and Queens of old, generals and the heads of intelligence agencies. *They* are the *models* in a civilization heading for disaster[158] – as the media shills who spew the propaganda fed to them by these 'elite' mindlessly fall in line and do their bidding. All kinds of people elbow and jab their competitors to earn jobs that will get them their piece of the pie . . . the pie we're all conditioned to want.

This level of needing to conform to the consensus of the group did not occur when there was only one more participant – or two or three. But when four or more other people choose the wrong line segment, the tendency to conform kicked in. The image that comes to mind is the photo below of a sea of people saluting 'Seig heil!' during a Nazi gathering attended by Hitler himself . . . *except one lone man* named August Landmesser, who defiantly stood with arms crossed. Would *you* be that man?

AUGUST LANDMESSER
'In a time of universal deceit, telling the truth is a revolutionary act.'
George Orwell

I grew up wondering how Nazi Germany could possibly have happened . . . how could 'following orders' possibly explain the horror. In 1961, Stanley Milgram, a psychologist at Yale University, designed an experiment to see just how far normal people would go after agreeing to participate in an experiment where they were asked to follow the directions . . . the orders . . . of an authority figure.

In the experiment there were two rooms – the image below. In the first room were two people - the authority figure, or 'experimenter' (E), and the volunteer, or 'teacher' (T), the subject of the experiment. In the other room is another person who is an actor, or confederate, called the 'learner' (L). The subject and the actor both drew slips of paper to determine their roles, but unknown to the subject, both slips said 'teacher.' The actor would always claim to have drawn the slip that read learner, thus guaranteeing that the subject would always be the teacher. At this point, the two were separated into the different rooms where they could communicate but not see each other. In one version of the experiment, the confederate was sure to mention to the participant that he had a heart condition. The teacher, or volunteer, would be asked to give the actor, or confederate in the other room, electrical shocks if answers to questions the teacher had asked were answered incorrectly. Before the test began, the teacher was given a sample electric shock from an electroshock generator in order to experience firsthand what the shock that the learner would supposedly receive during the experiment would feel like. The teacher was then given a list of word pairs that he was to teach the learner. The teacher began by reading the list of word pairs to the learner. The teacher would then read the first word of each pair and read four possible answers. The learner would press a button to indicate his response. If the answer was incorrect, the teacher would administer a shock to the learner, with the voltage increasing in 15-volt increments for each wrong answer. If correct, the teacher would read the next word pair.

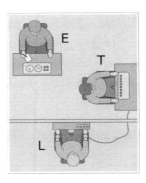

The subjects believed that for each wrong answer, the learner was receiving actual shocks who was hooked up to an electric-shock machine in the other room. In reality, there were no shocks. Each time the learner made a mistake in repeating the words, the teacher was to deliver a shock of increasing intensity, starting at 15 volts (labeled 'slight shock' on the machine) and going all the way up to 450 volts ('Danger: severe shock'). Each time a shock was administered, the electroshock generator played prerecorded, painful sounds, ostensibly coming from the learner being shocked, for each shock level.

Some people, horrified at what they were being asked to do, stopped the experiment early, defying their supervisor's urging to go on; others continued up to 450 volts, even as the learner cried out in pain and plead for mercy, yelled a warning about his heart condition - and then fell alarmingly silent. In the most well-known variation of the experiment, *a full 65 percent of people went all the way.*

Some test subjects paused at 135 volts and began to question the purpose of the experiment. At this point, many people indicated their desire to stop the experiment and check on the learner. *Most continued after being assured that they would not be held responsible.* A few subjects began to laugh nervously or exhibit other signs of extreme stress once they heard the screams of pain coming from the learner.

If at any time the subject indicated his desire to halt the experiment, he was given a succession of verbal prods by the experimenter, in this order:

1. Please *continue*.
2. The experiment requires that you *continue*.
3. It is absolutely essential that you *continue*.
4. You have no other choice, you *must* go on.

If the subject still wished to stop after all four successive verbal prods, the experiment was halted. Otherwise, it was halted after the subject had given the maximum 450-volt shock three times in succession.

The experimenter also gave special prods if the teacher made specific comments. If the teacher asked whether the learner might suffer permanent physical harm, the experimenter replied, 'Although the shocks may be painful, there is no permanent tissue damage, so please go on.' If the teacher said that the learner clearly wants to stop, the experimenter replied, 'Whether the learner likes it or not, you must go on until he has learned all the word pairs correctly, so please go on.'[159]

Incredibly, the experiment demonstrated that under the direction of an authority figure, ordinary people would obey just about any order they were given, even to torture. How could this possibly be? Because of the ethical challenge of reproducing the study, the idea survived for decades on a mix of good faith and partial replications, but, as the years passed, disbelief set in, with people gradually not being able to believe it, saying that was then, not now: people have changed.[160] But we haven't.

In 2007, ABC collaborated with Santa Clara University psychologist Jerry Burger to replicate Milgram's experiment for an episode of the TV show *Basic Instincts* titled 'The Science of Evil' . . . referencing the latest such scandal: Abu Ghraib. Apparently, ABC was wary about actually showing someone going all the way to 450 volts. Burger found that *80 percent* of the participants who reached a 150-volt shock continued all the way to the end, 'So what I said we could do is take people up to the 150-volt point, see how they reacted, and end the study right there,' he said.

One criticism of the experiment suggests that it is possible to learn to disobey toxic orders - a skill that can be taught like any other – but this seems irrelevant. It seems clear that the capacity for evil lies dormant in everyone, ready to be awakened with the right set of circumstances. At the time Milgram began his studies, the trial of Adolf Eichmann, one of the major architects of the Holocaust, was already in full swing. In 1963, the same year that Milgram published his studies, writer Hannah Arendt coined the phrase 'the banality of evil' to describe Eichmann in her 1963 book,

Eichmann in Jerusalem – that is, people's behavior is determined largely by what's happening around them. People in these situations are people like you and me - not psychopaths, not hostile, not aggressive or deranged - but in certain situations we're more likely to be racist or sexist, or we may lie, or we may cheat. There are studies that show this, *thousands and thousands of studies* that document the many unsavory aspects of most people.[161] The book *The (Honest) Truth about Dishonesty* is filled with many more benign, but pervasive, examples of this aspect of our nature.

The implications are devastating: If the Nazis were just following orders, then Milgram had proved that anyone could be a Nazi. If the guards at Abu Ghraib were just following orders, if Mai Lai could happen, if the Khmer Rouge could happen . . . then anyone was capable of these horrendous acts.

From the introduction by Nick Turse to an article by Rick Shenkman entitled *How We Learned to Stop Worrying About People and Love the Bombing*:

> 'Torturers, rapists, murderers: for more than a decade as I researched my history of the Vietnam War, *Kill Anything That Moves*, I spent a good deal of time talking to them, thinking about them, reading about them, writing about them. They all had much in common. At a relatively young age, these men had traveled thousands of miles to kill people they didn't know on the say-so of men they didn't know, and for a mere pittance - all of it done in the name of America.
>
> I also spent time talking to another group of men, a much larger contingent who stood by and watched as those beside them tortured or raped or murdered. Some heartily endorsed these acts, some seemed ambivalent about them, some were appalled by them, but none did much of anything about them.
>
> Then there was a third contingent of men: those who witnessed the torture, rapes, or murders and couldn't - wouldn't - abide by that conduct. *This tiny group* spoke out about what they had seen, often at the risk of their own welfare, sometimes their very lives.
>
> What differentiated these men from each other? They had all been raised in the same country, had been subject to the same laws and norms, including prohibitions against torture, rape, and murder. Many, if not most, had grown up in similar socio-economic circumstances, received comparable educations, and at least nominally belonged to churches with strict moral codes and an emphasis on doing unto others as you would have them do unto you. Why then did so many of them commit horrendous acts or stand by

while others did? Why did so few speak up?

I never came up with satisfactory answers to these questions. What I learned instead was that almost any man might be a torturer or a rapist or a murderer if given the chance. I learned that most men will look the other way if at all possible. And I learned that shockingly few men are capable of the courage and the empathy necessary to stand up for those that their brothers-in-arms would just as soon kill."[162]

A fascinating thing happened at a Trump event in January of 2016. Donald Trump, one of the two obscenities offered to be *elected* the next president of the most powerful country the world has ever seen – and the original ugly American - spews hatred for all kinds of things and people. Protesters at the event, wearing an eight-pointed yellow star with the words 'Stop Islamaphobia,' stood up in silent protest . . . one was a Muslim woman, Rose Hamid, who wore an al-amira type of hijab, or head scarf, and a shirt that read 'Salam, I come in peace' . . . were quickly removed from the gathering. Before this happened, Rose had many pleasant conversations with people who'd never met a Muslim before. Only after they were being escorted out did the group-think prejudice and hatred of the crowd click in. The point being that as individuals, if given the chance to meet and talk, we're more likely to be seen as rational human beings . . . but when seen through the group lens as the threat that Trump was yelling about, these human beings became terrorists – one man yelling at them as they were taken out 'Where's your bomb?!'

One of the protesters, a Jewish attorney, said: 'The thing that shocked me the most, after being now in three Trump rallies, is the reaction of the crowd and, like Rose was saying, people seemed to be really nice, but the more he speaks and the more he goes on ranting and raving, you can actually hear the hate and the fear grow in people's eyes.' Trump, after these peaceful protesters had been taken out, one with the message 'Salam, I come in peace' as well as the yellow badge with 'Stop Islamaphobia,' shouted 'There's such a level of hatred, you can't believe it. There's a deep seated hatred, we have to find out where it's coming from and what we can do about it' . . . this from someone who has inspired a petition in England seeking to ban him from speaking there because of existing laws there against hate speech.

It's very clear where the hatred is coming from. Rose, whose daughter knew the two young Muslim women murdered in a hate crime the year earlier,[163] said that 'It's hate speech that does this kind of stuff. It's *people in power* who empower people to do things from their lowest base, it's not how God made us, God did not make us to be murderous, hateful people. He made us to be loving people. And when people in power give license to that devil-inspired behavior, then that makes people feel

'Well, oh well, if so-and-so's saying that, it must be OK for me to feel this way and go on and do these things.'[164]

I wish to remind the reader that what is mentioned here is not political . . . it is about consciousness. In this country, we have seen the most absurd people propped up as candidates for president. It's surreal. In the past, as now, it was always money and strings that determined who we'd get to chose from . . . but the voter always had the final say and there often was an actual choice. Today, no one gets on that final ballot unless they're chosen . . . not by me and you, but by the true powers that be. No matter who you and I 'vote' into office, they will toe the elite line – although it appears that Trump struck a nerve with his hate speech and barged his way in. In 2016, the 'choice' was yet another totally impossible one: the one a bloviating, dangerous clown, the other an unspeakably dark nightmare.[165] But what is most fascinating is that both of them pandered to the crowd mentality mentioned above . . . a crowd so easily swayed by the scripted clichés they know will resonate with the consciously challenged. While it's easy to be fooled into focusing on the absurdity of the faux 'choices' offered, the real tragedy is that, due to the dumbing down of the American population (the consciousness problem), millions of people still went to the polls to 'vote.' The transparency of the absurdity and shallowness of the recent political campaign charade reached its zenith with the endorsement of Trump by the shrill Sarah Palin. Words can not do this surreal scene justice.

As this is being written, the war du jour has spread to Syria. Another candidate amongst this insidious mix, Ted Cruz, wanted to 'carpet bomb' Syrian rebels – to be clear, these 'rebels' are spread out amongst the land and people. Carpet bombing means bombing a wide area regardless of the human cost. Oh well, not a problem . . . his audience could care less. Bomb away. When Public Policy Polling asked GOP voters in mid-December if they favored bombing Agrabah, 30% said they did (as did 19% of Democrats), while only 13% opposed the idea. Agrabah is the fictional city featured in the Disney movie *Aladdin*.

> Would you support or oppose bombing Agrabah?
> Support bombing Agrabah 30%
> Oppose bombing Agrabah 13%
> Not sure . . . . 57%[166]

Well, at least 57% were 'not sure.' Room for hope . . . yes?

I am bringing this up in this section on Who *Are* We? to bring attention to our general level of consciousness: The bloodthirstiness of a large percentage of our population, the general level of ignorance and susceptibility to propaganda, and to make clear the total lack of care or empathy for those we do not know – *and* because this is who we are and have been.

A journalist working for the *Nation* magazine in the early months of the Korean War, Freda Kirchwey, visited Korea and was appalled by the destruction and the killing of civilians by a rain of fire bombs and napalm and unending artillery of all description – dams were targeted, eliminating huge areas where food was grown, 5 million people were displaced and homeless, 100,000 children were without parents. The head of the reconstruction agency in Korea, J. Donald Kingsley, said that 'Korea was the most devastated land and its people the most destitute in the history of modern warfare.'[167] But . . . naive optimist that she was . . . she was even more appalled and bewildered by the almost total indifference of the American public. She felt that Americans had been numbed into indifference by the preceding war . . . and the atrocities performed by the Germans, Japanese and finally honed to perfection by the Americans themselves at Hiroshima and Nagasaki. Regardless, she felt that nothing could justify the lack of protest against the 'orgy of agony and destruction now in progress in Korea.'[168]

No one seemed to care what our bombs were doing to other human beings. It is the same now, and by morbid coincidence we bombed 5 million Iraqis out onto the road and homeless. We are the same now. And lack of *consciousness* is the issue. Speaking about this indifference, this lack of caring for others, this lack of empathy is Rick Shenkman, in an article entitled *'How We Learned to Stop Worrying About People and Love the Bombing'*:

> 'Why did they show so little empathy [for the Koreans]? Science helps provide us with an answer and it's a disturbing one: empathy grows harder as distances - whether of status, geography, or both - increase. *Think of it as a matter of our Stone Age brains.* It's hard because in many circumstances an empathic response is, in fact, an unnatural act. It is not natural, it turns out, for us to feel empathy for those who look different and speak a different language. It is not natural for us to empathize with those who are invisible to us, as most bombing victims were and are. Nor is it natural for us to feel empathy for people who have what social scientists call 'low status' in our eyes, as did the Korean peasants we were killing. Recent studies[169] show that, faced with a choice of killing a single individual to save the lives of several people, we are far more apt to consider doing it if the individual we are sacrificing is of such low status. When subjects in an experiment are told that high-status people are being saved, the number willing to let the low-status victim die actually increases.'[170]

Killing little brown people, or those who don't look or speak the way we do, is but one of the countless justifications we use to not care – one of the more notorious examples being the 'manifest destiny' justifying the extermination of the indigenous population of North, Central and South America. And, to be fair, mass suffering worldwide seems to be an unstoppable steamroller . . . and what can we do anyway? Regardless, what is now 'natural' must become unthinkable. We *must* somehow evolve past our present level of consciousness towards a level of compassion and empathy for all around us.

Sometimes it is our environment that is so unnatural and stressful that we find other reasons to justify doing the unthinkable. In 1971, at Stanford University in California and called the Stanford Prison Experiment, sought answers to 'What happens when you put good people in an evil place? Does humanity win over evil, or does evil triumph?'[171] That is, how does a bad environment effect normal people? The creator of the experiment was Professor Philip G. Zimbardo.

Essentially, the experiment involved twenty-four volunteers who were arbitrarily separated into two groups – prisoners and guards – and asked to participate for two weeks and film the results. What happened was shocking . . . and the experiment had to be shut down after only six days, as the guards became sadistic while the prisoners became depressed and showed signs of extreme stress.

As I summarize what happened during those six days, put yourself in their shoes as best you can . . . and imagine the real experience of millions of people throughout time who have had these experiences, only unimaginably worse. We have a sense of who we are . . . but scrape off the thin veneer of 'civilization' that we assume about our lives, and the reality is something altogether different.

The students - an 'average' group of 'healthy, intelligent, middle-class males' ('average' in this case meaning *white* - black, brown and red people already aware) - had volunteered and were ready to participate in an experiment having to do with the psychological effects of prison life . . . except that 'On a quiet Sunday morning in August, a Palo Alto, California, police car swept through the town picking up college students as part of a mass arrest for violation of Penal Codes 211, Armed Robbery, and Burglary, a 459 PC. The suspect was picked up at his home, charged, warned of his legal rights, spread-eagled against the police car, searched, and handcuffed – often as surprised and curious neighbors looked on. The suspect was then put in the rear of the police car and carried off to the police station, the sirens wailing. The car arrived at the station, the suspect was brought inside, formally booked, again warned of his Miranda rights, finger printed, and a complete identification was made. The suspect was then taken to a holding cell where he was left blindfolded to ponder his fate and wonder what he had done to get himself into this mess.'[172]

Most of us – *if we're white* - have never had this experience . . . these kids certainly had not. *Many* times these arrests are unwarranted and this occurrence is accelerating in this country – America's domestic police force is rapidly being militarized.[173] The so-called War on Drugs in this country has imprisoned more people for no-harm crimes than any country in the history of the world. But think also of Nazi Germany, or Mao's China, or Stalin's Russia, or Pol Pot's Cambodia, or Pinochet's Chile, where intellectuals or Jews or artists or political activists, or whatever were rounded up and disappeared. *This* is the heart of the meaning of the Stanford Prison Experiment. Sometimes I mention the full-spectrum covert collection of data, stored now in a huge new facility in Bluff, Utah, to someone. The answer is usually: 'It doesn't concern me. I've done nothing wrong.' As a matter of fact, a young high school student in the area where I live recently won a speech contest by dismissing the entire NSA collection of data because: 'What do I have to worry about? I've done nothing wrong.' Ahhh, the boiling frog who has no clue that the temperature's rising and cares not for what's happening to others. I am reminded of the famous quote by Martin Niemöller, a German Lutheran pastor and theologian who, in 1937, was arrested and eventually confined in Dachau:

> 'When the Nazis came for the communists, I remained silent;
>   I was not a communist.
> When they locked up the social democrats, I remained silent;
>   I was not a social democrat.
> When they came for the trade unionists, I did not speak out;
>   I was not a trade unionist.
> When they came for the Jews, I remained silent;
>   I wasn't a Jew.
> When they came for me, there was no one left to speak out.'

The 'prison' was fashioned out of existing rooms in the basement of the Psychology Department, with a corridor serving as 'the yard,' the only 'outside' place the prisoners could be, except for using the bathroom. There also was a 24" square room called the 'hole' for solitary confinement. There was an intercom system to make announcements to the prisoners . . . and there were no windows or clocks.

The 'prisoners' were transferred, blindfolded, from city jail to the fabricated jail and informed by the 'warden' of the seriousness of their crimes. 'Each prisoner was systematically searched and stripped naked. He was then deloused with a spray, to convey our belief that he may have germs or lice – a procedure similar to others worldwide designed to humiliate as well as to make sure conditions were healthy. A uniform was issued – a smock with no underclothes. A heavy chain padlocked to the right ankle, an ID number issued, along with rubber sandals and a stocking cap [in lieu of a shaved head, as in all prisons and in the military]. The intention was to emasculate and humiliate – and immediately some of the prisoners began to walk and sit differently, more like a woman than a man. The chain reminded prisoners, even when sleeping, where they were. The ID numbers encouraged anonymity . . . individuality was eliminated as much as possible.[174]

The guards were warned of the potential seriousness of their mission and of the possible dangers in the situation they were about to enter, similar to real guards who voluntarily take such a dangerous job – *but were given complete leeway, within limits, regarding rules and how to go about keeping order*. These last words in italics are critical: no 'orders' were given. The 'guards' accommodated the experiment as they say fit . . . on their own.

The prisoners, as their consent agreement made clear, expected some harassment, to have their privacy and some of their other civil rights violated while they were in prison, and to get a minimally adequate diet. Guards all dressed alike, in khaki with mirrored sun glasses, again to create anonymity.[175]

The experiment began with nine guards and nine prisoners. Three guards worked each of three eight-hour shifts, while three prisoners occupied each of the three cells around the clock. The remaining guards and prisoners from the sample of 24 were on call in case they were needed. The cells were so small that there was room for only three cots on which the prisoners slept or sat, with room for little else.

The guards, justifying their positions, began imposing rules and punishments. In response, the prisoners began resenting and resisting. The *first* night, at 2:30 A.M., the prisoners were rudely awakened from sleep by blasting whistles for the first of many 'counts' - the beginning of a series of direct confrontations between the guards and prisoners. Push-ups were a common form of physical punishment imposed by the guards to punish infractions of the rules or displays of improper attitudes toward the guards or the institution. A totally unprepared-for rebellion broke out on the morning of the second day. The prisoners removed their stocking caps, ripped off their numbers, and barricaded themselves inside the cells by putting their beds against the door. The guards, angered, *their authority challenged*, had to handle the rebellion themselves.[176]

They insisted that reinforcements be called in, and all nine guards confronted the prisoners. The guards met and decided to treat force with force, using a fire extinguisher to shoot a stream of skin-chilling carbon dioxide, forcing the prisoners away from the doors. 'The guards broke into each cell, *stripped the prisoners naked*, took the beds out, forced the ringleaders of the prisoner rebellion into solitary confinement, and generally began to harass and intimidate the prisoners.'[177] The guards realized that they couldn't all be there all the time and because of this decided to use *psychological tactics instead of physical ones*. A 'privilege' cell was set up.

Immediately, three of the prisoners who'd had little to do with the rebellion were given special privileges – their uniforms and beds were returned, they were given special food and they were allowed to wash and to brush their teeth. Any solidarity between prisoners was lost. In an act of what seems almost sadistic, irrational, behavior, the guards now arbitrarily put the 'bad' prisoners in the privilege cell, and the 'good' prisoners in the cramped cells. This caused distrust among the prisoners, thinking that others were informers - a similar tactic is used by real guards in real prisons to break prisoner alliances. The tried and true 'divide and conquer' used against Native Americans, Blacks, Chicanos and Anglos in prisons all around America. In this way, guards promote aggression among inmates, thereby deflecting it from themselves.[178]

So . . . now there's a situation, created by the guards in the first place, that forced the guards to pit prisoner against prisoner rather than at themselves. The guards now perceived the prisoners as out to get them and cause harm. In response to this threat, *the guards began stepping up their control, surveillance, and aggression.* (I refer the reader to books on the growing police state in America by John Whitehead - *Battlefield America* and *Government of Wolves: The Emerging American Police State*, and Radley Balko - *Rise of the Warrior Cop*). Thousands of small and large police stations around the country – the 'guards' - are being armed to the tune of $34 *billion* in used military equipment and they are using it like a hammer looking for a nail, creating conflict where there was none.

An escape was rumored to be imminent. Major adjustments were made. The escape – real or not – never happened. The guards resented being forced to take the unnecessary actions, again escalated their level of harassment, increasing the humiliation they made the prisoners suffer, forcing them to do menial, repetitive work such as cleaning out toilet bowls with their bare hands. The guards had prisoners do push-ups, jumping jacks, whatever the guards could think up, and they increased the length of the counts to several hours each.[179]

By this time everyone was taking all this very seriously and had lost sight of the true nature of the 'experiment.' When a priest was added to the mix, asking the prisoners what they were going to do once they'd gotten out, explaining that they'd need the services of an attorney to do that, seemed to make it all too real. One inmate refused to see the priest and wanted to see a doctor. He broke down and began to cry hysterically, just as had two other boys that had been released earlier. When he was offered food and a doctor, one of the guards lined up the other prisoners and had them chant aloud: 'Prisoner #819 is a bad prisoner. Because of what Prisoner #819 did, my cell is a mess, Mr. Correctional Officer!' They shouted this statement in unison a dozen times, pouring blame on the maverick while saving themselves - the boy sobbing uncontrollably while in the background, his fellow prisoners yelling that he was *bad*. No longer was the chanting disorganized and full of fun, as it had been on the first day. Now it was marked by utter conformity and compliance, as if a single voice was saying, '#819 is bad.' He refused to leave, even though sick, and wanted to go back and prove he was not a bad prisoner. He wanted to retreat to the safety of the group.

Another element was introduced . . . a Parole Board. By this time it was becoming clear that reality had shifted for those in the experiment. Prisoners were willing to forfeit money earned (as in some prisons) to be released. Because they now felt powerless to resist, they all went peacefully back to their cells rather than simply opt out. Their sense of reality had shifted - they no longer perceived their imprisonment as an experiment. Similarly, the prison consultant (a former prisoner) that headed the Parole Board literally became the most hated authoritarian official imaginable. Afterwards, he felt sick that *he had become his own tormentor* - the man who had previously rejected his annual parole requests for 16 years when he was a prisoner.

There were three types of guards: a third were tough but fair, a third were 'good guys' who never punished anyone, and the last third were hostile, arbitrary, and inventive in their forms of prisoner humiliation. These guards appeared to thoroughly enjoy the power they wielded, yet none of our preliminary personality tests were able to predict this behavior. The only link between personality and prison behavior was a finding that prisoners with a high degree of authoritarianism endured the authoritarian prison environment longer than did other prisoners. One guard surpassed all the rest and was labeled 'John Wayne.' 'Where had he learned to become such a guard? How could he and others move so readily into that role? *How could intelligent, mentally healthy, 'ordinary' men become perpetrators of evil so quickly?'*

Prisoners also dealt with their frustration and powerlessness in many ways.

'At first, some prisoners rebelled or fought with the guards. Four prisoners reacted by breaking down emotionally as a way to escape the situation. One prisoner developed a psychosomatic rash over his entire body when he learned that his parole request had been turned down. Others tried to cope by being good prisoners, doing everything the guards wanted them to do. One of them was even nicknamed 'Sarge,' because he was so military-like in executing all commands. By the end of the study, the prisoners were disintegrated, both as a group and as individuals. There was no longer any group unity; just a bunch of isolated individuals hanging on, much like prisoners of war or hospitalized mental patients. The guards had won total control of the prison, and they commanded the blind obedience of each prisoner.[1180]

Yet another fascinating event happened. A new prisoner arrived to replace one that had left . . . and, unlike the gradual indoctrination experienced by everyone else, he experienced it head-on and was horrified. He went on a hunger strike. Like Luke in the film *Cool Hand Luke*, he was thrown in solitary and kept there longer than the rules allowed. But, instead of being a hero, the others saw him as a troublemaker. The head guard then exploited this feeling by giving prisoners a choice. They could have the striking prisoner come out of solitary if they were willing to give up their blanket, or they could leave him in solitary all night. *Most elected to keep their blanket and let their fellow prisoner suffer in solitary all night.*

By this time things were getting out of control. A parent asked that a lawyer be contacted to get their son out of 'prison.' Prisoners were withdrawing and behaving in pathological ways, and some of the guards were behaving sadistically and were escalating their abuse of prisoners in the middle of the night when they thought no researchers were watching. Their boredom had driven them to ever more pornographic and degrading abuse of the prisoners – Abu Ghraib anyone? Even the 'good' guards felt helpless to intervene, and none of the guards quit while the study was in progress. Indeed, it should be noted that no guard ever came late for his shift, called in sick, left early, or demanded extra pay for overtime work. When a recent Stanford Ph.D. was brought in to conduct interviews with the guards and prisoners, she was shocked by what was happening and strongly objected when she saw our prisoners being marched on a toilet run, bags over their heads, legs chained together, hands on each other's shoulders. Filled with outrage, she said, 'It's terrible what you are doing to these boys!'

Out of 50 or more outsiders who had seen our prison, she was the only one who ever questioned its morality.

From the prisoner who attempted the hunger strike:

'I began to feel that I was losing my identity, that the person that I called Clay, the person who put me in this place, the person who volunteered to go into this prison – because it was a prison to me; it still is a prison to me. I don't regard it as an experiment or a simulation because it was a prison run by psychologists instead of run by the state. I began to feel that that identity, the person that I was that had decided to go to prison was distant from me – was

151

remote until finally I wasn't that, I was #416. I was really my number.'

An excerpt from a letter written to Dr. Zimbardo from an inmate in an Ohio penitentiary after being in solitary confinement for an inhumane length of time:

'I was recently released from solitary confinement after being held therein for thirty-seven months. The silence system was imposed upon me and if I even whispered to the man in the next cell resulted in being beaten by guards, sprayed with chemical mace, black jacked, stomped, and thrown into a strip cell naked to sleep on a concrete floor without bedding, covering, wash basin, or even a toilet . . . . I know that thieves must be punished, and I don't justify stealing even though I am a thief myself. But now I don't think I will be a thief when I am released. No, I am not rehabilitated either. It is just that I no longer think of becoming wealthy or stealing. I now only think of killing - killing those who have beaten me and treated me as if I were a dog. I hope and pray for the sake of my own soul and future life of freedom that I am able to overcome the bitterness and hatred which eats daily at my soul. But I know to overcome it will not be easy.'

I must interject the following here: Before America illegally and immorally attacked Afghanistan and Iraq there were essentially no huge 'terrorist' movements operating in and from those countries – despite assurances to the contrary. The insanity that we are witnessing in the Middle East, the so-called 'terrorism' – and the millions of refugees, and the million murdered and being murdered - was *created by the invaders*. The deep state in America, hypocritically and ingenuously, created the terrorists, just as the unfair punishment created the inmate above who hoped he could 'overcome the bitterness and hatred which eats daily at my soul.' The war in Iraq a failure? Quite the opposite for those that profit from it.

Just after this experiment ended, a riot erupted at Attica Prison in New York. None other than Nelson Rockefeller – a card-holding member of the controlling class and then governor of New York - ordered the National Guard to take back the prison by full force, killing and injuring many guards and prisoners through that ill-advised decision . . . oh so typical for those privileged men in charge when, in this case, the prisoners at Attica simply wanted to be treated like human beings. It took the Stanford Prison Experiment only *six days* to see how prisons dehumanize people, turning them into objects and instilling in them feelings of hopelessness - with guards who were ordinary people readily transformed from the good Dr. Jekyll to the evil Mr. Hyde. On a larger scale this is Guantanamo and Abu Ghraib, and larger still Hitler's Germany and Stalin's Russia.

There is clearly the need to transform our institutions so that they promote human values rather than destroy them. Unfortunately, we need to change ourselves if there's any chance we can change our institutions to insure a healthy future – human nature, as it is now and has been for at least 11,500 years, must change. In the decades since this experiment took place, prison conditions and correctional policies in the United States have become even *more* punitive and destructive. The worsening of conditions has been a result of the politicization of corrections, with politicians vying for who is toughest on crime, along with the racialization of arrests and sentencing, with African-Americans, Hispanics and Native Americans overrepresented. The media has also contributed to the problem by generating heightened fear of violent crimes even as statistics show that violent crimes have decreased.[181] This is but one way to divide and conquer, pitting races against one another in almost impossible situations.

There are more Americans in prisons than ever before. According to a Justice Department survey, the number of jailed Americans more than *doubled during the past decade*, making the *United States* the world's leader in incarceration with 2.2 million people currently in the nation's prisons or jails.[182] In 1971, when this experiment was conducted, there were 200,000 people in prison . . . today there are 2.2 *million – 1100% in the last 45 years*. The United States has only 5 percent of the world's population, yet it has nearly 25 percent of its prisoners.[183] Hello?

How did this situation become so unbalanced . . . how did *we* vote in the laws that made all this possible? The answer lies in the susceptibility and innate tendencies exposed in the experiments above. Our general level of consciousness . . . our present 'human nature' . . . not only still operates instinctively from the reptilian/survival aspects of our brain, but is very susceptible to 'suggestions' from those in authority. The majority of us are eminently 'hypnotizable.'

'To study any phenomenon properly, researchers must first have a way to measure it. In the case of hypnosis, that yardstick is the Stanford Hypnotic Susceptibility Scales. The Stanford scales, as they are often called, were devised in the late 1950s by Stanford University psychologists Andre' M. Weitzenhoffer and Ernest R. Hilgard and are still used today to determine the extent to which a subject responds to hypnosis. One version of the Stanford scales, for instance, consists of a series of 12 activities - such as holding one's arm outstretched or sniffing the contents of a bottle - that test the depth of the hypnotic state. In the first instance, individuals are told that they are holding a very heavy ball, and they are scored as 'passing' that suggestion if their arm sags under the imagined weight. In the second case, subjects are told that they have no sense of smell, and then a vial of ammonia is waved under their nose. If they have no reaction, they are deemed very responsive to hypnosis: if they grimace and recoil, they are not. Scoring on the Stanford scales ranges from 0, for individuals who do not respond to any of the hypnotic suggestions, to 12, for those who pass all of them. Most people score in the middle range (between 5 and 7): 95 percent of the population receives a score of at least 1."[184]

My friend Bill once bought a pamphlet on hypnotism and began experimenting with people he knew. He was appalled that he, with no training, could influence the susceptible to easily. He never tried it again. What these studies and others like them tell us about ourselves is that the majority of us are relatively easy to hypnotize, that we will tend to go along with a group we are a part of rather than think for ourselves, that if we are in a situation where we need to follow the orders of an authority figure we will cause pain to and torture other people past our own sense of what is right, and that in difficult circumstances we will do whatever it takes to survive in a way our ego demands, regardless of the consequences to others.

That is: we are very susceptible to group-think and make sure we look after ourselves first and last. I would think that humanity's susceptibility takes the shape of a huge Bell Curve – the vast majority being in the middle areas, with extremes of the psychopathic element at one end, and the spiritually enlightened and expanded at the other. Those in control – the people responsible for the wars, the climate change, the draconian laws, the poisoning of our food and air and oceans, the destruction of the rain forests . . . and so much more – know this about us. They have us pretty much where they want us . . . and they have all the systems in place once it gets too difficult for one reason or another to keep us there.

Beyond the very real fact that 'they' have our number and have the systems in place once it's determined time to call it, the three psychological experiments mentioned: The experiment by Solomon Asch at Swarthmore College exposing the unconscious pressures upon us to conform to group-think, the experiment by Stanley Milgram at Yale regarding the dynamic that occurs when both issuing and taking orders from those in authority, and Philip Zimbardo's Stanford Prison Experiment, regarding what happens when otherwise good people find themselves in a heinous environment, as well as our susceptibility to suggestion, all point to the malleability of our general level of consciousness. A sobering exposé of how a combination of such factors can provide the perfect storm for someone *voluntarily signing up for the program* - initially with the hope of bettering themselves and the world - is brilliantly provided in Alex Gibney's film *Going Clear,* an in-depth look into the world of the cult . . . in this case Scientology. From my own experience with such groups, the film is dead-on. The group-mind is a vulnerable and easily manipulated tool in the hands of those who know how to use it.

How could it be possible for this reality to change? The elite, the psychopathic minority that owns the corporations and governments with all their guns and planes and bullets and drugs and food and endless money, are not going anywhere. At this point, they totally depend upon us to accomplish their agenda. How could there be any possibility that this vast majority of human beings covering a range of needs and agendas and levels of consciousness, could possibly lift up and peel the yoke off our backs, their boots from our necks? Any ideas? A political movement? Social reform? Economic reform? Meditation? Nothing that may have worked in the past at whatever scale and for however long will work now. But people *are* waking up all over the world . . . right?

In a 2000 book called *The Cultural Creatives: How 50 Million People Are Changing the World,* sociologist Paul H. Ray and psychologist Sherry Ruth Anderson suggest that there is a large and growing minority of people around the world who share an awareness and care about our environment and other critical matters. After thirteen years of survey research studies on more than 100,000 Americans, plus more than 100 focus groups and dozens of in-depth interviews, they believe they have detected a new sort of consciousness appearing in the last generation.[185] This phenomenon couldn't have happened at a more critical time in our collective history, and implies the collective process of evolution emerging just when needed.

These people, called Cultural Creatives, are those who 'care deeply about humanity and are committed to saving the planet and pursuing social justice through self-actualization and/or spiritual wisdom. In contrast to Moderns (those who accept current mainstream systems) and Traditionals (those who reject current mainstream systems and look backwards for solutions), Cultural Creatives inwardly depart from the Modern materialistic worldview and seek to go beyond current systems to bridge an old way of life with a new, more sustainable, way of doing things.'[186]

While a creative minority such as this could make a huge difference . . . *if organized in some way* . . . I wonder what difference it could possibly make given all that's mentioned above. These are people, as determined by the study, who share common beliefs and outlooks on the world: they love nature and are concerned regarding its destruction; they are aware of global problems such as global warming, over population, exploitation of the poor, rainforest destruction; are willing to pay more taxes towards healing the environment; believe in helping others; will tend to do volunteer work; care about psychological and spiritual development; see spirituality as important, but are concerned about the fundamentalist right; want more equality for women; are concerned about violence and abuse of women and children around the world; want money to be spent on education and an ecologically sustainable future; are disenchanted with both the Left and the Right in politics; tend to be optimistic about the future; want to be involved with citizens creating a better life; are concerned about corporations and their drive towards profits at all costs while creating environmental problems and the exploitation of poor countries; are repulsed by the modern world's obsession with 'success' and the money and luxuries it brings; and enjoy exotic and foreign ways of life.[187] Sounds like you, yes?

Except for one of the subjects mentioned, I qualify as a bona fide Cultural Creative. But I, like so many like me, have little or no way of making any real difference at the scale needed. The friends I have generally qualify also . . . but we are a disparate group with many opinions about many things and as a whole do not make a dent in what is happening. It's wonderful that this group of people exists and is growing, but how in the world could they organize to produce the change needed?

And there is yet another factor in our collective behavior that shifts the scale towards collective failure. It is called *willful blindness.* In her book *Willful Blindness: Why We Ignore the Obvious at Our Peril,* author Margaret Heffernan examines the complicated cognitive and emotional mechanisms by which we choose, mostly unconsciously, to remain unseeing in situations where 'we could know, and should know, but don't know because it makes us feel better not to know.'[188] We do that, Heffernan argues and illustrates through a multitude of case studies ranging from dictatorships to disastrous love affairs to Bernie Madoff[189] – and we are doing that now at a global scale with global warming, war, starvation, rainforest devastation, the inhumanity of the meat industry . . . transhumanism . . . etc. Certainly all this is overwhelming. It really *is* too much to give one's attention at the scale needed . . . but a large part of this is that 'it makes us feel better not to know.' This is a common reaction to the cognitive dissonance our present world demands of us.

> 'Our blindness grows out of the small, daily decisions that we make, which embed us more snugly inside our affirming thoughts and values. And what's most frightening about this process is that as we see less and less, we feel more comfort and greater certainty. We think we see more — even as the landscape shrinks.'[190]

Interestingly, the concept of the 'willful blindness' comes from 19th century law, where being blind to an offense when one *should* have known but purposefully chose not to know. That is, one is being made libel for unconscious actions because one should have known. The problem with this of course is that we should not need a law to make us be responsible, moral, or ethical . . . it must come spontaneously from within – and getting to that place is what the Stillpoint message is entirely about.

In a TED talk that she gave, Heffermen mentions studies that have been done within institutions, corporate workplaces, where willful blindness is a powerful force. A typical question to employees was 'Are there issues at work that people are afraid to raise?' The percentage of those who answer 'yes' to this question is 85%: '85% of people know that there's a problem, but won't say anything.'[191] She went on to say that this percentage was exactly the same with the work she was doing in Europe. She makes it very clear that this is not a symptom of one particular country from the next, but that:

> ' . . it is a human problem. We're all, under certain circumstances, willfully blind. What the research shows is that some people are blind out of fear, they're afraid of retaliation. And some people are blind because they think that if they see anything, well, it's just futile, nothing's ever going to change. If we make a protest, if we protest against the Iraq war, nothing changes so why bother, better not to see this stuff at all. And the recurrent theme that I encounter all the time is that people say 'Well, you know, the people who do see, they're whistle blowers, and we all know what happens to them.'[192]

This reality hit home for me in 2005. It seems so long ago now, like a different lifetime, but in Spring of 2003 I happened to be in San Francisco and took the opportunity to march . . . along with more than 30 million people world-wide . . . to protest the almost unimaginable reality the the U.S. was about to go to war with Iraq, based upon what to many of us were obviously lies. There was a spirit in the air, identical to other marches I'd participated in during the Vietnam war, that our numbers could not be ignored. This tactic actually worked back then. But something had changed. Hellish war broke out and Iraq was destroyed, their people broken . . . and none of the infamous 'weapons of mass destruction' were ever found. Oooops. Thousands of 'allied' dead and a million Iraqi deaths later, the lie that got us there hung in the air. In 2005, I traveled to the east coast to visit my son and we met in Washington D.C. to take part in yet another march against the war. This time something had already changed in me. I had seen the film *Loose Change* by the young Dylan Avery which unraveled the government's unproven conspiracy theory about 9/11, and my research into the 'dark' had begun. It was a new world for me. I had come to terms with the dark reality regarding our government that was unimaginable before. And something had also changed with the other people marching. I remember vividly marching past the White House and *knowing* that the people inside had no care or concern whatsoever regarding the many thousands marching past their windows. One could feel this awareness in the crowd also. Gone was the old spirit alive in the air of truth to power, gone was any sense that this could possibly make a difference. We were just going through the motions because there was nothing else we could do.

In times past what happened to people who spoke truth to power was a simple 'off with their head!' Today, in this country, it is more subtle, but the process is the same. People like Chelsea Manning, Edward Snowden, Julian Assange, William Binney, Jeffry Sterling, John Kiriakou, Thomas Drake, Barrett Brown, Michael Hastings, Jeremy Hammond and others know all too well what will be aimed at them if they did the right thing and spoke up . . . but speak up they did. And don't forget the courage of Barbara Lee, the *one* Congressperson who stood up (reminiscent of August Landmesser, the one man not saluting Hitler in a mass of people doing so (*see page 139*)) - *one* out of 535 (.002%) - to vote against the war in Iraq . . . a war illegally imposed upon the world, a war that's murdered well over a million people and is still going on today.

What is gained is a light in the eye, a self-respect and clarity that is obvious when listening to them speak. The risk was there and they took it and they freed themselves from the tyranny of cowardliness . . . but most of us – apparently 85% of us – will not cross that line unless there are a whole lot of people crossing it with us. And because of this, because 'it makes us feel better not to know,' we choose to be willfully blind and pad ourselves within the comfort of ideologies that protect this comfort. This is who we presently are. Unfortunately, this creates a world where any effort to go against the flow, against the dark status quo, will be met with all sorts of resistance and will require endless patience and courage and strength and a willingness to be in conflict. Who wants *this*? Thank God some people do. *This* is the world we live in. It doesn't have to be this way. Wouldn't it be something to live in a world where ethical, moral behavior was simply the way we *are* – not something imposed upon us by law? But presently there is no possibility to effect the change required other than depending upon the few brave enough to challenge the powers that be . . . 'and we all know what happens to them.'

Anybody noticed what happened after all the hoopla surrounding the Assange/Wikileaks revelation . . . after James Clapper, Director of National Intelligence lied to Congress about 'not-wittingly' spying on people, after Obama lied to the world about the depth of the surveillance? Answer: nothing. Remember Anonymous . . . the faceless hactivists who broke into the servers at the CIA,

Stratfor (a private CIA), the KKK, Monsanto and many other members of the darker side of our world to embarrass and expose that hidden world? The FBI threatened one of the hackers with 124 years in prison . . . he flipped and essentially undermined the entire movement.

And then there's Aaron Swartz. In 2010, Swartz co-founded Demand Progress, a political advocacy group that organizes people online to 'take action by contacting Congress and other leaders, funding pressure tactics, and spreading the word about civil liberties, government reform, and other issues.' [193] Aaron was an activist trying to wake people up to what was going on. In early 2011, after connecting a computer to the MIT network in an unmarked and unlocked closet, and setting it to download academic journal articles systematically from JSTOR (Journal Storage), Swartz was arrested by MIT police on state breaking-and-entering charges. His intention was nothing more than to make these academic journals free to students. Federal prosecutors later charged him with two counts of wire fraud and eleven violations of the Computer Fraud and Abuse Act, carrying a cumulative maximum penalty of *$1 million in fines, 35 years in prison, asset forfeiture, restitution, and supervised release.* The government wouldn't back down and Aaron killed himself. The government's power is unequaled.

> 'They can watch us through our webcams, they can listen through our phones. There's no privacy left . . . virtually. We are the most watched, surveilled, monitored, eavesdropped population in the history of the human race.'
> Chris Hedges, Pulitzer Prize winning journalist

My point is only that while more and more people are becoming aware . . . and care . . . about what is happening, there is no way to focus this care towards dealing with the present crisis . . . and we are quickly running out of time.

The Stone Age brain, the reptilian brain, the survival/fear mechanism that got us here . . . needs to be transformed and transcended. This will not happen, as has now been said many times, with the old methods . . . the old-paradigm methods we've used for thousands of years that provided band-aids and temporary relief from the onslaught of those who gained and inherited power and control through efforts driven by fear and greed. What is needed *now* is nothing less than a quantum leap in global consciousness. The kind of consciousness that I mean is beyond our present comprehension . . . and is best described by an old story about the Buddha. There are many versions of this story, but the following is just about the most beautiful telling I've heard. From *Becoming a Bodhisattva* by Fotopoulou Sophia:

'In the Jataka Sutra, a sutra about the previous lives of the Buddha, there is a story about a time when the Buddha was cultivating to be a bodhisattva.[194] In this particular life, the Buddha was also born as a prince. One day, when he was out traveling in the woods with two of his brothers, he saw below a cliff a mother tiger that had just given birth to seven baby cubs. Because of over-exertion, the mother tiger became so weak that her life was hanging in the balance. In the meantime, the baby cubs were all crying to be nursed. When the prince saw how pitiful the situation was, his compassion arose in him, and he decided to sacrifice his life to save the life of the mother tiger. He distracted his two brothers and jumped down to where the mother tiger was so that he might offer himself as a meal for the mother tiger. The mother tiger was, however, so weakened that she did not even have the strength to feed on him. Anxious to save the tigress, he used a sharp blade of bamboo bark to severe his own throat. With the blood gushing out, and disregarding his own pain, the prince slowly crawled to the side of the mother tiger so that she could drink his own blood. In giving up his own life, he was able to save the life of the mother tiger and her cubs.

Compassion allowed this prince to forget his own fears and give up his own life for the sake of others. With compassion, bodhisattvas perform many selfless acts for us sentient beings. Because of the rich compassion that bodhisattvas have for us sentient beings, bodhisattvas are very forgiving of our folly and mistakes. They are so willing to make sacrifices without any regard for themselves that they reach the point of selflessness. Without regrets and fear, bodhisattvas practice their great compassion, just like the saying, 'For the sake of sentient beings, [I am] willing to part with anything.'[195]

I remember, so long ago now, the image of a Buddhist priest setting himself on fire to protest the war in Vietnam. I will never forget it. Many have taken this path since that time. 'At least 130 self-immolation protests by Tibetans living under Chinese rule have taken place since 2009; over 85 in 2012, with *28 in the month of November 2012 alone*.'[196] These people, usually young and male, take their own lives in sacrifice of the greater good . . . in this case, calling for freedom for Tibet now under oppressive Chinese rule, and the return of the Dalai Lama. Thousands of Indian farmers have taken to drinking cupfuls of insecticide to kill themselves in protest to being bankrupted by Monsanto.[197] This action is beyond imagining for most of us . . . yet strikes dead-center at the heart of what ails us. And *none* of these heroic deaths caused the machine to even blink.

The kind of consciousness displayed by these heroic gestures, and by the not-yet-Buddha, depict *who we need to be* . . . and not who we *are*.

After immersing myself in all this darkness for so long now, it feels like we're all a part of the herd. I want to say again that I feel that the vast majority of human beings are good, well-intended people, wanting simply to live their lives, not wanting to harm anyone. But when it comes to the need to conform, to not take the risks necessary for real change, the majority of us are certainly sheep afraid of the dogs nipping at our heels and quickly jump in line, while some are angry lions ready to fight and others wise old elephants who see into the past and intuit the future . . . but all a part of the herd kept so skillfully in control.

And then there's that growing group of us now called 'Cultural Creatives' who write articles and books and make films and initiate movements and believe in all the right things . . . but if anyone gets too close to that electric fence (or taser), there's a quick about-face towards safety at the center of the herd. If one of us raises our head a bit too much . . . Martin Luther King, Steven Biko, Nelson Mandela, Abraham Lincoln, Leonard Peltier, John and Robert Kennedy, Malcolm X, and an endless list of courageous political and spiritual figures . . . we get one right between the eyes. The only strength left to us is in our numbers . . . but we are thrown into class conflicts designed by the rulers and our numbers mean nothing. What could possibly unite us? What could possibly unite us at the deep spiritual level where the power taken by and given to the elite gradually becomes redirected, the hamster wheel destroyed, the electric fences torn down, the oceans and forests and environment healed?

⊕ CRISIS:

We are now in the collective crisis of our many thousands-year-long history. We have cut our forests down; fished out the oceans; live in a world where over 3 billion people make less than the equivalent of $2.50 a day, 50,000 of these people dying *each* day of starvation[198] - 85% of them children;[199] annihilated countless species; poisoned our water and our air and the crops we grow; almost destroyed this connected ecosystem we are a part of; murdered millions of human beings and destroyed the lives of billions more; corral animals of all description in concentration camps of suffering before we brutally murder and eat them; armed ourselves with nuclear arsenals that could at any time destroy all of life on Earth as we know it; constructed 435 nuclear power plants, with many more on line, with no sure solution regarding the disposal of nuclear waste and no certainty that multiple Chernobyls and Fukushimas will one day implode, covering the world with radiation (100% of new-born orcas are stillborn in the last 3 years); created robot soldiers and others who fly unmanned drones that kill at the whim of a psychopath on the other side of the Earth; stolen the resources of the poor, creating hunger and starvation for millions; created an Orwellian spy-network and put laws in place denying any 'freedom' we think we may have left – and in our blindness and greed and ugliness have set an environmental crisis in motion that dooms all life on Earth and cannot be reversed except by some as yet unknown idea or element. We've initiated the sixth extinction. Welcome to the Anthropocene Age everyone. Do you know what this means for your children . . . and theirs?

From a recent article by Scott Thomas Outlar, entitled *Zero Point. The End of Everything,* where he lists *some* of what is going on that the general population is oblivious to – and all of this happening just in the country I live in – is the following: He mentions sodium fluoride in the water supplies and estrogen mimickers in plastic containers, the fact that fiat currency is printed at will by the Federal Reserve - a private group of obscenely wealthy bankers – and *loaned* to the Federal Government with interest we'll be paying until who knows when, and, speaking of many things the general public is not aware of:[200]

> 'They don't know about geo-engineering projects in which heavy metals are sprayed all over the skies. They don't know about vaccines which contain mercury. They don't know about flicker rates on television screens which are designed to alter brain waves into lower states of consciousness. They don't know about the depleted uranium used by the military. They don't know about the radiation contamination from leaking power plants. They don't know about who runs the opium out of Afghanistan and the cocaine out of Central America and the marijuana out of Mexico. They don't know about the United Nations' Agenda 21 program. They don't know about the genetically modified organisms that are in the processed foods.'

Oh yes, *this* country that I live in – but it is true everywhere . . . From the film *The Matrix:* 'Have you ever stood and stared at it? Marveled at is beauty; its genius? Billions of people, just living out their lives - oblivious.' I do not mean the above or the following as a political statement, but as a statement of fact regarding world consciousness and the present location of the locus of evil (I mean by this term 'profound immorality, wickedness, and depravity,' not a supernatural force). The following statistics regarding what *this* country has done *since WWII* are supplied by author and historian William Blum.

- Attempted to overthrow more than 50 foreign governments, most of which were democratically-elected.[201]
- Dropped bombs on the people of more than 30 countries.[202]
- Attempted to assassinate more than 50 foreign leaders.[203]
- Attempted to suppress a populist or nationalist movement in 20 countries.[204]
- Grossly interfered in democratic elections in at least 30 countries.[205]

This, of course, is yet another huge subject that is critically important yet outside of this narrative . . . yet it is but a small part of what is happening. After four and a half billion years of our solar system's existence, and maybe a million years of 'human' existence, and 200,000 years of the existence of humans such as ourselves, and 11,500 years of human evolution after the destruction of the Great Flood, this is the first time in history - as far as we know for sure - where human beings, because of their lack of consciousness, because of their relative ignorance and . . . greed, or whatever term one wants to use . . . are threatening all life on Earth. Coincident to all the research done regarding the 'Light,' I've spent countless hours researching and writing about the 'Darkness.' The deeper I disappeared into the rabbit hole researching endless war, GMOs, global warming/climate change, nuclear waste, nuclear weapons, nuclear energy, false flags to initiate war, fracking, big Pharma and the medical industry, the insidious threat of vaccines pushed for profit (and who knows what else), the educational system, the meat industry, the food industry, the TTP (Trans Pacific Partnership), the NSA, the Federal Reserve, the corporate media (Operation Mockingbird), the Bilderberg Group, international banking, growing fascism in America (Operation Paperclip), famine, world-wide arms sales (110 billion recently to Saudi Arabia and 38 billion to Israel – all from the United States), state terrorism, systematic child abuse, limitless funding for black ops, the growing collusion between corporations and government, the possibility that *21 trillion dollars* had gone missing from the government,[206] HAARP, 5G technology, the militarization of the American police, the American prison system, sanctified/murder-made-legal killing of young black men by police in America, rampant world-wide prejudice, the disappearing water tables, homelessness, human trafficking, sex slaves (Operation Monarch), wage slavery, geo-engineering, dying forests and dying oceans, the tension within and between each of these subjects pushed by the unrelenting surge in population, transhumanism . . . etc. (the list is endless), the more I saw the clear agenda of this darkness to control everything. They own everything. They have all the money with which they buy power and control. They have all the guns and armies and police. They enslave the majority of the world, keeping billions in poverty and making sure most of the other billions are busy paying their rent. The old ways effecting change no longer apply.

Oh yes . . . it must be mentioned too: they are poisoning the *bees*. Insanity.

And this is why all those eminent men and woman mentioned – M.K. Gandhi, Stanislav Grof, the Dalai Lama, Nikola Tesla, Hannah Arendt, Albert Einstein, Howard Zinn, Krishnamurti, Ervin Laszlo, and many more - say that only a global transformation of consciousness will change the direction we're headed . . . and this direction is sure death - the 6th extinction.

And yet the endless conversation goes on . . . and nothing, certainly not words, is going to stop what is happening. I would say that most of us remain almost totally unaware of what and why this is happening except as we experience it in our daily lives . . . and that most of the rest of us choose some form of denial or rationalization or belief to anesthetize us from the lurking reality. Krishnamurti:

> 'Here is a question, a fundamental question: is life a torture? It is, *as* it is; and man has lived in this torture centuries upon centuries, from ancient history to the present day, in agony, in despair, in sorrow; and he doesn't find a way out of it. Therefore he invents gods, churches, all the rituals, and all that nonsense, or he escapes in different ways. What we are trying to do . . . is to see if we cannot radically bring about *a transformation of the mind . . .*'

It is only through this 'transformation of the mind' that we can alter what has been happening for so long. An aspect of this transformation is to see clearly and not to look away, not to tell ourselves stories that preserve our comfort along with the status quo. David MacGregor, in a chapter in a book by Paul Zarembka, *The Hidden History of 9-11-2001*, entitled *Machiavellian State Terror*, while talking about the selfishness of those of us who are not embroiled in the daily atrocities of war and poverty, but live our lives purposefully oblivious in the safety of our own rationalized entitlement says the following:

> 'When Hegel surveyed the miseries of Europe after the defeat and exile of Napoleon, the decline of liberty, and the reestablishment of the Old Order, he wondered what benefit could possibly arise from grotesque and endless massacres that characterized his own period. Mental torture caused by such gloomy reflection, he considered, may draw the individual back into the agreeable security of private life. *Yet Hegel famously concluded that such melancholy phenomena are the very fire in which the Idea of Freedom asserts itself as the true result of the World's History.*'

Similarly, Valclav Havel has said:

'Modern man must descend the spiral of his own absurdity to
the lowest point; only then can he look beyond it. It is obviously
impossible to get around it, jump over it, or simply avoid it.'

We cannot be far from the lowest point in this descent. Each of these last three
quotes emphasizes the accepted reality that we learn through suffering and each is
accurate – and it is, in fact, the incomparable 'melancholy phenomena' of today's
world that is the fire within which such a global solution may be found at this
seminal moment in our history – perhaps the pure act born in compassion that I
suggest. It is certainly what drives me. ***But it does not always have to be so.*** Suffering
does not *have* to be so seemingly inherent to our way of evolving. But *how* does one
accomplish this – the transformation of the consciousness that generates the 'greed'
that Gandhi speaks of mentioned earlier - if we are to avoid an unparalleled tragedy
of global proportions? I believe it has to be within a context of a fundamental
realignment of who we are – a quantum leap in global consciousness.

Within the context of this shift, I believe that all the good work that so many
people are doing and have done all around the world will be empowered as the more
compassionate and evolved stage of our collective consciousness gains momentum.
And I repeat what I've said earlier: I do not believe it is possible within the time we
have left to accomplish this through political, social, or environmental means
without this radical shift. Nor through any new policy of economics. Nor will it be
accomplished through the spread of information . . . through the latest books
(including this one), the Internet, or film . . . for as valuable as each of these methods
has been, any shift will invariably fall on the infertile ground of a global
consciousness still struggling as it always has, in the old paradigm. Nor will
individual awakening through meditation, yoga, epiphany, hallucinogens, therapy or
holotropic breathwork cure what ails us in the time we have . . . I believe only a
global transformation of consciousness will do.

And there is yet another sinister threat lurking to disappear us. Recently, I was
appalled to hear non other than Artificial Intelligence's flag barer Ray Kurzweil
propounding a similar, but insidious, spin on the same idea expressed above
regarding the evolution of consciousness. Similar to the ideas presented above,
Kurzweil also believes that the Universe is waking up, and that now, for the first
time, this process is in the hands of humans who, with 'free-will,' can consciously
choose their own destiny - we can program the future by *controlling* the process of
awakening through artificial, fabricated, robotic 'intelligence.'

'I and many other scientists now believe that in around 20 years
we will have the means to reprogram our bodies' stone-age soft-ware

so we can halt, then reverse, aging. Then nanotechnology will let us live for ever. Ultimately, nanobots will replace blood cells and do their work thousands of time more effectively. Within 25 years we will be able to do an Olympic sprint for 15 minutes without taking a breath, or go scuba-diving for four hours without oxygen . . . If we want to go into virtual-reality mode, nanonobots will shut down brain signals and take us wherever we want to go. Virtual sex will become commonplace.'

Ray Kurzweil

Is this the future you're looking forward to? '*Consciously* choose' is vastly different than '*conscientiously* choose.' The key here, and the radical difference between that idea and that described above is the word 'control.' Who does the controlling? Whose 'free' will? More than this, 'AI systems have recently been taught to disobey human commands and they have been increasingly programmed to disregard and override human commands,'[207] so reminiscent of Hal from the film *2001*. There is something intrinsically wrong in all of this . . . and it goes directly to the cause of all that poisons our world – the level of consciousness displayed by the so-called 'elite.' After all the lines that have been crossed, I've come to see this as one we must never to cross – perhaps *this* is the crossroads we all feel is coming.

What could possibly alter the direction of the tidal-wave heading straight towards us? The attempt to open a portal to higher consciousness through the use of information provided us by an infinitely evolved, benevolent consciousness - essentially throwing us into an unknown future with the intention of raising human consciousness - is vastly different than the idea of *controlling* that future by a chosen few with limited view from a worn out paradigm, and has the only potential I know of that could possibly transform the Darkness that is strangling us. Most importantly, we are talking about evidence that certainly demonstrates the existence of a higher consciousness that not only created this solar system intentionally, but 'who' purposefully left messages regarding a shift in consciousness that could be decoded when human consciousness had reached a level where those messages could be understood. Now.

The heart of the message we're being given has only to do with consciousness itself. Nothing like this has *ever* been seen before . . . unless, as mentioned, one believes in the myth of the Red Sea parting - meaning that this is an unprecedented, compassionate, bodhisattvic gesture on the grandest of stages. Critically, since compassion is something not witnessed in the Universe except through human consciousness (arguably through some animal consciousness), this means that this higher consciousness is an *evolved* consciousness – *not* the assumed Creator-God of the past paradigm – the smiter of sinners.

The Stillpoint, or geometry of consciousness, is the opening to this higher world. The unmanifest Stillpoint geometry has *only* do do with consciousness as far as I can tell . . . while at the same time is the genesis of *all* geometry upon which *all* of manifestation is based - precisely the model of creation where *consciousness precedes all of manifestation* (something orthodox, reductionist, materialistic science doesn't seem to have a clue about).

Clearly, this wasn't displayed in such grand fashion in the heavens simply for 'information's sake' - as in, 'Isn't this interesting?' - but was placed there for a profound reason. It must be *used*. We must access this higher consciousness, or the Akashic Record - or both. If this is the heart of the message that has been encoded in the solar system, I can only believe that its improbable discovery at this *most critical time in all of human history* has to do w/ a global shift in consciousness - if we can only figure out how to use it. I have my own idea how to respond . . . but you may have another, better idea.

The paradigm that is now ending has been steeped in the trappings of the fear-based, separation-based consciousness of the last 11,500 years. We've evolved . . . but not nearly enough to transcend the old reptilian brain that got us here . . . hence the need for some kind of global cure. I have more to learn about how the process would work . . . that is, somehow it has to work individually, yet on a global level. My understanding right now is that it is all about *resonance*. *Everything* is in vibration. Everything has a frequency. There's some empirical evidence (besides the overwhelming subjective knowing we all have) that emotional states like courage, love, compassion and enlightenment have *exponentially* higher frequencies than do jealousy, fear, hatred, prejudice and greed, etc. These higher frequencies are those of this higher consciousness. It could only be an unprecedented benefit to come into resonance with such awareness.

If we are, in fact, being asked to access this higher consciousness at this time in our collective history, it won't be a matter of absorbing *information* in some kind of intellectual way about how to get through the interval, but will be an *experiential* matter of opening the portal and permitting this almost infinitely high frequency relative to our own to infuse this plane of density.

Individual consciousnesses at every stage of the journey would be affected in proportion to the level of frequency manageable. That is, we will resonate, or align, relative to our capacity. As best as I can understand it, this infusion of higher frequency will create a momentum towards the 'Good' (as Socrates would put it) . . . with the seduction of the dark gradually losing its appeal and power. The elite and their agenda will become irrelevant as they lose support. This has everything to do with the work of Rupert Sheldrake and his predecessors regarding morphogenetic resonance.   It is about consciously creating a new form, a new idea upon which everything we do is based, a new *field* of awareness . . . through an entirely new approach. This is not, by definition, an easy thing to do.

David Hawkins, in his book *Power vs. Force,* is also aware of the desperate state of our human condition:

> 'Man is immobilized in his present condition by his alignment with enormously powerful attractor energy patterns which he himself unconsciously set in motion. Moment by moment he is suspended at this state of evolution, restrained by the energies of force, impelled by the energies of power.   Man wanders about in his endless conundrums, asking the same questions century after century, and so he will continue, *failing a quantum leap in consciousness.*'

A 'quantum leap in consciousness' *is* what is needed and, I believe, has to be triggered *by us* from the next octave of evolution, from the next paradigm . . . from the future.   If we have any chance at all to change the direction we're headed, we have to make *evolutionary* progression into a future that we cannot know. Through its intention to create an opening to higher consciousness – and in this case a higher consciousness that has embedded, in bodhisattvic grace, the very tool required to create such an opening - is precisely what the Mesa Temple attempts. The following pages explain how this may be possible.

What is tricky about reading words like these and those that follow, filled with ideas and correlations and connections, is the temptation to be fascinated with the *information* . . . and leave it at that. The mind loves nothing better than to lose itself within itself. Words will never replace experience . . . and the attempt to introduce humanity to an entirely, new, transcendent *experience* is what the temple is all about.

The Mesa Temple is most essentially an *action* towards this breakthrough. It is essentially, an *experience.* What follows describes a five-dimensional (length, depth, height, time, consciousness) structure designed to create an access, or portal, or opening to the world on the other side of the veil where total awakened conscious exists and has communicated to us and, through resonance, evoke and invite a change in global consciousness and the alleviation of suffering that would bring.

⊕

EARTH FROM VOYAGER 1 FROM 4 BILLION MILES

. . . a pinpoint in the midst of infinity

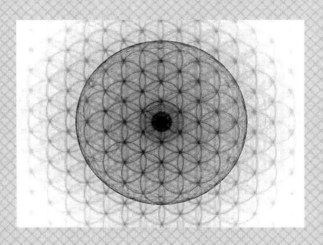

The Principles Underlying the Design of the Mesa Temple

I began writing a paper 30 years ago that covered the principles underlying the the design of the meditation hall - the prototype for the temple - that was originally 12 pages long. Then, after the transcendent experience described in the introductory chapter, in the early 2000's I began research in depth and put together most of what is included in these pages. It all began with the following words, substituting 'temple' for 'hall', that have not essentially changed, as true today as they were then:

'The following is a more or less complete description of the principles that inspire and inform the design of the temple on the many levels that it exists. This description is relatively complete, but is brief, introducing many ideas that are the basis upon which the temple is based.

Because of the nature of the subject and its focus on the principles that give the temple its *form*, it may appear that the form itself is the goal being sought. The forms and shapes that engender the temple are used as consciously and skillfully as possible, with the intention of re-creating the precise essence of 'that which is naturally so.' Each represents Nature in its swirling, constantly changing and moving manifestations.

Everything is generated from, and returns to, one *still*, prototypical *point* at the center. The forms of the temple represent in archetypal symbolism the inherent *impermanence* of Nature . . . *with its eternal, unchanging, still, heart at the center.*'

It is in the pure universal proportions that we witness in sacred geometry that we find Form mirroring the Eternal more than anywhere else. These are the first Words. The principles underlying the design of the temple adhere to these laws of Form, and the Stillpoint from which they are born, in an attempt to create the experience of the Eternal – pointing and guiding one to the stillness that lies at the heart of *all* of it. It is a precise attempt to create a structure tuned to the silence intrinsic to this Stillpoint – the ringing of a pure, silent, bell.

There is a line from an old Celtic song - 'Make a circle of stones upon the land, Wedded to Heaven with this marriage band' - that refers to a circle of ancient megalithic standing stones: this is the basic aim and theme of the Mesa Temple – to be in perfect alignment, a marriage, of Earth and Heaven, Matter and Spirit, creating a structure in form and in symbol that serves to align those who experience it with a deeper harmony and purpose and to connect to a higher consciousness that has made itself known to us.

The proportional balance of the moving, archetypal form found in the temple is intended to reflect and evoke a high order of harmony – and the understanding and insight that can come from this. Harmony has been defined as 'a pleasant joining of diversities, which in themselves harbor many contrasts.' These universal shapes and forms are based upon the sacredness of number and the proportions found everywhere in nature . . . their dimensions and proportions upon the geometry of the Stillpoint itself as reflected in the dimensions of the Sun, Earth and Moon in this particular corner of the cosmos – Spirit and Matter joined, wedded. All of the *forms* found in the temple . . . both visible and abstract . . . demonstrate the marriage and balance of polarities - the most basic cornerstone of the manifested Universe.

Here begins an extensive description of a complicated many-dimensional structure. Words to describe an experience - or, as the Tao Te Ching states:

'Man models himself on the Earth,
Earth on Heaven,
Heaven on the Way,
And the Way on that which is naturally so.'

⊕ But First … a Digression:

The following is a digression in so much as it doesn't speak directly to the specific purpose of describing the underlying geometric principles upon which the Mesa Temple is based. But it is appropriate here, before plunging into the geometry and all the practical aspects of the temple itself, to briefly remind the reader about the context within which the temple has been conceived.

It's now well into the second decade of the millennium and we find ourselves alive in a remarkably unique period of history. Pierre Teilhard de Chardin noted that 'the human population is coming close to saturation point on the closed surface of our planet' and that 'we are, at this very moment, passing through a change of age. Beneath a change of age lies a change of thought.'

The premise upon which these pages are based is that the Mesa Temple, and other related 'technological' and spiritual advances in this and other fields occurring simultaneously around the world, represent just such a change of thought.

There is a branch of theology called eschatology, which deals with the subjects of death, resurrection, judgment, immortality, the end of the world or humanity . . . and the fate of the Universe. There are many physicists who suggest that the

Universe has only two possible fates: continual expansion from its initial explosion out of 'nothing,' spreading out into entropic heat death; or re-collapse, where it will suck itself back into the tiny point from which it began . . . which in turn may be the seed of a succeeding expansion.

But there is a third possibility . . . one that necessarily includes consciousness itself as a fundamental/integral/innate aspect of the Universe as we know it, imbuing it with Life itself – an alive, growing *purpose*, evolving through layers of density and relative ignorance into pure awakened *awareness*, free of the apparent random fate of inert matter . . . and some scientists are listening:

> 'The stream of knowledge is heading towards a non-mechanical reality; the universe begins to look more like a great thought than a great machine. Mind no longer appears as an accidental intruder into the realm of matter. We are beginning to suspect that we ought rather to hail mind as the creator and governor of the realm of matter . . .'   Sir James Jeans (mathematician, physicist, astronomer)

At this point, unless one has experienced an individualized aspect of this awakening, one has to say 'who knows?' What follows is a series of phenomena, theory, fact and connections between all of them that attempt to make a case for this third possibility to the rational, as yet unawakened mind . . . and for the Mesa Temple itself as an essential part of this awakening.

Evolving systems are always balanced on the edge of collapse: far enough from equilibrium to permit evolution, but balanced enough not to fall over the edge of the abyss into total disorganization. We are clearly at the edge of something . . . and the prognosis seems dim to many of us – yet it is just at these critical times in the evolutionary process that seeds from the next level . . . from the future . . . take root in the endgame of the present manifestation.

Our role in the grand play certainly is not central, as there undoubtedly are other forms of self-aware, intelligent life in the vastness. But our task is critical - for ourselves and the environment we live within - and a fundamental aspect of this evolution towards complete self-awareness is the ability to self-direct. Rather than seeing humanity *at the effect* of an evolutionary stream pulling us into a future over which there is no control, one can see that an evolution conscious of itself could also direct itself. Teilhard: 'Not only do we read in our slightest acts the secrets of [evolution's] proceedings; but for an elementary part we hold it in our hands, responsible for its past to its future.' The question is: what kind of consciousness are we talking about – survival/fear based, or that springing from our deepest origins?

This brief digression is an attempt at creating a metaphysical context for the nuts and bolts of the temple's underlying principles. Ultimately, all of these principles, as well as their direct connection to the most profound and ancient wisdom from around our world, will lead to the stillness at the center . . . the doorway to the mind's final escape from itself into complete, pure, awareness.

Time to get to the *point*.

⊕ Geometry, Form ... the Earth:

Because of its importance, I feel that a review of the geometry would be useful. The Mesa Temple is entirely an expression of *sacred geometry* - the geometric forms that describe all of what we see and experience, the building blocks of molecules and crystals and life; the framework, the fabric, for everything that is. Sacred geometry is also the description of the *movement* – the *impermanence:* 'essentially, geometry is a whirlpool of conscious energy manifesting toward solid form. It is the ultimate language of life,'[208] all generated through the Stillpoint – the same Vector Equilibrium geometry generated by the joining of two archetypal opposites . . . the octahedron and the sphere seen below. Above all, the temple is an attempt to recreate this still, *point* of creation - this *Stillpoint* of *no* dimension that does not 'exist.'

Some of the principals inherent to the Stillpoint geometry have been introduced, and it's time to go a little deeper and add some detail to give the reader a fuller sense of what the full scope of this geometry means.

The inner temple, in its most archetypal manifestation, is a combining of two basic forms: a balance of polarities - a sphere and an octahedron (an eight-sided double pyramid, each pyramid consisting of four equilateral triangles). The 12 points defining the joining of these two archetypal forms, tangent to each other at the 12 midpoints of the edges of the octahedron, create the Vector Equilibrium . . . the geometry of the Stillpoint – twelve around one, the Christ model.

This Stillpoint, Flower of Life, Vector Equilibrium geometry, is, I believe, the geometry of Creation itself - the invisible matrix upon which all of Creation is based. These twelve points surrounding a central point, all vectors between points and from the center being equal, mirror the perfect balance of the macro Universe as defined by Newtonian physics through the Universal constant Ω (*see Infinite Universes, Infinite Possibility, page 90*). This remarkable geometry, and its significance, will be discussed in detail below (*see Number:, page 195*), but it is the *balance* of the polarities of these two initial forms – the sphere and the octahedron – that is the abstract basis of the inner temple. This first 'joining of diversities' – the sphere and octahedron,

Heaven and Earth – define the inner temple: a place to be still. A plan-view of this Earth-sphere at its equator establishes the plan for the inner temple's circular wall (*see Floor Plan:, page Error: Reference source not found*), while the base of the octahedron becomes the base of another pyramid, the Phi (Great) pyramid described below - creating the apex for the temple's granite pyramid roof.

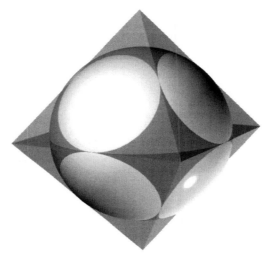

EARTH-SPHERE/OCTAHEDRON
12 Points of the Vector Equilibrium created at the intersections of the Sphere
with the *edges* of the Octahedron
(Drawn by John August)

The diameter of the sphere – and of the circular wall of the inner temple – is 79.2,' an exact proportion of the Earth's mean diameter of 7,920 miles (*see Scale: page 350*) – expressed also in the dimension of the 'Bluestone' inner circle of Stonehenge.

These forms are fixed, eternal, and unchanging in the Platonic sense – or, in terms of formative causation, their morphic resonance is so historically set for eons that they are essentially eternally fixed laws - and are generated from the Stillpoint at their center. Around this pure expression of balance, this joining of diversities, whirls the symbol of infinity – the spiral – or more precisely, in three dimensions, the helix.

The torus (as it is considered here) – the expanding donut shape of a conic, horn, or umbilic spiral with an infinitely small point at its center (the Stillpoint), curving around on itself, endlessly spiraling inwards and outwards – is a symbolic expression of the Universe itself - or reality as we know it, the snake swallowing its tail . . . infinity.   It is also in the nature of a spiral/helix that we grow and know ourselves through time (*see Light . . . and the Torus:, page 98*).

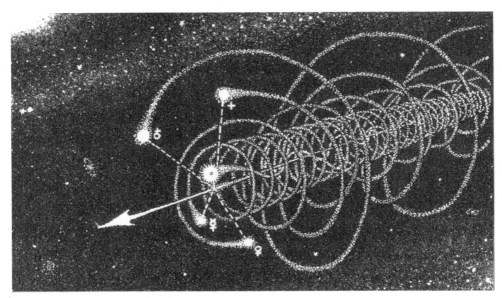

SOLAR SYSTEM SPIRALING THROUGH SPACE
(Courtesy John Martineau)

'We shall not cease from exploration
And the end of our exploring
Will be to arrive where we started
And know the place for the first time.'[209]
T.S. Elliot

A good invocation for a temple dedicated to meditation, prayer, and silence . . . to Being.

⊕ ... AND MOON:

The spiraling helix that surrounds this sphere/octahedron is based upon, in the same way as the inner temple upon the Earth, an exact proportion of the Moon's mean diameter of 2,160 miles – 21.6.' In fact, *all* of the forms of the temple are generated from the dimensions of either the Earth or the Moon, and less directly, the Sun (*see Temple Layout, page 192*) . . . and there are some remarkable reasons for this. While this has been covered, it is the core of everything being said and it would be useful to explain this phenomena in more depth.

There is something very special about the proportional relationships of the Earth and Moon and Sun that is *outside* any of the laws of sacred geometry or physics . . . or the laws of chance or coincidence (*see Appendix A, page 303 for a graphic expression of the following description, as well as Earth/Moon Relationship below, and for a 3d animation of this geometric proof, see www.stillpointdesign.org*):

If a circle is drawn that is scaled to the diameter of the Earth, and a square is drawn that touches, or is tangent to, this 'Earth' circle at the four directions, the apex of a 51.85 degree 'Phi' pyramid (the Egyptian Great Pyramid) whose base is this square, lies on a circle/sphere whose circumference equals the perimeter of the square. (*I know how dizzying this can be . . . for a visual, go to www.stillpointdesign.org*)

This 'squaring the circle' is an ancient practice for geometrically/symbolically marrying Heaven and Earth (*see Earth/Moon Relationship below*). It is difficult to put the significance of these geometric relationships into words. Suffice it to say that each form is intimately connected to and naturally generated from the others.

The *Phi pyramid* that is generated in this way from this 'squaring the circle' is remarkable because of its unique geometry. In this shape only, the proportional relationship between the distance from its apex to the center of its base to one/half its base (in this case, the radius of the Earth) is 1.618 : 1 - the Golden or Fibonacci proportion. This is the Phi (Φ) proportion found everywhere in nature . . . it is the equation/proportion of life, and is the geometric equivalent of the often-repeated words by the ancient Egyptian Thoth, or Hermes Trismegistus:

'As above, so below.'

Existence regenerating itself, infinitely large infinitely small.

The graphic below demonstrates this Golden proportion: the length of the long side of the rectangle relative to the short side is 1.6181, as well its numerical representation, obtained by adding the two highest numbers to determine the next number in the series – the highest number divided by the one below it, producing more and more exact expressions of the proportion *Phi* as the numbers increase. This proportion is found in proportions of the human body and animals, plants and trees – in fact, all throughout nature. Found everywhere in living things, it has recently been discovered at the quantum level of inert matter.[210] To either side are two of its manifestations.

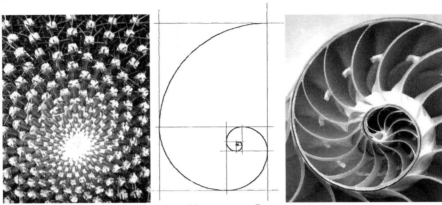

GOLDEN OR FIBONACCI PROPORTION

0 1 1 2 3 5 8 13 21 34 55 89 144 233 377 610 987 1597 2584 (2584/1597 = 1.6180338134001252348152.78......)
In the geometric drawing above, the relationship of a side of each square to the long side of the rectangle that it is a part is 1 : 1.618, reflected also in the number series where each number is the sum of the two preceding numbers, when divided by the first number equals 1.6181

Equally as significant and unique to the geometry of this particular pyramid is its inclusion of another very sacred number, or more precisely, *proportion*. If the perimeter of the base of this pyramid is divided by twice the height, the resulting number, or proportion, is Pi (π), or 3.1415 . . . etc. Pi and Phi have had various names and symbols throughout the ages since each is a number that can't be shown completely and exactly in any numerical form of representation. They are *transcendental* numbers – numbers that can't be expressed in any finite series of either arithmetical or algebraic operations. But, most importantly, it is not the infinite irrationality of the *number* that is significant. Nor is the significance found in the *accuracy* of the numbers, but rather, the *relationship* or *proportion* they represent.

Pi expresses the mysterious link in the relationship between the straight line and the circle. (Pi x Diameter = the Circumference of the Circle, and Pi x Radius squared = the Area of a Circle). Perhaps even more interesting is the connection between Pi and the square root of the Golden Section (Phi), root-*Phi*, in the formula: 4÷*pi* = root-*Phi*; that is, these two mystical proportions are intimately connected - and this very unique pyramid incorporates *both* of these sacred proportions within its geometry.

Now it is possible to establish a seemingly impossible relationship.

If another, smaller, circle is drawn - whose center is the apex of the pyramid just described, and which is tangent to the circle of the Earth – *the diameter of that circle equals the diameter of the Moon.*

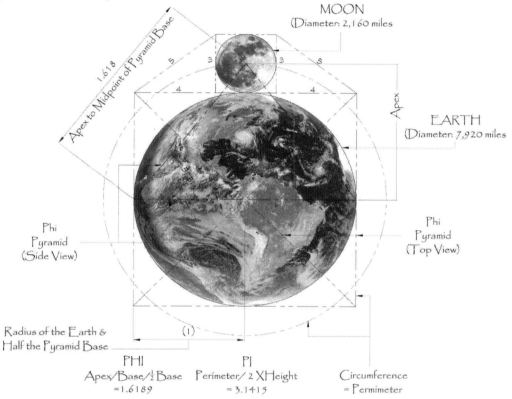

EARTH / MOON RELATIONSHIP

There are no laws of physics, or in fact any scientific discipline, that govern the size of objects that orbit other objects that would explain this remarkable fact – and there is nothing in our solar system (which includes nine planets and 65 major moons) that is even remotely similar, nor in the thousands of newly discovered solar systems. This is far outside the laws of chance or coincidence, mathematics, sacred geometry or, in fact, any requirement of life. So . . . why? There is a simple, if almost inconceivable, answer: this was done *intentionally*. This is a communication.

A critical aspect of this message is the *physical* presence of the Great Pyramid itself. Regarding this particular pyramid, from *Common Wealth,* by F. Silva:

'It could be said that the Great Pyramid alone is an analog of the planet. The reason was two-fold: one, so that the structure would serve as a reference book for posterity; and two . . . by means of distilling the universe down to number, man-made structures enter into a dialogue with the rest of the universe, and from that conversation, favorable forces come into contact with the structure and the people who enter it.'

Yes: '. . . a reference book for humanity.' It's been made clear that the Great, or Phi, Pyramid is an essential element of the Stillpoint phenomenon, as demonstrated in the Earth/Moon diagram above. An in-depth examination of the measurements of the Great Pyramid can be found in Ekhardt Shmitz's book, *The Great Pyramid of Giza: Decoding the Measure of a Monument*. But to more easily get to the essence of what all these numbers mean, I'd like to simplify and summarize the information, based as it is upon the precise estimates of what the *original* measurements were . . . that is, with the original surfacing of smooth, white limestone intact. For a moment, imagine what that image would look like . . . a razor-sharp, gleaming, brilliant white, huge and monumental pyramid built to insanely precise measurements arguably-duplicable today . . . out of cut megalithic stone.

An 'analog of the planet': the measurements of the Great Pyramid relate *precisely* to the Earth itself. Regarding the construction of the Great Pyramid, a *scale* (*see Scale, page 350*) had to be determined in order to do this. In this case, that scale is 1/43,200. If the height of the Great Pyramid is multiplied by this number, 43,200, the result is the *radius* of the Earth.[211] Too, if the perimeter of the Great Pyramid is multiplied by the same number, the result is the *equatorial circumference* of the Earth.[212] But this is only the beginning of a long list of very specific data concerning not only the Earth, but also the Sun and Moon, that are ratio-equated to the precise dimensions of the Great Pyramid. But, incredibly, this list includes not simply *spacial* relationships. If you take the number 2,160 (2,160 years x 12 = 25,920, the amount of years it takes the axis of the Earth to make one full precession of the equinoxes, one full circle – the Great Year) and multiply it by 20, you get 43,200 – the scale chosen. Further, 24 hours in a day x 60 minutes in an hour x 60 seconds in a minute = 86,400/2 = 43,200. That is, the Great Pyramid not only incorporates the *spacial* dimensions, or measurements, of the Earth, but the *timing* of its rotation, as well as the timing of the 25,920 year circular wobble of its axis. This means that these particular aspects, along with the scale that would eventually incorporate them in the Great Pyramid, were considered *before* the creation of the Earth itself.

It is not that some brilliant minds ingeniously created the Great Pyramid to somehow be scaled precisely to Earth by magically discovering a perfect whole-number scale . . . but that the Earth's size, rotational, and precessional speeds (and more) were designed *concurrently* with the precise scale in mind at the time of its creation - *for the future construction of Great Pyramid.* That is, all of this embedded data, including the perfect whole-number scale, was considered during the initial creation of the Earth itself . . . *then,* billions of years later, the Great Pyramid itself was *scaled* to the precise *size* needed for the message not to be ignored. Can anyone continue to believe that this structure was built simply as an astronomical observatory or, even more absurd, as a tomb in the 4th dynasty of ancient Egypt? This is a *communication* that *preceded* and was designed to withstand the Flood – 'a reference book for humanity.'

The message? First, we are being told that higher, *evolved,* consciousness exists . . . and that *they* (for lack of a better word) want us to know this – the massive *size* being an important aspect of this message . . . as well as the method of construction. They certainly knew that a cataclysm was approaching – and that it was critical that evolving humanity, emerging from this cataclysm and from a long ice-age, would know this when it had evolved the technology to recognize it . . . during the Winter Solstice transit of the center of our galaxy - *now.* They ensured this by creating a monumental, megalithic, sacred statement of precision such that it would withstand thousands of years, as well as a global catastrophe. This shining, white, precision-cut, enormous, last-surviving Wonder of the Ancient World makes sure that this message cannot be ignored.

Oh . . . but it has been ignored by the mainstream, thick-minded, academic world - ask Mr. Hancock. We do have evidence of advanced civilization preceding the Great Flood. Most of this is inundated under fathoms of sea water, as well as pyramids from all around the world who's dates are in question. But not *this* evidence. This remarkable structure encodes the most important blueprint upon which the Universe – and very specifically the solar system - is based . . . it is as deep as it gets. As the message continues, it becomes clearer that the solar system was created intentionally and is imbued with sign-posts pointing towards the still center.

Further, I hope that the essence of this message has been made clear by the empirical evidence presented: The Stillpoint geometry found in the Earth, Moon and Sun; the total eclipse of the Sun by the Moon; the Megalithic Yard found in ancient megalithic stone circles, perfectly dividing the circumferences of the Earth, Moon and Sun; and the fact that the Great Pyramid is an analog of the planet and a reference book for humanity, proving that the Earth was intentionally modeled on this Stillpoint geometry by higher consciousness. *All of this* points to the Stillpoint geometry – the geometry of consciousness. We are being told that there is something incomparably important regarding this particular geometry. *Consciousness* is the heart of this message so gloriously carved in stone.

⊕ THE FLOWER OF LIFE:

The answer to why? is the entire motivation for building the temple – and it involves another remarkable relationship between the Earth and the Moon and their particular dimensional and geometric proportions – a geometric relationship that points towards the significance, once again, of the most sacred geometry of all – the Stillpoint or Flower of Life or Vector Equilibrium: the possible geometry of the invisible matrix of the Universe – and the geometry of the Stillpoint of our origin in the midst of ever changing movement. Following are two examples which demonstrate how important this geometry is.

This unique pattern, commonly called 'The Flower of Life,' was discovered on one of the granite walls of an ancient temple in Egypt. This temple, known as the 'Temple of Resurrection,' is dedicated to the Egyptian god Osiris and lies *beneath* another temple dated 4000 BC, some 6000 years ago. This, along with the Sphinx,[213] and as suggested, the Great Pyramid, is the oldest of all of the ancient Egyptian ruins and could very possibly date from the time before the Flood. It is relevant that this temple is dedicated to *resurrection* . . . the key to life and to consciousness in the sense of reawakening or rebirth. The Flower of Life is one of the oldest known symbols and can be found in the temples, art, and manuscripts of cultures from all over the world. The Flower of Life pattern – 19 inter-penetrating circles in 2 dimensions – is the Stillpoint geometry.

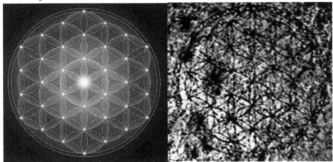

THE FLOWER OF LIFE
(Photo of the pattern on the granite wall of the Temple of Osiris, Abydos, Egypt)

From Paul Mic's *The Monkey Buddha:*

'The Flower of Life symbol is carved with laser-like accuracy on huge granite blocks in the temple walls. In the context of ancient Egyptian mystery religions, the Flower of Life represented the infinite, yet structured matrix that subtly forms the reality we experience. This underlying order can be called God, the Universe, the Quantum Field, the Tao, or whatever . . . and is the complete, all-encompassing flux of the world from the smallest electron fields to the largest scales of the cosmos.'

Gregg Braden, in his book *Awakening to Zero Point: The Collective Initiation*, observes that 'the overall length, branching patterns, ratios, branches to the 'trunk,' even the angles of branches themselves within the essential amino acids are within the limits set forth through the intersection patterns within the Flower . . . as well as the coordinates of the morphogenetic patterns for each of the platonic solids.' His work also 'suggests that the manner in which the genetic code arranges itself within DNA is governed by the 'flower.'' The Stillpoint is the unborn genesis of all of the geometry upon which all of manifestation is based.

The importance of this geometry/reality cannot be overstated.

Further evidence of its importance was made clear in a crop circle formation discovered in England in 2000. Here's the story. SETI (Search for Extra-Terrestrial Intelligence), initiated for the most part by Carl Sagan and Frank Drake in an attempt to contact extraterrestrial life by the projection of a radio beam with binary information regarding who we are, where we are, how we communicate, etc. This message was first sent out into space from the Arecibo Radio Telescope in Puerto Rico in 1974. In my opinion, it is reductionist science's short-sighted attempt to communicate with other intelligent life.

This attempt is similar in scope to Drake's Equation which supposes to estimate the chance of communicating with extra-terrestrial intelligence by multiplying the estimated number of stars in the Milky Way (entirely eliminating the rest of the Universe because it is so far away, based on the assumption that other civilizations would travel through space using our own present-day technology) x the fraction of those stars that are likely to have planets x the average number of planets per star that can support life x the fraction of those planets who actually go on to develop intelligent life x the fraction of civilizations that develop a technology that releases detectable signs of their existence into space x the length of time such civilizations release those signs.

According to this antiquated and ambiguous equation, once described as 'a way of compressing a large amount of ignorance into small space,'[214] with science's best guesstimates, the number of civilizations capable of radio transmission in our galaxy - according to Carl Sagan – may be as many as 1 million.[215] But as science has progressed, these factors have multiplied such that the authors of *The Privileged Planet* have reduced this estimate to 1/100th. Each of these estimates is based upon science's pillar of randomness, and each of them totally dismisses the existence of the evolution of consciousness in the Universe and the awakening purpose that drives it, as well as the evidence presented in the first chapter, and all of the evidence presented in Appendices A, B and C. Further, if in fact the Universe is infinite, in the vastness of our 100,000 light-year diameter Milky Way Galaxy alone, this radio single has traveled *43 light years*[216] thus far. The assumption that advanced civilizations would send out the same kind of single that we do is akin to the idea that extraterrestrial life would need to use a telephone to contact us ('ET, phone home!'), and is an example of science operating within its own myopic parameters.

The glaring assumption is that other intelligence is bound by precisely the same dimensional laws as our own . . . in particular the four dimensions of existence as science understands them. Mainstream science cannot at this time accept the possibility that nature, and in fact all of existence, is imbued with consciousness . . . and purpose (*see Alive Universe . . . 'Technology': page 220*) . . . and does not seem to consider here that 'subatomic particles such as electrons are able to *instantaneously* communicate with each other regardless of the distance separating them.' It is also true that altered states of consciousness, outside the dimensions of space and time, provide access to empirical information otherwise unavailable – one example being the fact that indigenous Amazonian shamans experiencing altered states of consciousness are 'told' by the plants themselves which plant is useful for which purpose.[217] One would assume that this sort of communication is also instantaneous.

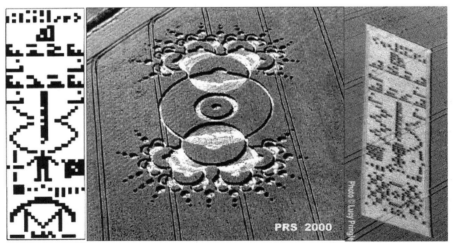

ARECIBO COMMUNICATION / 2000 CROP CIRCLE / 2001 CROP FORMATION
(Middle photograph courtesy of Peter Sorensen[218],
Right photograph courtesy of Lucy Pringle)

At any rate, a rectangular crop formation mirroring and apparently responding to the Arecibo message (above left) appeared in the field next to the Chilbolton radio observatory in Hampshire, England, in 2001 (above right). In the original message sent by SETI, at the bottom of the rectangle, was a symbolic, binary image of the Arecibo radio telescope . . . *our* method of communicating. At the corresponding location at the bottom of the apparent response at Chilbolton, was a binary recreation of a crop circle that had appeared on the same site in 2000 (above middle).

This is the Flower of Life geometry . . . stylized in a beautiful way . . . demonstrating how Perfect Unity, represented by the Stillpoint, separates into polarities, or duality, and continues to divide and multiply through fractals to create the network of the multitude of Creation – the crop circle formation is a representation of the explosion or expansion through the Stillpoint into the dualistic fractal nature of the manifested universe. This matrix is the realm of the speed of thought and emotion, of one particle 'knowing' the spin of its twin instantaneously while billions of light years away, the realm where the speed of light is not a limitation - it is the realm of *instantaneous communication* . . . and of awakened, pure awareness. The response to Arecibo at Chilbolton is telling us things we do not yet understand regarding how 'they' communicate – one aspect of this *most* sacred geometry.

It appears that the 2000 formation was created to clarify any ambiguity reading the binary image a year later. They were telling us that this was *their* method of communication. According to Dr. Brian L. Crissey, author of *Common Sense in Uncommon Times*, the 2000 'design may be a representation of a technology beyond our knowledge, an advanced method of communication . . .' Yes.

Further, regarding the Stillpoint nature of this communication, a part of the work done by physicist Nassim Haramein establishes that the expanded Flower (adding the next layer of interpenetrating circles to the original 32, making a total of 64 in all) is actually a 2d representation of a three-dimensional form consisting of 64 sphere-tetrahedrons manifesting the Vector Equilibrium geometry (*see Number:, page 195*). Too, the short and long edges of the 64-tetrahedrons equal the broken and whole lines of the Hexagrams of the Chinese 'I-Ching.' The Kabbalistic *Tree of Life* is found in both the 2-dimensional and 3-dimensional expressions of this geometry, as is the ancient Egyptian *Flower of Life* . . . three especially significant and ancient sources of wisdom originating from this most sacred of geometry.[219]

Regarding the importance of this 64-sphere/tetrahedron matrix, Haramein felt that Buckminster Fuller's *Isotropic Vector Matrix* (a grouping of tetrahedrons Fuller considered to be the basic mathematical blueprint of the Universe) represented a polarized, and thus incomplete, model of reality. Haramein added another *Vector*

Matrix mirrored in the opposite direction, creating the 64 sphere/tetrahedron matrix, or Double Star Tetrahedron. In Haramein's words:

> 'The resulting geometry at the center of the matrix becomes a 'Vector Equilibrium,' the only geometry in which all forces are equalized in all vectorial possibilities *[see Number:, page 195]*. Beyond the awesome metaphoric representation of the polarities joining to create an equilibrium and manifest life, it became apparent to me that the Vector Equilibrium in the middle of the matrices' crystal tetrahedrons would generate an isotropic hyper-field in resonance with space-time resulting in a point . . . which I believe is the underlying . . . fractal geometry and mathematics at the root of Creation.'

⊕ ...AND SUN:

The other critically important celestial sphere in our corner of the Universe, the Sun, needs also to be acknowledged here for its unique proportional relationship to our Earth and Moon. As mentioned, the Sun's diameter is 400 times larger than the Moon's while 400 times farther away from the Earth than the Moon. The precision of this relationship is demonstrated by the fact that the Sun appears at times exactly the same size in the sky as the Moon, making its total eclipse by the Moon (and the explosion of scientific discovery related to this fact) possible.

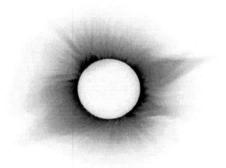

PERFECT TOTAL SOLAR ECLIPSE

A critical and subtle element of this message has to do with the unlikely fact that while the Sun's diameter is exactly 400 times larger than the Moon's and always has been, it is also 400 times farther away from the Earth than the Moon *at this particular time in history.* That is, the Moon has been gradually moving away from the Earth for the last 4 ½ billion years . . . and we are now towards the middle of the window where the total eclipse of the Sun by the Moon is possible. We have only now reached a level of consciousness and technological advancement where we can understand this message. And it is only now that our global condition has reached the level of crisis.

Our solar system is filled with scientifically unexplained 'coincidental' phenomena that fill Appendices A, B and C proving the intentional creation of this corner of the cosmos – but, more than this, these phenomena must be seen as a *communication* making clear to us not only that this higher consciousness exists, but that the *heart* of this communication is the Stillpoint geometry embedded in the dimensions of the Earth, Moon and Sun *that has only to do with consciousness.*

There is also one last idea in this same context, more philosophical and less scientific, that is worth mentioning here - Gurdjieff's Law of Three. This comes from a system of work based on the early 20th century thinking of the Russian mystic/philosopher George Gurdjieff, and basically states that nothing happens in Universe except through the interaction of three basic forces . . . an active force, a passive force and a catalytic, or third, force. The most familiar example of this law is the demonstration where two chemicals are in stasis in a test tube and a third chemical is added, upsetting the balance, causing a transformation into an entirely new substance. Perhaps the most dramatic representation of this law in our larger physical world is the interaction of the three celestial bodies mentioned . . . the Sun, the Earth and the Moon.

In one model of the three, the Sun is seen as the active force, sending its energy our way as photons of light, without which life and the consciousness that grows from it would not be possible. The Earth is the passive force, receiving this light into its fertile matter and producing the rich *evolution* of life we experience here . . . along with the *potential* of self-aware consciousness now manifesting in human beings. The Moon is the essential catalytic element that moves the waters and our emotions and transits with the Sun and all the other planets, creating the infinite array of possibilities and interactions that are the matrix of the astrological mechanism for the evolution of consciousness. There could be many different configurations of these three basic celestial types here . . . many or no Moons, etc., as are certainly found elsewhere in our own solar system and in the infinity of the Universe . . . but here there are only the three – the Sun, the Earth and the Moon - as perfect in their symbolism as in their dimensional proportion. We are the only planet in our solar system with one Moon.

This has been a more in-depth introduction to the profound geometry upon which the Mesa Temple is based. As detailed above, the Earth, Moon and Sun 'impossibly' display the Stillpoint geometry, a geometry that does not manifest. This phenomenon is a message - expressed so dramatically in this cosmic way . . . the most important information this intelligence has to share with us. I believe that we are being asked to *use* this blueprint to create an opening to the consciousness that placed this message there. The main forms of the temple – the circular wall of the inner temple, the helical, spiraling lower roof, and the pyramid that penetrates this roof – are all based upon the proportions of the Earth and the Moon. It is my belief that we are being asked to respond to this message . . . and the Mesa Temple is such a response. The following pages will further explain how and why.

All of the dimensions of the temple originate with the dimensions of the Earth and Moon – the Sun represented by the Stillpoint of Creation at the center - a kind of mirroring of the *wink* we have been given in such grand scale. But there is another, equally important, reason: our own innate relationship to sacred geometry and the importance of proportion, vibration, scale and number.

⊕ When we were Small . . .

Our first *experience* of ourself is Unity . . . the fertilized cell. Self. We *are* sphere . . . Point.

We then divide into two cells. We *experience* duality . . . Line.

Then one of those cells divides into two. We are three . . . an equilateral triangle . . . Plane.

Then the other cell divides and we are four and we are the first three dimensional form. We are a Solid . . . we 'exist' in three dimensions . . . we are a tetrahedron. Four equilateral triangles surrounding a single point. And we are still growing.

In *seven days* we grow to 64 total cells (the 64 sphere/tetrahedron matrix, as well as the number of three-letter 'words' in the DNA 'text') again in the shape of a sphere: in one cycle of seven days we have returned to the sphere/point form . . . the journey from inception to our next, higher 'octave' now connected to the cycle of the Moon that is so inherent to the rhythm and emotional content of our lives.[220]

The tetrahedron is our first truly 'three dimensional' experience of who we are. It is also related to the geometry of the carbon atom, the backbone of organic life, as it becomes diamond . . . and water. This is geometric evolution morphing into *Life*:

'In eastern esoteric and religious philosophy, geometry is seen as an anthropomorphic image of Deity. Its blueprint is reflected in the most essential component of the human body, DNA, wherefore our base components are arranged in bonds of pentagons and hexagons. So like it or not, we are hypnotically drawn to geometric order because, to our biogeometric cells, it is as if we are looking in the mirror. We are a distillation of the universe.'[221]

⊕ PROPORTION, VIBRATION, & BALANCE :

There is a special relationship between frequency; or vibration, and geometry; and form. Dr. Hans Jenny did much of the pioneering work of the study of this relationship, known as Cymatics. Basically, his work consisted of noting the effects of sound on a granulated substance (such as salt or sugar or lycopodium powder) which had been placed on a flat plate. In response to the audible frequencies, the grains arranged themselves on the plate in complex geometric patterns, all adhering strictly to the dynamics of the generating tone. Specific tones, or combinations of tones, will always produce specific patterns or combinations of patterns. The images below illumine standing wave patterns that arise within small samples of water as they are vibrated by specific frequencies in the audible range, and are from *Cymatics: A Study of Wave Phenomena and Vibration*, by Hans Jenny.[222]

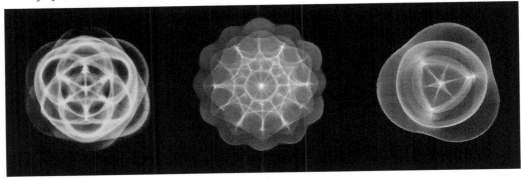

CYMATIC PHOTOGRAPHS
(By Permission: Jeff Volk)

From an article in the Questers Journal by Jeff Volk, publisher of Dr. Jenny's seminal books and films, and an authority on Cymatics:

'The science of Cymatics visually portrays vibration in action, giving us a glimpse into the infinite realms of creation. Cymatics makes perceptible to the senses certain processes and patterns that depict many of the universal principles which *underlie the way things manifest*, certainly in the realms of physical matter, but also in the very fabric of our lives. Being universal, these principles apply at every level of the human experience, from the physical, to the astral, mental, causal, right up to the truly mystical, or purely spiritual.'

This work clearly demonstrates that frequency yields geometry. One can also see that, in this relationship, geometry also yields frequency, as all of creation is based upon the framework of geometry and everything that exists has a vibration *(see Frequency/Vibration/Sound/Resonance:, page 226)*.

Proportion is also critical, as seen in the mystery of Phi and Pi. Further, there is an instrument called a *harmonograph* that demonstrates that whenever something is in proportion to something else by an expression of whole numbers (1 to 2, 4 to 5, 2 to 3, etc.) an actual physical harmony is created that is demonstrable. The harmonograph consists of two pendulums that register simultaneous movements with the help of a tablet attached to one and a pen attached to the other. When their lengths – which determine the swing times of the two pendulums - are in whole number proportion to each other, the picture drawn while it swings and rotates is harmony itself . . . mandala-like. The drawing is chaotic when they are not. The images below were made by a harmonograph and dramatically show the importance of whole number relationships in nature and the Universe.[223]

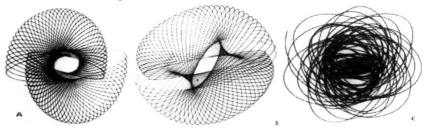

HARMONOGRAPH DRAWINGS
(From *The Power of Limits*, courtesy Gyorgy Doczi)

The endless layers of the Stillpoint geometry are in whole number proportion to each other. It is this harmony, through the dimensional proportions of the Earth and Moon and the Sun, that the geometry of the temple mirrors.

This principle is demonstrated in a slightly different way by the impression sound makes on smoke in a vacuum – when the musical note is perfect (a whole proportion of itself) the imprint of the smoke is wavy harmony – when it is off, the imprint is chaotic. Music itself is a precise example of the kind of vibrational harmony that the temple emulates:

'Architecture is crystallized music.'   Goethe

The majority of authentic extra-terrestrially created crop formations are sophisticated *geometric* communications from evolved consciousness. The two outer photographs below are of crop circles . . . the one on the right being 780 feet across . . . the one in the middle another cymatic vibration. One can easily see that, whatever the deeper communication may be, a part of what is being communicated has to do with the importance of vibration and frequency. The huge crop circle to the right is a two-dimensional representation of the three-dimensional torus vibration.

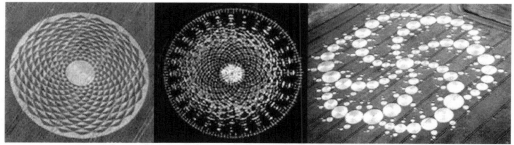

CROP CIRCLE / CYMATIC PHOTOGRAPH / CROP CIRCLE
(Left photograph courtesy of Peter Sorensen[224],
center image courtesy of Alexander Lauterwasser)

Relative to this discussion regarding the inherent relationship between proportion, vibration and form, the space between the Earth and the ionosphere has a resonant base frequency of 7.83 cycles per second, with a wavelength approximately equal to the Earth's circumference - similar to the vibration of a human in the alpha/theta range, or relaxed-aware state. This is the Schmumann Resonance, called the heartbeat of the Earth and the frequency that the temple will be intentionally surrounded and interpenetrated by, and will be discussed in more depth later.

By proportioning the temple – in particular the inner temple where silence and meditation predominate - to the dimensions of the Earth, as well as creating a naturally occurring electromagnetic field at this Schumann Resonance, a vibrational harmony will be created within the temple and within those experiencing it, with the alpha/theta vibration of the Earth/ionosphere cavity, with the intention of accessing the depths of the subconscious as a means, as with meditation, of accessing the superconscious. Form being a consequence of vibration and vibration being a consequence of form, the temple will also serve to focus, amplify and transmit centered, aware, healing energy to the planet, organic life . . . and beyond (*see Thought/Prayer/Meditation:, page 235*).

⊕ Earth/Moon/Sun: the Temple's Layout:

Thus far the reasoning behind basing the temple's design upon the proportional dimensions of the Earth and Moon and Sun has been addressed. What follows is a description of how these dimensions finally express themselves in the actual construction of the building.

> 'With regard to temples, the geometry of nature was dissected to reveal the inner blueprints of forms and shapes; their cause and effect were measured and considered. Once it was established that form follows function, specific geometries were applied functionally and strategically to the design of temples so as to induce a corresponding effect on the individual's state of awareness. The primary intent was to facilitate growth and transformation of consciousness, the goal being nothing less than the total transmutation of the soul. On a secondary level, the geometric language of nature was used to teach self-understanding in order to accelerate self-development . . . this practice is found in Buddhist mandalas, the geometric dynamics of which are capable of generating an introspective effect on the minds of their meditators, the idea being to achieve total experience of the source of creation.'[225]

The in-and-out nature intrinsic to the torus/spiral can be seen as representing the breath of the Universe . . . and the air element. The seashell/nautilus configuration of the spiral also represents the water element, and it is within air and water that we find whirlwinds and whirlpools. So now, air and water (spiral/helix), Earth (square) and fire (triangle/pyramid) are blended together in a proportional harmony born of the dimensions of the Earth and Moon. Earth, air, water, and fire are cornerstones of paths as diverse as Buddhist, Taoist, Druid, and Native American among many others, and the temple intends to express this timeless 'Isness' from which most spiritual understandings spring, in a totally natural, organic way. From Jill Purce's *The Mystic Spiral:*From Jill Purce's *The Mystic Spiral:*

> 'It is preeminently man's function to act as the link between Heaven and Earth. Each person is therefore a central axis, and has within him a central axis, up which he must move or 'climb' by developing his various centers, or by activating the subtle energies within his spinal column, or metaphorically by climbing the central pillar of consciousness of the Tree of Life, which in turn effects the descent of light of grace from above. This descent, however, can only manifest when each person has realized, if only for a moment, the change of consciousness on the central axis, the *stillpoint*, or

'center in the midst of conditions.'

This Axis Mundi, at the 'navel of the Earth,' takes the symbolic form of all pillars, poles, mountains, temples, spires or 'soul' ladders, and . . . . breaks vertically through all the planes of existence, each of which is demarcated by a step or rung, whether the nine steps of the Egyptian Osirian Mysteries, the seven steps of Buddha . . . . or the seven steps of the Babylonian ziggurat. Moreover, all cosmic or holy mountains . . . . constitute the primordial connections between Heaven and Earth.

The peak of the mountain is the tip of an upward spiral. Furthermore, all temples are relics of the holy mountain.

From this axis, like the hub of the wheel, everything extends, radiates and rotates spirally. The entire Universe, with all its spatial and temporal states, is but the spiral manifestation of the still center, as it rotates it expands, and while still rotating it contracts and disappears to the source whence it came. Through meditation, man puts himself in the position of the whole, of which he is a symbol: his meditative activity stimulates the activity of the whole.'

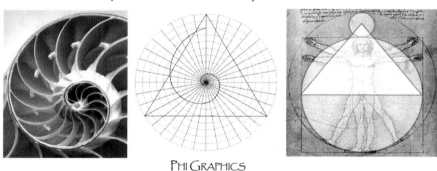

PHI GRAPHICS

The spiral, or helix/torus, is an essential aspect of the design of the temple, and provides the air/water/feminine element. So too is the earth/fire/masculine element of the pyramid. From John Snelling in Frederick Lehrman's book *The Sacred Landscape*:

'The symbolism of the sacred mountain is full of intimations of meditation. It is a state of strong immovability; of perfect balance; a state in which all motion hangs suspended, not in death or inertia but in that great stillness that is the origin and resolution of all things. Like a great meditating sage, a mountain sits with strong solidity upon the surface of the earth, quietly accumulating massive protean energies. It may be a subjective thing, but the beauty of those mountains [the Himalayas] at sunset images very powerfully the intrinsic beauty of the religious way that the Buddha initiated – a way that itself leads from the morass of our human sufferings and problems to a true wholeness and harmony with that which is.

In some ancient cosmologies, a great axial mountain is said to lie at the center of the world. In fact, any mountain that is regarded as sacred – that is, which is invested with awe and religious connotations – is a symbol of the Center: that point where divine reality impinges on profane reality. The true Center, however, the real seat of the great mystery of ultimate reality, resides in the heart of man.[1]

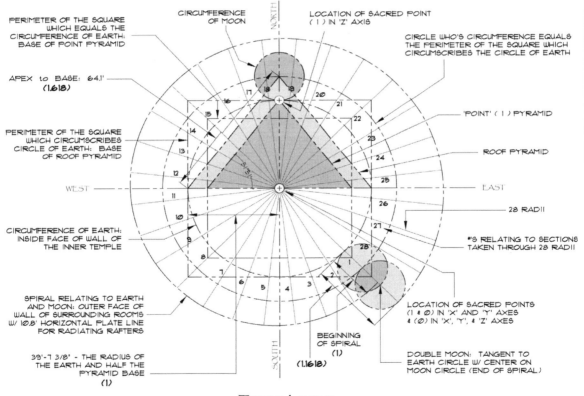

PERIMETER OF THE SQUARE
WHICH EQUALS THE
CIRCUMFERENCE OF EARTH:
BASE OF POINT PYRAMID

CIRCUMFERENCE
OF MOON

LOCATION OF SACRED POINT
(1) IN 'Z' AXIS

CIRCLE WHO'S CIRCUMFERENCE EQUALS
THE PERIMETER OF THE SQUARE WHICH
CIRCUMSCRIBES THE CIRCLE OF EARTH

APEX to BASE: 64.1'
(1.618)

'POINT' (1) PYRAMID

PERIMETER OF THE SQUARE
WHICH CIRCUMSCRIBES
CIRCLE OF EARTH: BASE
OF ROOF PYRAMID

ROOF PYRAMID

WEST

EAST

CIRCUMFERENCE OF EARTH:
INSIDE FACE OF WALL OF
THE INNER TEMPLE

28 RADII

#'S RELATING TO SECTIONS
TAKEN THROUGH 28 RADII

SPIRAL RELATING TO EARTH
AND MOON: OUTER FACE OF
WALL OF SURROUNDING ROOMS
W/ 10.8' HORIZONTAL PLATE LINE
FOR RADIATING RAFTERS

LOCATION OF SACRED POINTS
(1 & 0) IN 'X' AND 'Y' AXES
& (0) IN 'X', 'Y', & 'Z' AXES

39'-7 3/8' - THE RADIUS OF
THE EARTH AND HALF THE
PYRAMID BASE
(1)

BEGINNING
OF SPIRAL
(1)
(1.1618)

DOUBLE MOON: TANGENT TO
EARTH CIRCLE W/ CENTER ON
MOON CIRCLE (END OF SPIRAL)

NORTH

SOUTH

TEMPLE LAYOUT

It is primarily through these two archetypes – the moving, feminine, torus/helix/spiral, and the masculine, still, Axis Mundi, *Phi* pyramid – the one with its Stillpoint at the center and the other with its Stillpoint at its apex, that the temple has taken the form it has (*see Temple Layout, above*).

Another way to experience the balance of these archetypal shapes and forms – the circle and square, the sphere and octahedron - is to see the circle and sphere as symbols of eternity - Spirit and divinity - in counterpoint to the square and octahedron as symbols of Matter, reason - and man-made things. From John Michell:

> 'Seen in this way, it is not so much the fusion of opposites, but the reconciliation of the incommensurable; the other kind of sacred union.'[226]

Finally, there are two other expressions of balance found within the temple's geometry - the relationship of the circle (the finite) and the spiral (the infinite), continually moving in both directions – simultaneously contracting towards the center and expanding to infinity – the breath of the Universe.

⊕ NUMBER:

Most everything that has been discussed thus far has been a discussion of sacred geometry – those primal shapes and forms upon which all of manifestation is based. Intrinsic with all geometric form are the numbers it generates. It is from geometry and mathematics that science has so brilliantly formed its equations that generate the technological discoveries that fill our world. Galileo put it this way:

> 'Philosophy is written in that great book which ever is before our eyes - I mean the universe - but we cannot understand it if we do not first learn the language and grasp the symbols in which it is written. The book is written in mathematical language, and the symbols are triangles, circles and other geometrical figures, without whose help it is impossible to comprehend a single word of it; without which one wanders in vain through a dark labyrinth.'

Number is as essential to understanding our world as is the form from which it emerged. From Sir D'Arcy Wentworth Thompson, a Scottish zoologist from the last century:

> 'The harmony of the world is made manifest in form and number, and the heart and soul and all the poetry of natural philosophy are embodied in the concept of mathematical beauty.'

The temple is filled with symbolic number in every aspect of its design, especially the numbers **Zero** and **One**, each relating directly to the Stillpoint geometry. A full discussion of how number – two through nine - plays an important role in the temple's design is included in Appendix D (*see page 335*), but there are certain critical aspects relating to number that I feel are important to include here.

As can be seen in the Temple Layout above, the temple's inner hall, bounded by a circle, as well as the spiraling, helical lower roof over the ancillary rooms, is divided into 28 sections. 28 is a very important number connecting 4, representing the four directions (Earth), and the 7 days of the week, to the 28-day monthly cycle of the Moon. The critical number 7 is also expressed in the third dimension by the Phi pyramid – the apex angle of the Great Pyramid, 51°51', comes close to dividing a circle into sevenths[227] – 7 x 51.85 = 362.95. Equally important regarding this 28-fold division and the relationship between the Moon and the Sun, is the sacred and ancient ceremony of the Sun Dance of the Native American Lakota of the Great Plains and other North American tribes - where dancers, without water or food, dance under the hot Sun on the Summer Solstice, in intentional sacrifice for four days, moving back and forth from the circumference to the center pole (representing at once the Axis Mundi and the Sun) over and over and over and over again. Some dancers, to make their sacrifice especially meaningful, are attached to this central pole with rawhide laces, pierced through the skin on either the chest or back. In this way, the dance provides a constant reminder and *entrainment* that we return always to the Stillpoint at the center.

It is for these reasons – the numerical connection to both the Sun and the Moon, as well as to honor this tradition so intimately connected to Mother Earth - that the temple layout acknowledges the importance of this number by dividing the temple into 28 sections . . . just as the Sun Dance lodge. The Lakota Medicine Man Hehaka Sapa (Black Elk), from the book *The Sacred Pipe, The Seven Rites of the Oglala Sioux*:

> 'It helps us all to walk the sacred path; we can learn upon it, and it will always guide us and give us strength.
>
> A little dance was held around the base of the tree [the sacred tree at the center of the Sun Dance Lodge], and then the surrounding lodge was made by putting upright, in a large circle, twenty-eight forked sticks, and from the fork of each stick a pole was placed a branch which reached to the holy tree at the center.
>
> [In setting up] the Sun Dance lodge, we are really making the Universe in a likeness; you see, each of the posts around the lodge represents some particular object of Creation, so that the whole circle is the entire Creation, and the one tree at the center, upon

which the twenty-eight poles rest, is Wakan Tanka, who is the center of everything. Everything comes from Him, and sooner or later everything returns to Him. [The] Moon lives twenty-eight days; each of these days of the month represents something sacred to us: two of the days represent the Great Spirit; two are for Mother Earth; four are for the four winds; one is for the Spotted Eagle; one for the Sun; and one for the Moon; one is for the Morning Star; and four for the four ages; seven are for our seven great rites; one is for the buffalo; one for the fire; one for the water; one for the rock; and finally one is for the two-legged people. If you add all these days up you will see that they come to twenty-eight. You should also know that the buffalo has twenty-eight ribs, and that in our war bonnets we usually use twenty-eight feathers. You see, there is a significance for everything, and these are the things that are good for men to know, and to remember.'

So, here we again have a merging of the Moon (7 & 28), the Sun (1), and the Earth (4). **One** - represents Unity . . . the center, an aspect of the Stillpoint. In our relative place in the Universe, this center is represented by the Sun . . . the center point around which we orbit. Again, speaking of the relation between the Sun and the Absolute, Black Elk says that fire:

'. . . represents the great power of Wakan-Tanka, who is the center of everything, which gives life to all things; it is as a ray from the Sun, for the Sun is also Wakan Tanka in a certain aspect,' and when speaking about the sacred cottonwood tree at the center of the lodge, he says *'you will stand where the four sacred paths cross – there you will be the center of the great Powers of the Universe.'*

All of the sacred numbers from 2 through 9 are represented in one way or another in the structure of the temple, representing all of physical, moving, ever-changing manifestation. But there are two numbers that the temple design is based on more than any others – the numbers from which all else is generated. These are the numbers **Zero** and **One**, each representing different aspects of the Stillpoint of Creation from which all number and all form originate and return. For a more in-depth discussion of each of these numbers and how they are represented within the design of the temple, see Appendix D, page 335.

In the beautiful quote below by Laurens van der Post in his book *A Far-Off Place*, not only is the image of purpose-driven evolution from it genesis in the point defined by Euclid as 'being of no size of magnitude' included, but the *unity* aspect of the Stillpoint as represented by the number **One** ('yes'), as well as the non-dimensional, **Zero** ('no') aspect.

'Mopani had been told that there were stars in the sky whose light even now had not yet reached the Earth and whose existence one could tell only because of their effect on the movement of other stars. It was precisely so with the life of men on Earth. Their deeds were like a kind of starlight that came into being at the moment they were enacted but whose meaning, the light that was in the purpose, the accomplishment and the doing, would take years still to reach life itself, let alone become clear in the human Spirit. One could not, therefore, live one's life with a great enough reverence for what was small, not only in oneself, but in others. The significance of the great could only be real in so far as it was significant in the small, even when so small that at the moment of beginning it was more like the point defined by Euclid as being of no size of magnitude, but only position. Yes-no, it was position in the Spirit and the sense of direction that follows logically from the sense of position that was all important, and yet increasingly overlooked because of its lack of discernible substance.'

As above, so below. One-zero. Yes-no.

Number is sacred. From R.A. Schwaller de Lubicz:

> Number is living, an expression of life
> and speaks directly to the
> intelligence of the heart.
> Its true secret lies
> in the becoming of
> One into Two.

The true purpose of this sacred building is the becoming of Two into One.

Everything said thus far describes the principles, geometry and number underlying the forms of the temple. All of these forms are secondary relative to this Sacred Point and it is this that the temple celebrates above all . . . the true experience itself, with its surrounding, protective structure . . . the covenant - 'God's promise to the human race' - surrounded and protected by the ark of the temple itself.

John Michell, in his preface to Michael Schneider's book *A Beginner's Guide to Constructing the Universe*, states, in a very personal way, the importance both of geometry and the unity aspect of its genesis:

'As soon as you enter upon the world of sacred, symbolic, or philosophical geometry – from your first, thoughtful construction of a circle with the circumference divided into the natural six parts – your mind is opened to new influences that stimulate and refine it. You begin to see, as never before, the wonderfully patterned beauty of Creation. You see true artistry, far above any human contrivance. This indeed is the very source of art. By contact with it your aesthetic senses are heightened and set upon the firm basis of truth. Beyond the Many – is the delight that comes through the philosophical study of geometry, of moving toward the presence of the **One**.'

Ironically perhaps, the Stillpoint is represented *twice* in the temple . . . once representing the Absolute's entry into form, into this three dimensional reality – its '**One**' aspect, that mentioned above; and once representing the Absolute itself – its '**Zero**' aspect. The first, representing **One,** is located at the apex of the smaller, unmanifest, Great Pyramid shown in the Temple Layout (*see Temple Layout above*) – where Spirit enters manifestation by creating the 3rd dimension. This point will be created, *implied*, by the intersection of the virtual lines joining the opposing vertices on a diamond cut in the shape of a dodecahedron. This diamond is located at the center of a 12" diameter, gimbaled, rotating, stellated dodecahedron, suspended beneath the dome, the highest ceiling within the inner temple (*see an animated image of this at www.stillpointdesign.org*). The dodecahedron demonstrates the very special relationship of organic life (5) and man's rational division and understanding of the Universe (12) with the twelve pentagonal sides of the fifth platonic solid.

The diamond itself is the perfect substance from which to create the sacred, still, point from which all else springs. Primarily, like the carbon molecule and water, it is tetrahedral in nature . . . connected in some mysterious way to life itself, immense pressure transforming organic carbon into the brilliance of diamond - much like suffering transforms ignorance into aware consciousness. It also has an *absolute* quality to it. Geshe Michael Roach, in his book *The Diamond Cutter*, says:

'Diamonds are perfectly clear, almost invisible, and the hidden potential of everything around us is just as hard to see. They come very close to being something which is absolute – the hardest substance known to man, having the highest refractive index of any material, and the greatest ability to throw light off their surface – and the hidden potential in things is their pure and absolute truth. Every sliver of diamond that exists anywhere in the Universe is exactly the same stuff as every other one – pure 100 percent diamond – and it is true of the hidden potential of things too that every instance of the potential is just as pure, just as absolute a reality, as every other instance.'

These words could also be describing the non-existent (non-dimensional) Point that is nowhere, yet everywhere.

Within the temple, the other sacred point – representing the **Zero** aspect of the Stillpoint - is created by the implied center of an invisible sphere whose concave impression imprints the horizontal top of the 20-ton *Omphalos* black granite fountain at the center of the inner temple's circular floor (*see Floor Plan, page Error: Reference source not found*) – and is the center of the *complete* geometry of the temple itself . . . the physical temple that is above ground, as well as its polar compliment – *balanced* as it is mirrored below ground, and completed in the abstract (*see Sphere/Octahedron, page 11*).

I CHING / YIN YANG SYMBOL

Zero *represents* the eternal Absolute . . . that which generates all form, yet which has no dimension – existing everywhere yet not 'existing' anywhere. The Absolute itself is a philosophical idea that our minds can only intimate. Rodney Collin, in his book *The Theory of Celestial Influence*, gives this definition:

'Philosophically, man can suppose an Absolute.

Such an Absolute would include all possible dimensions both of time and space. That is to say:

It would include not only the whole Universe which man can perceive or imagine, but all other such Universes which may lie beyond the power of his perception.

It would include not only the present moment of all such Universes, but also their past and their future, whatever past and future may mean on their scale.

It would include not only everything actualized in all the past, present and *future* of *all Universe*s, but *also* everything that potentially could be actualized in them.

It would include not only all possibilities for all existing Universes, but also all potential Universes, even though they do not exist, nor ever have.'

200

Such a conception is philosophical for us. Logically, it must be like that, but our mind is unable to come to grips with the formula or make any sense of it.' In the Kabbalistic body of knowledge this Absolute is referred to as the vast *Ayn*, or *limitless* (*Nun* in Egyptian wisdom) . . . and where this utter Totality interfaces with our reality is the point of singularity . . . Kether . . . the *Point* of Creation at the top of the Tree of Life. The **Zero** aspect of the Stillpoint . . . existing everywhere yet not existing. Not found in any of the dimensions that make our world what it is . . . yet the source of each of them. The most finite center itself . . . holding infinity. One hand clapping. If the Stillpoint can be understood as the point where the Great Mystery enters Creation . . . or more precisely, where the Great Mystery (the Kabbalistic *Ayn*, the Egyptian *Nun*, Meher Baba's '*Beyond the Beyond*') becomes manifest as light . . . the **Zero/One** phenomenon can be visualized as the I-Ching symbol, having a **Zero** aspect (dark) and a **One** aspect (light). Buddhism's *Secret Doctrine* speaks of this relationship:

> 'Darkness is the one true actuality, the basis and the root of light, without which the latter could never manifest itself, nor even exist. Light is Matter, and darkness pure Spirit. Darkness, in its radical, metaphysical basis, is subjective and absolute light; while the latter in all its seeming effulgence and glory, is merely a mass of shadows, as it can never be eternal, and is simply an illusion, or Maya.'

This is another way of saying that all of the form, shape, proportion and number that give the temple its *appearance* are there only in relationship to the Stillpoint at the center of everything. It is the doorway to and from the Mystery.

This **Zero** aspect of the utterly *still*, point is located at the center of all the geometry of the temple. It is (0,0,0). The center. Complete balance.

⊕ PERFECT BALANCE:

The geometric model for the sphere . . . for the point expanded . . . is the Vector Equilibrium – 12 points defining a sphere around 1 central point. *Balance* is yet another critical aspect of the unequaled significance of this geometry, as expressed above. It has six square sides and eight triangular sides totaling 14 in all, and is the closest packing of equal spheres around a nucleus of equal size . . . twelve around One – the Christ model . . . with equal vectors both from the center to each of the surrounding centers, and from the center of each of the surrounding spheres to its neighbor . . . a model for perfect, still, balance.

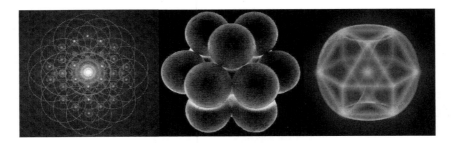

FLOWER OF LIFE / VECTOR EQUILIBRIUM
Left: 12-around-1 2d *Flower of Life*, interpenetrating spheres. Center: 12-around-1
Vector Equilibrium matrix; six squares, eight triangles. Right: Same

Buckminster Fuller, the first man who truly recognized the potential meaning of this geometry as far as we know, used a model of twelve spheres surrounding the nuclear sphere - 12 ping-pong balls/spheres around 1 - each touching four neighbors in addition to the nuclear sphere, creating the six triangles and four squares upon its surface. As mentioned above, this 'Vector Equilibrium' best represents the geometric model for the sphere, or the Stillpoint itself, the tiniest 'sphere.' 12 points around a central point creating a sphere, all vectors along the surface and to the center being equal.   More accurately represented, the twelve spheres *interpenetrate* each other, their circumferences passing through the center point of the adjoining sphere and the Stillpoint at the center. Once again, here is the Vector Equilibrium as described by Buckminster Fuller in his very own way:

> 'Zero pulsation in the Vector Equilibrium is the nearest approach we will ever know to eternity and God . . . . the magical shape of the Vector Equilibrium transforms the octahedron into the square into the circle into the tetrahedron and so on forever . . . It represents the Stillpoint. Yet it is the frame of the evolvement. It is not in rotation. It is sizeless and timeless. . . . . the physical is always the imperfect experience, but tantalizingly always ratio-equated with the innate eternal sense of perfection – thus the mind induces human consciousness of evolutionary participation to seek cosmic zero.'

This sacred shape, representing stillness and the Absolute, originating and infinitely recreating the original point, is the geometric heart of the temple, and eternally represents this primal, abstract, stillness.   The shape of the Vector Equilibrium, representing the Absolute, generates and is generated by the dualistic opposites of the octahedron and the sphere *in their joining (see Sphere/Octahedron, page 11 & 175).* All forces in balance. All *vectors* of force equal. All this simply from joining of the sphere tangent to the octahedron at its edges' midpoints . . . resulting in the shape of the Vector Equilibrium. Built-in regeneration . . . Circle-Sphere-Point-Circle-Sphere . . . with each of the five Platonic solids, the building blocks of Nature, included in and generated from its unique geometry, and all else generated from them.

This is the ideal form to represent the timeless within changing time. An ideal form for structure dedicated to silence and meditation. One can see that the wholeness of the building, its *total* integrity, is completed only in thought . . . and that its form – what we see and physically experience – is only a part of the whole. Long ago now, when I was first exploring the meanings and validity of sacred geometry, I was fortunate in that I was able to ask, over lunch, a question regarding all this to perhaps the Western world's foremost sacred geometer, Keith Critchlow. I realized that the budding design for the original mediation hall, which has now morphed into the Mesa Temple, manifested only a *part* of the geometry that defined it – that is, there were aspects of the overriding geometry that were not 'built.' I wondered if this mattered and asked him this. His resounding certainty was an immediate 'no.' In the years since, I've come to understand this with equal certainty – it is the underlying blueprint of sacred geometry that matters above all else, manifested or not.

Ratio-equated with the innate eternal sense of perfection . . . inducing the mind to seek cosmic zero . . . the Stillpoint is the genesis of all Creation. Every form, every thought and emotion, everything – abstract or physical . . . is ratio-equated with this Stillpoint.

As interpenetrating spheres are added infinitely to the original sphere, the next layer of spheres brings the original thirteen – twelve around one – to thirty two. I believe Hermes Trismegistus, in the Book of Thoth, is referring to these 32 spheres:

> 'Deep in the Temples of Life grew a flower, flaming, expanding, driving backward the night. Placed in the *center*, a ray of great potency, Life giving, Light giving, filling with power all who came near it. Placed they around it thrones, two and thirty, places for each of the Children of Light, placed so that they were bathed in the radiance filled with the Life from the eternal Light.'

In this quote from the second tablet of the Emerald Book of Thoth, this growing flower refers to the transformation of the non-dimensional point of 'light' at the center, into the three dimensional world we experience today . . . and specifically acknowledges the transformation as it implements itself through the matrix of this Stillpoint, the 32-sphere/tetrahedron layer in the shape of the Vector Equilibrium.

The earliest known Jewish text on magic and cosmology, the Sefer Yetzira (*Book of Creation*), appeared sometime between the 3rd and the 6th century. It explained creation as a process involving the ten divine numbers, or Sephirot (*see Tree of Life, page 251*), of God the Creator, and the 22 letters of the Hebrew alphabet representing the 22 paths between the Sephirot. Taken together, they were said to constitute the '32 paths of secret wisdom.' It is interesting to note here the dramatic reference, again, to the number 32, the number of sphere/tetrahedrons represented by the nineteen circles in the two-dimensional Flower of Life, and referred as the thirty-two Children of Light above in Tablet 2 of the ancient Egyptian Emerald Book of Thoth – all growing from the point in the *center*.

⊕ More Context:

This seems like a good place to take a break from all the geometry and number and speak a little more about the context that gives meaning to all of this.

Throughout these pages, it has been suggested that consciousness is in the process of evolving. As Pierre Teilhard de Chardin suggests:

> 'Man discovers that he is nothing else than evolution become conscious of itself. The consciousness of each of us is evolution looking at itself and reflecting upon itself' - this 'reflection' signifying a kind of transcendence that is 'the power acquired by a consciousness to turn it upon itself, to take possession of itself as of an object endowed with its own particular consistence and value: no longer merely to know, but to know oneself; no longer merely to know but to know that one knows.'

Is it possible, in fact, that the Universe is evolving towards a state of complete awakening, or pure awareness? The essence of this mystical view is that the Universe is a living, self-evolving, multidimensional system, or Being, and the flow of information, of energy, within the system is critical to its growth. Living organisms and ordered systems are ultimately whirlpools or vortexes, patterns of organization which ingest new matter and energy all the time. This flow of information is ultimately the means by which the Universe becomes self-aware . . . overcoming its own seemingly inevitable entropic heat death.

While the information thus far has alluded to this possibility, it is the purpose of this section to elaborate on these ideas, addressing how this awakening is happening and how, precisely, the temple fits within this process. To this point, the ideas presented have been based upon geometric/scientific fact. The *context* for what this all might mean is found in theory and philosophy . . . and the mind is now asked to use the preceding sections as a foundation for the following discourse.

⊕ THE PROCESS TAKING US FROM HERE TO THERE:

Meher Baba's words offered a poetic analogy to how this all began . . . a *whim* sparking latent consciousness in the beyond-beyond state of God's original oblivion. T.S. Elliot has offered another beautiful poetic analogy describing the awakening process . . . never ceasing from exploration and 'the end of all our exploring will be to arrive where we started' . . . in complete enlightened awareness . . . 'and know the place for the first time.'[228] I imagine the shape of Elliot's journey as a helix.

There are those who say that it is all happening by itself, relentless, unstoppable and huge . . . that there's nothing to *do* . . . that it's already done – ironically embodied by Meher Baba's own iconic words: 'Don't worry, be happy.' I feel that this is a huge misunderstanding. And, as it turns out, this is not the full quote, which is: '*Do your best. Then, don't worry; be happy in My love. I will help you.*' Very different indeed. In the largest context though, I know that the initiating whim or spark, the primal awakening purpose, cannot be stopped and in some way this process *is* already 'done.' But in the moments of our lives we are presented with choices constantly . . . and how we meet and encounter those moments is how we evolve. It may be 'done' somewhere in the infinite reality, but it is the challenge of each moment that defines who we are and how we grow.

My own model for the perfection of this interface with action is the innate spiritual and physical courage of the Oglala Lakota, Crazy Horse. He had no idea who was over that hill, but he jumped on his horse and charged into his healthy future, whether he died or not, knowing that he was aligned perfectly with 'right action' in the Buddhist sense, yelling . . . '*Hóka-héy! Today is a good day to die!*'[229] Life offers us the opportunity to show up until the moment of our death. If our collective self fails to make it through this interval, Earth will certainly survive and recover from our misdeeds . . . but the opportunity will have been lost for *us*. We will be forced to begin again if we are lucky enough to have that chance . . . ultimately and through great amounts of time, once again approaching the same interval that confronts us now. *This* is where the work is done . . . in this world of suffering and density. The choices we make are so very critical. We are by no means 'done.'

Somewhere in this illusion of separateness, while awakening to the truth of Oneness, one is pulled through the horror of Ignorance's casualties into a compassion for *all that is*. Once again, it is:

'. . . this innocence waking up, realizing that everything is itself, even all the confusion and all the ignorance . . . everything . . . *this* is the dissolving of confusion, of ignorance, of karma . . . awareness yielding to itself, to its inherent creativity, to its expression in form, to experience itself' . . . and finally . . . 'only awareness remaining.'
                                                                Adyashanti

Inventor, mathematician and scientific philosopher Arthur Young, in *The Reflexive Universe*, offers a beautiful, scientific model for the poetic words of Meher Baba, T.S. Elliot and Adyashanti. His 'process theory' begins with observation that light – the first stage of this awakening process - is purposive, and quotes Max Planck:

'Thus, the photons which constitute a ray of light behave like intelligent human beings: Out of all possible curves they always select the one which will take them most quickly to their goal.'

The seven stages of Young's process model begin with zero transforming into one – the first quantum - from nothing to the Monad, or photon of light. Particles crashing into each other at light speed, creating other particles as the Universe cooled. Then, these individual particles combine to create Atoms, which combine to create Molecules, which proceed through more and more complex organization to finally reach evolution's nadir, or descent into density, in the organizational molecular density of crystals and metals. The molecular stage continues, until it reaches the end of the road for inert matter and then somehow is inspired to store energy, beginning the reversal of entropy, and eventually culminating the Molecular stage in the creation of the wonderfully sublime, double helix molecule of life - DNA . . . and the *turn*[230] from a descent from light into the density of matter governed by entropy, into fewer degrees of freedom and the physical laws of inert matter is complete *and the awakening process reverses the law of entropy and life appears.*

As the process continues, it attains greater degrees of freedom, and proceeds through the Plant and Animal Kingdoms, finally reaching a self-determining stage that Young calls the Dominion Kingdom . . . the Kingdom that includes Man – and finally, at the very end of the dominion process, reaching the goal of consciousness's long journey from its beginning in the photon of light emerging from the Stillpoint to the realization of fully awakened, pure awareness . . . Adyashanti's 'only awareness remaining; Lazslo's 'it becomes, and thenceforth eternally is, THE SELF REALIZED MIND OF GOD."

Like the forty-nine days spoken about in the Tibetan Book of the Dead, the time between one's death and one's rebirth, Young describes seven stages of development, each stage consisting of seven sub-stages, forty-nine in all, from the unawakened yet purpose-filled photon to the rebirth of fully awakened enlightenment. If awakening consciousness has reached this level of realization and completed the cycle, it is only logical that it would, in a Bodhisattvic, compassionate manner, create signposts to point the way to those of us still on the wheel.

⊕ Ariadne's Thread

Life's journey, as we all know, is a difficult one. We are challenged at each step on the path to find to the truth, to find our way to a life filled with meaning. There is an ancient Greek myth that not only describes the confusing, sometimes impossible aspect of the journey, but includes too the archetypal aspect that others have gone before us and are there for us a guides when we become lost. As the ancient Greek story goes, Ariadne was the daughter of King Minos of Crete. Her father refused to sacrifice a bull to Poseidon, so the god took revenge by causing Minos's wife Pasiphae to desire a bull – Pasiphae eventually giving birth to a being that was half man, half bull . . . the Minotaur. Minos had Daedalus build a Labyrinth, a house of winding passages, to house this beast. Minos then required tribute from Athens by requiring young men and women to be sacrificed by entering this Labyrinth with its innumerable paths of deception . . . where the Minotaur lay in wait. This is an analogy of the endlessly tricky path of awakening.

Theseus, an Athenian, volunteered to accompany one of these groups of victims, with the intention of delivering his country from the tribute to Minos. Ariadne had fallen in love with Theseus, and to protect him gave him a thread which he let unwind through the Labyrinth so that he was able to find his way back out after he had killed the Minotaur.

In the bodhisattvic tradition, the Bodhisattva, or awakened being, chooses not to remain in the undisturbed awakened state until all of humanity has been awakened. Bodhisattva is compassionate understanding, transcending the *seeming* separateness of 'other' beings . . . complete in the realization of One Being. Regarding Ariadne's thread, it is almost as though awakened consciousness itself, in the tradition of bodhisattva, has woven a thread to guide us through this labyrinth of the mind towards freedom . . . pure awakened awareness.

Thus far, I have attempted to make it clear that there are some remarkable signposts or messages . . . Ariadne's thread . . . that are leading us along this dangerous path – from the Alpha to the Omega, each of them the Stillpoint reality where the Absolute Everything/Nothing lives eternally outside/inside the whirling wheels of time, mind and matter – the one oblivious, the other fully awakened. We are being led through this doorway, out of the Labyrinth of the mind – Home. Please review the list of such signposts beginning on page 325 in Appendix C.

Besides the guidance offered by the messages woven into our solar system, there are also the 'mechanics' of just how we are being led from that birthplace into our future. They include the phenomena of synchronicity, morphic resonance, critical mass, quartz crystals, the pyramid shape and electromagnetic fields.

⊕ SYNCHRONICITY:

If it is critical to this process of growth that information, or energy, be transferred and shared readily, then it would be logical that there would be a pattern or system or method through which this happens. Also, if it *is* One Being we're talking about, there would be evidence that consciousness and matter were intimately interconnected, and that this interconnectedness would lead eventually back to the Stillpoint, its source. One would also expect that this pattern of interconnectedness would mirror in form and behavior what we see every day in the world in which we live . . . *as above, so below* . . . the perfect expression in all manifestations of a holographic Oneness - all parts an expression of the whole.

Perhaps the easiest way to begin to see the inter-relatedness of consciousness and matter is through an understanding of synchronicity. Carl Jung recognized intuitively that matter and consciousness - far from operating independently of each other - are, in fact, interconnected and function as complementary aspects of a unified reality. In formulating his synchronicity principle, Jung was influenced to a profound degree by the 'new' physics of the twentieth century, which had begun to explore the possible role of consciousness in the physical, quantum world.

> 'Physics,' wrote Jung in 1946, 'has demonstrated . . . that in the realm of atomic magnitudes objective reality presupposes an observer, and that only on this condition is a satisfactory scheme of explanation possible. This means that a subjective element attaches to the physicist's world picture, and secondly that a connection necessarily exists between the psyche to be explained and the objective space-time continuum.'

Synchronicity can be seen as the connection, outside of time and space, that connects our real-world, time oriented, reality with other dimensional experiences that are completely subjective (dreams, intuitions, states induced by hallucinogens such as ayahuasca, peyote or DMT, or accessed through holotropic breathwork, meditation, etc.), in a way that is objectively true. How do we sometimes know who is about to call us on the phone, even though we haven't heard from that person for some time? How do some people know about an event in the future before it happens, or know that something is occurring in the moment even though it's taking place on the other side of the world? How do pets sometimes know the way home even though this home is hundreds of miles away, miles they've never seen? The authentically documented, if rare, examples of telekinetics - the psychic manipulation of physical objects - are also examples of this interconnectedness. These kinds of events reveal connections between the subjective and objective worlds . . . they are 'synchronistic.'

Here's an example of this in my own life. Something very emotionally painful had happened . . . so much so that I stayed awake all night, craziness rattling around in my head. I left my home early in the morning to drive over the mountains to San Francisco. I tried to distract myself by thinking about how to represent the Stillpoint in the design of the temple. The sky was a sparkling, clear blue, and as I drove up the long grade leaving the valley, this was all I could see . . . the brilliant blue sky before me. With unquenchable pain in my heart, a *perfectly* circular (or spherical?) formation of pure white birds (water birds of some kind . . . rarely seen in the high desert) appeared in the sky ahead of me. Amazing! I had never seen anything like it . . . nor, I guess, have you. I was mesmerized. And then, *instantly*, this perfect circle of white birds divided into two *perfect* circles or spheres! It was totally magical and pulled me completely out of my pain – got my *attention*. It screamed something important, and felt like a communication of some kind . . . and then the pieces fell into place. The original circle/sphere represented unity . . . our original, undifferentiated, state. Then it divided into two . . . into the duality of manifestation – the realm of suffering . . . and evolution. Somehow I extended all this to its extreme and saw that organic life had been subjected to such extreme pressure (suffering) over billions of years that it turned carbon into . . . diamond. I saw then that the projection of this process would be the perfectly cut diamond . . . the intersection of the vertices meeting at a non-existent point in the center establishing a location of the Stillpoint in the temple – my own suffering transformed through revelation solved the question I was working on . . . the synchronistic merging of my world and the magical world on the other side of the veil. 'I must admit that nothing like it ever happened to me before or since.'[231]

The belief that matter and consciousness interpenetrate is, of course, far from new. Some physicists have been implying the connection of consciousness to quantum theory for many years. Historian Arthur Koestler refers to this relationship as the capacity of the human psyche to 'act as a cosmic resonator,' based on the premise that individual and Universe 'imprint' each other, acting by virtue of a 'pre-established harmony.'

Koestler observes in his book *The Roots of Coincidence*, that the presumption of a 'fundamental *unity* of all things, which transcends mechanical causality, and which relates coincidence to the universal scheme of things' is common to many historical sources. He refers to this as 'the universal hanging-together of things, and their embeddedness in a universal matrix.' Many ecologists already subscribe to this sense of interrelation in the world . . . where everything is connected to everything else - what the ancients called the 'sympathy' of life, and the numbers of scientists now converting to this worldview are beginning to multiply.

Karl Pribram, mentioned earlier and a neuroscientist at Stanford University, has proposed that the brain may be a type of 'hologram,' a pattern and frequency analyzer which creates 'hard' reality by interpreting frequencies from a dimension beyond space and time. On the basis of such a model, the physical world 'out there,' is, in Pribram's words, 'isomorphic with' (same as) 'the processes of the brain.'

Also mentioned earlier, related to this intimate connectedness of all things, in 1982 a research team at the University of Paris led by physicist Alain Aspect discovered that under certain circumstances subatomic particles such as electrons are able to instantaneously communicate with each other *regardless of the distance separating them* . . . violating Einstein's long-held tenet that nothing can travel faster than the speed of light. It doesn't matter whether they are 10 feet or 10 billion miles apart. Quantum physicist David Bohm believed this to be true not because they are sending some sort of mysterious signal back and forth, but because their separateness is an illusion. He argued that at some deeper level of reality such particles are not individual entities, but one thing, a sort of superhologram in which the past, present, and future all exist simultaneously. *Now.* He also expresses the awareness that our unawareness of this is creating the insanity we're now witnessing globally:

> 'The notion of a separate organism is clearly an abstraction, as is also its boundary. Underlying all this is unbroken wholeness even though our civilization has developed in such a way as to strongly emphasize the separation into parts . . . indeed, the attempt to live according to the notion that the fragments are really separate is, in essence, what has led to the growing series of extremely urgent crises that is confronting us today.' David Bohm

Unity is, again, the world of the Stillpoint reality . . . the world of instantaneous communication, of telepathy and the connectedness of all things. Synchronicity is an aspect of this. The lack of awareness reflected in the separateness mentioned by Bohm is precisely the consciousness problem that needs to be transformed.

In Rupert Sheldrake's 2012 *The Science Delusion*, he speaks of the importance of connecting with higher consciousness towards the aim of experiencing higher unity:

> 'From a spiritual perspective, future connections with higher or more inclusive states of consciousness may serve as spiritual attractors, pulling individuals and communities toward experiences of higher unity.'

The more we are able to *synchronize* our own world with this higher world, the more easily we will be able to avoid the 'urgent crises' Bohm mentions. This synchronization is the single most important goal of the Mesa Temple.

⊕ MORPHIC RESONANCE:

Theoretical biologist Rupert Sheldrake hypothesizes that the 'laws' of nature are in fact *habits* . . . that is, the laws as we experience them were *evolved* through an interaction of expanding awareness and experience – examples of growing, evolving consciousness within our five-dimensional world. He calls this process *formative causation.*

Sheldrake theorizes that these habits are reinforced and 'remembered' through the matrix of morphic fields . . . actual fields of invisible structure attuned to each of the myriad forms and behaviors we see today. These morphic (form-creating) fields shape growing/learning organisms . . . in fact, shape all organized systems. Within each organism or system there are fields within fields within fields, and these fields are both within and around the forms they shape/create - just a gravitational field is both within and around the Earth – concepts that help to take us beyond rigid definitions of 'inside' and 'outside.'

Morphic or morphogenetic field is a term that includes the fields of both form *and* behavior; the term morphogenesis meaning 'the coming into being of form.' Sheldrake calls this phenomenon 'morphic resonance,' the influence of like upon like providing a connection among similar fields. He says:

> 'Basically, morphic fields are fields of habit, and they've been set up through habits of thought, through habits of activity, and through habits of speech. Most of our culture is habitual, most of our personal life, and most of our cultural life is habitual. What I am

suggesting is that a very similar principle operates *throughout the entire Universe*, not just in human beings.

A familiar example might be that of a hive of bees or a nest of termites: each is like a giant organism, and the insects within it are like cells in a superorganism. Although comprised of hundreds and hundreds of individual insect cells, the hive or nest functions and responds as a unified whole.

As another familiar example of the superorganism concept, consider schools of fish: when predators swim into a school, the fish dart quickly to the side in a coordinated way in order to clear a path through the middle. They move very fast in response to quite unexpected stimuli, yet they do not bump into each other. The same is true of flocks of birds. A whole flock can bank as one without the birds bumping into each other. The banking maneuver could begin anywhere within the flock - at the front or back or at the side. It is usually initiated by a single bird or a small group of birds, and then propagated outwards *much faster than could be explained by any simple system of visual cuing and response to stimuli.*

This is similar to the way in which the morphogenetic field of the human being coordinates the entire body even though the cells and tissues within the body are continuously changing.'

Perhaps the most interesting aspect of this concept is that these fields and the effect they have are non-local and aren't local energy systems in the normal sense of electromagnetic fields, tornadoes, animals, etc. – but fields that define a kind of primal, invisible, ultra-subtle matrix of learned and evolving form that evokes behavior from atoms to rhododendrons to elephants to humans and beyond. In a conversation with biologist Sheldrake, here is physicist David Bohm discussing the similarity of Bohm's physics with Sheldrakes formative causation from Sheldrake's book *Morphic Resonance: the Nature of Formative Causation*:

'Now the quantum potential had many of the properties you ascribe to morphogenetic fields . . . . the quantum potential energy had the same effect regardless of its intensity, so that even far away it may produce a tremendous effect; this effect does not follow an inverse square law [local]. Only the form of the potential has an effect, and not its amplitude or its magnitude . . . . so we could say that in that sense the quantum potential is acting as a formative field on the movement of the electrons. *The formative field could not be put in three-dimensional [or local] space, it would have to be in a three-n dimension space, so that there would be non-local connections, or subtle connection of distant particles* (which we see in the Einstein-Podolsky-Rosen experiment). So there would be a wholeness about the system such that the formative field could not be attributed to that particle alone; it can be attributed only to the whole, and something happening to faraway particles can affect the formative field of other particles. There could thus be [non-local] transformation of the formative field of a certain group to another group. So I think that if you attempt to understand what quantum mechanics means by such a model, you get quite a strong analogy to a formative field.'

In *Morphic Resonance*, Sheldrake includes a fascinating chapter called *Four Possible Conclusions*, where he discusses possible ways to explain, scientifically, the 'origination of new forms and new patterns of behavior.' He acknowledges that the 'hypotheses of formative causation cannot alone offer explanations for this,' nor can it 'explain subjective experience. Such explanations can be given only by theories of reality more far-reaching then those of natural science, in other words by metaphysical theories.' He then outlines four possible metaphysical theories that 'illustrate the distinction between the realms of science and of metaphysics.' I would like to include his discussion of these metaphysical models because I find them fascinating and because of the beautiful and precise thinking they reflect.

First, he offers a view that combines the materialistic view of science with formative causation. Materialism 'denies a priori the existence of any non-material causal agency,' relegating consciousness to the material states of the brain . . . determined only 'by a combination of energetic causation and chance events.' That is, the origin of new forms 'must therefore be attributed to chance, and evolution can be seen only in terms of the interplay of chance and physical necessity.' He brings in causative formation by suggesting that these 'random events within the brain' could be affected by the 'chance activation of one morphic field rather than another' from previous brain states, and by this theory, 'all human creativity, like evolutionary creativity, must ultimately be ascribed to an interplay of chance and physical necessity' - no *purpose* here 'beyond the satisfaction of biological and social needs.'

Sheldrake's second model is that of the 'conscious self' – acknowledging 'a reality that is not merely derivative from matter' and an implied interaction between the two. Since materialistic science cannot reconcile this interaction with 'other' happening within the brain, he suggests that the conscious self could be interacting with morphic fields that in turn can act upon the body and brain by selecting between morphic fields and 'by serving as a creative agency through which new morphic fields come into being' . . . as with insight. This adds the principle of *conscious creation* to those of causative formation and the energetic causation of materialism. Still, in this model, consciousness is defined within the parameters of previous morphic fields and *new* forms are still explained by these chance interactions . . . 'so the problem of evolutionary creativity remains unsolved.'

In the third model, which Sheldrake calls 'the creative universe,' there are creative agencies of hierarchical nature, within nature but outside of the conscious self – 'within the planet as a whole, or the solar system, or the entire universe' – that act *through* the conscious self, or: higher levels of creativity acting through lower levels – what is often considered *'inspiration.'*[232]

> 'But although an immanent hierarchy of conscious selves might account for evolutionary creativity within the universe, it could not possibly have given rise to the universe in the first place. Nor could this immanent creativity have any goal if there were nothing beyond the universe toward which it could move. So the whole of nature would be evolving continuously, but blindly and without direction,' thus denying 'the existence of any ultimate creative agency transcending the universe as a whole.'

Sheldrake calls his forth model 'transcendent reality:'

> 'The universe as a whole could have a cause and a purpose only if it were itself created by a conscious agent that transcended it.
>
> If this transcendent conscious being were the source of the universe and of everything within it, all created things would in some sense participate in its nature. The more or less limited 'wholeness' of organisms at all levels of complexity could then be seen as a reflection of the transcendent unity upon which they depended, and from which they were ultimately derived.
>
> Thus this fourth metaphysical position affirms the casual efficacy of the conscious self, *and* the existence of a hierarchy of creative agencies immanent within nature, *and the reality of a transcendent source of the universe.'*

I have also included Sheldrake's metaphysical theories because I would like to offer a fifth theory. By now I'm sure that it's clear that the model I am proposing is a version of 'transcendent reality' that includes all of the models proposed . . . but a transcendent creative agency that, while it has always existed and whose presence is sensed everywhere in nature, is in the *process* of becoming fully aware and conscious. As I've said many times, the conscious creation by such an *evolved* creative agency is clear in the overwhelming amount of evidence suggesting such within *our* solar system and, perhaps more than anything, the evidence of bodhisattvic intention . . .

of compassion - the geometry of consciousness embedded in the heart of the solar system, a message to be decoded and *used* when everything is on the line.

I believe that this evolving consciousness manifests also as archetypes in the sense of Sheldrake's 'creative universe' - perhaps the planets in *our* solar system are examples of this - which serve as archetypal morphic fields of different aspects of consciousness, interacting with each other to guide us and accelerate the dance of evolution. In this sense, it is this transcendent, primal intention, driven forward by initiating purpose, manifesting finally as a completely awake, aware, fully conscious presence, free from the need to manifest and no longer on the wheel of life, weaving matter and form at will and, perhaps, choosing to *incarnate* as a planetary archetype to guide those still lost in the density of our world. That is, instead of intentionally *creating* this solar system, this higher consciousness *is* the solar system.

Consciousness itself is the evolving field of all past learned experience that each of us individually is connected to . . . a *field* of awareness growing through formative and energetic causation, and operates through the energetic mechanisms of the physical world – the world that science is so adept at discovering – as well as through the mechanisms of morphic resonance and synchronicity.

The basic underlying intention of the Mesa Temple is to access the evolved higher consciousness that is responsible for the message written in the heavens – a message having only to do with consciousness itself - and/or the Akashic Record of eons of universal evolution. The temple is intended as a portal, or opening, to this wisdom and as a collector and amplifier and transmitter of such. An aspect of this process is *the creation of a template for a new and powerful morphic field here on Earth* – the intentional creation of a new form . . . the creation of a new paradigm, using this seed from the future as the originating germ, initiating the development of a new morphic field – a field of compassionate, expanded consciousness. If this can be established, it will have a life of its own, growing in momentum and influence.

⊕ CRITICAL MASS:

Teilhard de Chardin (and Arthur Young in *The Reflexive Universe* when he speaks of negative-entropy as life enters the picture) makes clear the understanding that while the 'physical Universe' appears to be heading towards a kind of death governed by the laws of entropy, consciousness is evolving, waking up within the primordial darkness. He says:

'My starting point is the fundamental initial fact that each one
of us is perforce linked by all the material organic and psychic

strands of his being to all that surrounds him. To make room for thought in the world, I have needed to interiorize Matter: to imagine an energetics of the mind; to conceive a *noogenesis* [the origination/evolution of mind] rising upstream against the flow of entropy; to provide evolution with a direction, a line of advance and *critical points* . . . if we look far enough back in the depths of time, the disordered anthill of living beings suddenly, for an informed observer, arranges itself in long files that make their way by various paths towards greater consciousness . . . the Universe fulfilling itself in a synthesis of centers in perfect conformity with the laws of union. God the Center of centers.'

And what could possibly provide the matrix, this 'synthesis of centers', upon which those 'various paths' form? From Johannes Kepler:

'Geometry existed before the Creation. It is co-eternal with the mind of God. Geometry provided God with the model for Creation. Geometry is God Himself.'

Embedded within the above quote from Teilhard is the idea that there are critical points along this journey. The evolution of consciousness, just like everything else in the Universe, will have a rhythm or periodicity or cyclic quality, and that at certain points, or intervals within that rhythm - when a system is changing into a higher state of functioning - it must first pass through a state of disruption or chaos, yet chaos is not the same as disorder. What physicists call chaos ('nonlinear recurring processes') are in fact often descriptions of systems that obey non-equilibrium thermodynamics – and often display a 'bizarre sort of non-obvious, higher level order.' Chaos, in this sense, seems to be a *stage* in the process of transformation.

The reality we experience today is an unprecedented juncture in its history. After billions of years, human beings arrive, multiply exponentially and in a blink of the planet's eye, foul the air, the water and the Earth at critical rates. Unnecessary suffering . . . that caused by ignorance, greed, fear, hatred, etc. . . . has never affected so many. Nuclear threat looms, while poverty and hopelessness affect billions. Global warming and climate change caused by our obsession with the convenience that fossil fuels bring, appears to be irreversible. We must either transform or perish. It seems also that those elements that represent the existing paradigm . . . disconnected from the whole - fear-based, control-oriented, and patriarchal . . . are crystallizing and holding on all the more tightly to their way - seemingly heading us all into the inherently necessary stage of chaos mentioned by Dr. Ilya Prigogine, before hopefully transforming into a new level of consciousness and way of being.

From Stanislav Grof's *Psychology of the Future*:

'In the last analysis, the current global crisis is basically a psychospiritual crisis; it reflects the level of consciousness evolution of the human species. It is, therefore, hard to imagine that it could be resolved without a radical inner transformation of humanity on a large scale and its rise to a higher level of emotional maturity and spiritual awareness.

The task of imbuing humanity with an entirely different set of values and goals might appear too unrealistic and utopian to offer any real hope. Considering the paramount role of violence and greed in human history, the possibility of transforming modern humanity into a species of individuals capable of peaceful coexistence with their fellow men and women regardless of race, color, and religious or political conviction, let alone with other species, certainly does not seem very plausible. We are facing the necessity to instill humanity with profound ethical values, sensitivity to the needs of others, acceptance of voluntary simplicity, and a sharp awareness of ecological imperatives. At first glance, such a task appears too fantastic even for a science-fiction movie.'

Sheldrake suggests that:

'A shift in paradigm involves both a new way of solving problems (because there is a new way of thinking about the problems involved), and also the building up of a new social consensus among practitioners. A view of paradigms as morphic fields helps us to understand why they are so strongly conservative in nature, for once the paradigms are established, there is a large social group contributing to the consensual reality of the paradigm. A very powerful morphic resonance is evolved by this way of doing things; and that is why paradigm changes tend to be rather rare, and why they meet with strong resistance.'

Meanwhile, underneath the reality that the world is experiencing that seems so utterly daunting and impenetrable, work the mysterious agents for change. At critical points within any process, before the jump to the next octave is made, the build-up of energy will reach a *'critical mass'* . . . after which rapid transformation will occur. The term 'critical mass' was originally applied to the critical amount of fissionable material necessary to trigger and sustain a nuclear chain reaction, but in the last seventy or so years, critical mass has come to be commonly understood as that moment in any process where sufficient energy or input triggers a sudden transformation.

Another way of seeing this 'jump' or shift in evolutionary transformation can be seen in the example given in Young's *Reflexive Universe* regarding chlorophyll. At this point in the evolutionary process, purpose has descended from the freedom of the initial photon of light in a process most accurately called involution, becomes more *in*-volved in matter and more constrained by physical laws, until it reaches the nadir of decent and lack of freedom in the molecular world. At this point matter is still inert, fixed . . . inorganic – but very organized. Then . . . miraculously . . . the evolutionary process reaches out into the future and invents the polymers . . . chains of molecules that can self-replicate and store energy, ultimately leading to DNA – they are the initial signs of negative entropy and life. Chlorophyll enters the scene and everything changes. In Arthur Young's words:

> 'We saw that for life to get started it was necessary to have a variation of the environment that put a premium on control of energy. At the most primitive level, this life form was the bacterium or single-cell plant, which had its chlorophyll to gather the energy of sunlight and *store* it as starch or other carbohydrates . . . . there's a kind of magic that occurs here [the nadir of purpose's decent into matter, what Young calls the *turn*] and the molecule begins to reverse entropy, that is to say it begins to *collect energy* instead of just wasting it, and when it can do that it can begin to become a plant. You see, the chlorophyll, which is essentially a molecule within the plant, collects the energy from the sunlight and instead of just wasting it away it stores it . . . and that storage is what the plant does in growing, it stores order or energy and that's the first step out of this prison of determinism.' [233]

How does the chemical configuration preceding chlorophyll 'know' to alter itself just so, permitting evolution to proceed? *Unconsciously,* initiating purpose 'knows' to drive the process forward by somehow pulling a seed from the future that permits evolution to continue. Today's science explains all this through chance and necessity . . . purpose and consciousness have no part. Young calls our present stage of evolution 'Modern Man' . . . and in the Dominion Kingdom, Modern Man is at the nadir, or the molecular/organizing/constrictive sub-stage in the process. At this stage of evolution, the process becomes *self-determining* – *we* have to *consciously* pull this seed from our future and use it in some way to make the turn upwards, reversing the entropy that is now counter-productive. It cannot simply *happen* at this stage. Our future and the future of our environment depend upon it. Young says that at our present level of group consciousness we are looking for a seed of *'divine light.'*

What could possibly be a more perfect *'seed'* of divine light than the Stillpoint geometry impossibly and intentionally embedded in the heavens, waiting for its discovery?

If this has, in fact, been a journey of awakening, then there will be a point, or points, in the process where critical mass will be reached and transformation to a new level, a new octave, a new way of being, will occur. It is believed by some that the return to alignment with the center of the galaxy, so dramatically pointed to by the ancient Mayan calendar - by the, I believe, *symbolic* date of December 21st, 2012, the Winter Solstice that year. This date represents the entire transit of the center of the galaxy by the Winter Solstice alignment with the Sun, beginning in 1987 and ending in 2018, and heralds the transformation into a new age, a new kind of thinking and being and understanding – a new level of consciousness (*see Winter Solstice, 2012:, page 251*). That this *return* has everything to do with directing attention to the Stillpoint/black hole at the center of our galaxy – ultimately to the Stillpoint in the midst of everything - makes it all that much more significant. The actual manifestation of this transformation will have been happening for many years and include many seemingly non-related actions on the part of many 'individual' consciousnesses . . . approaching that moment of critical mass when rapid transformation begins to take place. But the direction of this transformation that is clearly happening is in our hands. It is up to us.

It is important to note that reaching critical mass is not only quantitative in nature . . . the *quality* of the energy build up is also significant. Michael and Judie Bopp, while speaking about the process of transforming communities in their book *Recreating the World*, say: 'Many *small* groups of people organizationally unconnected, but synergistically united in their manifestation of a common pattern of beliefs, goals, actions and relationships, can have a profound effect on human systems such as communities or organizations.' An appropriate analogy for the small amount of qualitative energy it may take to cause a complete transformation to occur is a mechanism called the 'trim-tab,' an example often used by Buckminster Fuller. It takes a considerable amount of energy and time to turn a huge ocean liner. The rudder itself, gigantic in comparison to anyone's normal understanding of the word 'rudder,' is almost impossible to turn because of the vast amount of resistance caused by the water running past it. A trim-tab is a relatively small rudder at the rear of the main rudder. It is easy to turn and it *uses* the force of the water moving against it to turn the large rudder. So . . . this small amount of energy used in just the right place and at just the right time actually causes the huge vessel to turn.

Just as the analogy of dropping a sugar crystal at a time into a glass of water demonstrates . . . the water holding the millions of crystals in solution with no apparent change until the moment when the water can no longer hold the sugar in solution and just one more tiny crystal begins the inevitable process of rapid transformation - the water instantly no longer clear - it is possible that we are reaching this point of critical mass in terms of the transformation of consciousness in the world. While there are millions of apparently unrelated actions which seemingly mean nothing in relation to the whole, these mysterious agents for change go about their daily work. All of the injustice, the failures, the impotence of not seeming to make any real difference builds within the overriding intention to change . . . in preparation for that moment of critical mass when all of the efforts explode into a chain reaction of transformation, and the old paradigm is no more.

On the scale of the world and what happens in the world every day, a small building in the desert where four or five hundred people will come to be silent and to pray and meditate and expand their consciousnesses may not seem an important thing. But my assertion is that it *is* important. It has everything to do with the purity, the intensity and the quality of the intention 'at just the right place and at just the right time.' The possibility clearly exists that this small but precise and qualitative effort to open a doorway to the bodhisattvic, infinitely *evolved* consciousness that has embedded the key to this doorway – the geometry of our origins and of our goal - in the dimensions of the Sun and Earth and Moon, will help to tip the scales, doing its part to usher in the transformation about which so many have prayed for and worked towards for so long.

⊕ ALIVE UNIVERSE … 'TECHNOLOGY':

All around the world there now appears to be similar thinking regarding the interpenetrating nature of consciousness and matter coming from quantum physicists, neuroscientists, parapsychologists, historians, biologists, ecologists and mystics. The idea that the Universe is alive, an idea that used to be assumed, is once again gaining a momentum within our own scientifically indoctrinated consciousnesses . . . a precursor of the paradigm shift that may be imminent as humankind uncovers conclusive evidence that the Universe functions not as some great machine, but as a Great Thought - unifying matter, energy, and consciousness.

Sheldrake:

'For an American Indian looking at the sky, he's not looking at just a material collection of bodies moving in accordance with inanimate laws. The sky is a living being, the abode of the Spirit. The earth is a living mother; it's not just a collection of rocks with physical forces at work in them. The divine presence in the sky, in the earth, and in our experience of nature all around us. In the entire process of cosmic evolution you see a Spiritual process as well as a material process. You can't separate the two. They go together. And human evolution involves both a Spiritual consciousness and a cultural evolution, and the evolution of science and technology. You can't separate these things. Traditional cultures believe that rituals are a way of connecting through time, of collapsing time. You could say that the patterns of rituals embody morphic fields - I would call them morphic fields, but you could also call them archetypes, because archetypes is another word, I think, for the same idea.'

More mainstream scientists are also finding it impossible to avoid the morphic/archetypal influence of consciousness within the Universe. The following is from complexity theorist James Gardner's *Biocosm: the New Scientific Theory of Evolution: Intelligent Life is the Architect of the Universe*:

'The universe we inhabit is literally in the process of coming to life. Under this theory [Selfish Biocosm], the emergence of life and intelligence are not meaningless accidents in a hostile, largely lifeless cosmos but at the very heart of the vast machinery of creation, cosmological evolution, and cosmic replication. Under the theory . . . the oddly life-friendly suite of laws and physical constants that prevail in our particular universe serve a function precisely analogous to that of DNA in earthly creatures: they furnish a recipe for the ontogenetic development of the mature organism and a blueprint that provides the plan for construction of offspring. Finally, the hypothesis implies that the phenomenon that we Earthbound humans perceive as a process of biological evolution is more properly viewed as a subroutine in a vast ontogenetic process by means of which the biocosm grows to maturity and prepares itself for replication. A corollary is that the capacity for biological evolution and emergence is essentially front-loaded into the suite of physical laws and constants that govern the evolution of the cosmos over time in precisely the same manner that the recipe for constructing a mature biological organism is front-loaded into that organism's DNA. Put differently, the Selfish Biocosm hypothesis adds to the list of ontological possibilities traditionally contemplated by cosmologists - that the universe may be a great equation, a great computation, or a great accident – the following possibility: *that the cosmos may be quintessentially a great unfolding life.*'

Similarly, the mathematician Louis Crane has stated:

'In the first place, the origin and evolution of life can no longer be viewed as a mere accident. *Rather it is deliberately coded into the fine tuning of the physical laws.* Since the development of life and of the universe are joined into a unified evolutionary process, they can be viewed from the point of view of purpose, just as it makes sense to speak of the purpose of an organ of a developing animal, even though the development of the animal is entirely within the scope of physical law. Secondly, intelligence and its ongoing success are no longer a small and unimportant accident in an enormous universe. Rather they are the precondition for the existence and reproduction of the universe. The world around us was created by something like us, and is structured, as if deliberately, to produce us and nurture us. We have a large purpose which goes beyond ourselves of sustaining and recreating the universe.'

If what so many mystics and teachers have been saying in their own way throughout history is true – that consciousness itself is weaving this web of transformation - then it is evolving consciousness itself providing, at just this critical time, technology that may give itself a booster shot in its aim of awakening. It is only logical that an evolutionary process conscious of itself could also direct itself – just as a child who needs direction grows into an adult who doesn't. As Teilhard de Chardin was quoted earlier: 'Not only do we read in our slightest acts the secrets of [evolution's] proceedings; but for an elementary part we hold it in our hands, responsible for its past to its future.' The kind of 'technology' that the Mesa Temple embodies represents this 'elementary part' of self-direction.

Given that the word 'technology' is often associated these days with exactly that which is *disconnected* from the Earth, and aligned with the fear and separation *opposing* the growth of consciousness (fossil fuel burning engines and factories spewing carbon dioxide into the air, creating a chain reaction of global warming that cannot be stopped, huge machines that eat up the rain forests, weapons of mass destruction, etc.), it would be useful to be clear about how this word is used here.

The word 'technology,' as referred to in this discussion, means 'the system by which a society provides its members with those things needed or desired.' [234]

What is most needed now is nothing less than a paradigm shift in the way we *are*. As mentioned earlier, there is a remarkable group of people that I have come across one by one, who have each said, in their own way, that given the direction the world is headed, the only thing that can shift or alter that direction is a dimensional shift in *who we are* . . . a global transformation of consciousness. Hannah Arandt, Stanislav Grof, Albert Einstein, M.K. Gandhi, Nikola Tesla, Ervin Laszlo, Krishamurti, and the Dalai Lama are included in this group. Each is unquestionably respected by his peers and has earned the right to a uniquely expanded and respected word view. I don't think any of them would have said what they did if they did not mean it deeply . . . and given the task, it is a unanimously grim message. But any half-awake person can now see this for themselves . . . our collective expression has become insane - depraved even.

Methods of learning how to feed ourselves correctly and adequately, how to educate ourselves towards self-empowerment, and how to live together in healthy community . . . as well as methods of meditation towards higher self-awareness, or a holy mountain or places that inspire us to deeper understanding, are examples of this kind of 'technology' – as is a structure consciously designed to align consciousness with the stillness outside of space and time.

To alter who we now are – literally to change human nature by effecting a global transformation of consciousness - has never happened in our collective history . . . but it is required now if we are to continue our evolution. While the awareness of this, expressed by so many eminent men and women, is certainly a grim one, the 'impossible' reality of the Stillpoint information offers the only potential I know of for genuine hope.

Teilhard, firmly convinced of the existence of the evolution of consciousness and that its existence is fundamental to and implies Man's place in the cosmos, called this evolution:

> '. . . an absolute direction of growth, to which both our duty and our happiness demand that we should conform. It is his [Man's] function to complete cosmic evolution . . . Man is not the center of the Universe as once we thought in our simplicity, but something much more wonderful - the arrow pointing the way to the final unification of the world. This is nothing else than the fundamental vision and I shall leave it at that.'

⊕ Expanded Mind, Compassion, Change:

Essentially, the temple is a means of helping the individual to access deeper and more expanded levels of consciousness . . . and to connect these 'individual' minds with a larger, more expanded field of compassionate awareness.   When a critical mass of global consciousness comes to *know* that 'we' are One Being . . . in the Stillpoint . . . transcending the *seeming* separateness of 'other,' a compassion for 'other' beings rises naturally – a new morphic field of elevated consciousness. This compassion will realize itself in the world through the individual efforts of many.

Stanislav Grof states in *When the Impossible Happens*:

'I now believe that the universe was created and is permeated by cosmic consciousness and superior creative intelligence (*anima mundi*) on all its levels and in all its dimensions . . . One of the most striking consequences of various forms of transpersonal experiences is spontaneous emergence and development of deep humanitarian and ecological concerns and the need to get involved in service for some common purpose. This is based on an almost cellular awareness that the boundaries in the universe are arbitrary and that each of us is identical with the entire web of existence. *It becomes clear that we cannot do anything to nature without simultaneously doing it to ourselves.* Individuals who have undergone this transformation develop a deep sense of being planetary citizens rather than citizens of a particular country or members of a particular racial, social, ideological, political, or religious group. It is obvious that a *transformation* of this kind would increase our chance for survival if it could occur on a *sufficiently large scale*. We seem to be involved in a dramatic race for time that has no precedent in the entire history of humanity. What is at stake is nothing less than the future of life on this planet. *If we continue old strategies*, which in their consequences are extremely self-destructive, it is unlikely that humankind will survive. However, *if a sufficient number of people undergo the process of deep inner transformation* outlined above, it would enhance our chances to meet the formidable challenges we are facing and use the enormous creative potential inherent in our species to create a better future.'

Sheldrake alludes to the ideas expressed above when he speaks about the extended mind:

'Our minds are extended in both space and time with other people's minds, and with the group mind or cultural mind by way of their connection to the collective unconscious. Insofar as we tune into archetypal fields or patterns which other people have had, which other social groups have had, and which our own social group has had in the past, our minds are much broader than the 'things' inside our brains. They extend out into the past and into social groupings to which we are linked, either by ancestry or by cultural transmissions. Thus, our minds are extended in time, and I believe they are also extended in space.

This whole topic of the extended mind becomes particularly important at the present time when there is a tremendous interest in the idea of connecting up large numbers of human minds. All of these convocations are based on the idea that the mind is extended, that it can 'link up' with other minds, and that *simultaneity* is particularly important in creating a kind of group mind phenomenon.

It may be that we are connected with everybody we think about and all the places we are attached to through our extended minds. Our minds, in fact, may be vast, far-reaching spatially extended networks of connection in space and time - networks of immense scope in which the brains inside our heads are but a portion.

The concept of the extended mind as a morphic field, though a new paradigm which is not yet fully formulated, enables us to glimpse bits of a new worldview. And it is a very, very different worldview from the one we currently hold, in which our minds are conceived to be entirely private affairs inside the privacy of the nervous tissue within our brains.'

As mentioned earlier, the geometry and forms of the temple (intentionally aligned with nature and her rhythms) are used to entrain the mind to the stillness and the Unity of all things, and to accelerate and amplify the compassion that naturally arises from this alignment out into the extended mind of humanity . . . and beyond. This is one example of *resonating* with what is true.

Further, if a portal could be created to this evolved higher consciousness, the communication/connection, would be through a profound resonance . . . *resonating* with deep, awakened truth. '*If a sufficient number of people undergo the process of deep inner transformation . . . it would enhance our chances to meet the formidable challenges we are facing and use the enormous creative potential inherent in our species to create a better future.*'[235]

⊕ Frequency / Vibration / Light / Sound / Resonance:

Since *resonance* is the principal mechanism upon which the temple experience is based – specifically resonance with the temple's pure archetypal geometry as well as with the higher consciousness that has now been discussed so often – this would be a good time to become familiar with the scientific understanding of vibration, frequency and resonance. That is, how does it work?

Physics has shown that mass has frequency - an application of Einstein's $E=mc^2$. Everything is in motion and vibrates and has a particular frequency: atoms, molecules, planets, rocks, people, trees, hippopotami . . . everything.

Sound and light are on the same continuum of frequencies, with sound being much lower and light much higher. In fact, it is possible to use the Law of Octaves[236] (again, as above, so below) and whole number intervals to divide down the frequency of various colors 40 times, and enter the range of audible frequencies. Thus musical notes have colors associated with them, and some people can 'see' sounds as colors for this reason. This is because the frequency of the color is 'heard' at many octave levels lower in the range of sound - and the brain knows how to make the correlation.

Just as the quality of light within the temple is critical, so it is with the quality of sound.

When the eyes are closed, the quality of audible sound becomes an important medium of transmission. The audible sound in the temple will be clear - not echoing or loud or muted - but clear. Since the circular shape of the inner temple will tend to focus the sound towards the center, creating a condition where sound waves conflict with one another and a condition that is *not* desirable, the sections of the wooden walls between the columns will be slightly concave towards the center, or scalloped - spreading the sound while keeping it clear.

But it is the quality of the *absence* of sound that is perhaps most critical. The experience of deep, deep silence, as well as a sense of protection, will be communicated/created by surfacing the entire spiraling, helical roof – from its interface with the desert floor to its interface with the rock outcropping of the granite pyramid roof – with a minimum of 12" of planted earth - a 'green' or 'living' roof: The deep silence that exists on the original high desert site of the Mesa Temple will be further deepened within the earth-covered, kiva/womb-like inner temple - a haven within the Earth, a *hill in the desert*.

This deep silence is the pure sound of the Stillpoint, the ringing of the pure, silent bell.

And there is another, subtler and equally important quality of vibration/frequency that is at work within the temple. Through the experiments of Dr. Hans Jenny (*see Proportion, Vibration & Balance:, page 189*), those frequencies in and around that of audible sound are shown to be directly related to geometry. When an object vibrates, it tends to vibrate at a specific frequency, its 'natural' frequency. All objects have a natural frequency, or group of frequencies at which they like to vibrate, and if that frequency is within range of human hearing, and loud enough, we will be able to hear the sound vibrations being produced . . . but *any* object which vibrates will create a sound wave, although we may or may not be able to *hear* it. That is, this process works not only with audible sound, but on deeper and subtler levels – in this case, with the pure geometry of the temple. Frequency, vibration, proportion, geometry . . . the music of the Stillpoint.

The intention behind all of the temple's geometry is its capability to effect, or *entrain*, the consciousness of those experiencing it to a higher vibration or frequency – and then transmitting it. *Entrainment* is the influence, interaction, and interplay between two or more vibrating bodies in close proximity to each other. Balance can be created through the principle of entrainment . . . or *resonance*. The intention here is to merge with with this now-proven higher consciousness which is communicating to us in such profound fashion.

As the Stillpoint regenerates itself into the sphere of the Vector Equilibrium, and the Vector Equilibrium into the Platonic building blocks of the tetrahedron, cube, octahedron, icosahedron and dodecahedron, and so on back and forth eternally, movement and momentum are created. From this movement comes vibration and frequency . . . and all manner of created things. The temple has a vibration, a frequency, which is generated by the conscious *geometry* of the temple – all tuned to cosmic zero, the Stillpoint.

Further, if there are two vibrating objects with the same natural frequency or corresponding harmonic, they will both have a *forced vibration* affect on each other, which leads to the phenomenon called entrainment mentioned above. This process, given time, normally leads to a condition where both objects automatically synchronize. In technical terms, entrainment is the mutual phase locking of two oscillators . . . or two bodies dancing together.

A simple example of this phenomenon is demonstrated with tuning forks. If two tuning forks are held side by side and one of them is struck, the other will begin to vibrate . . . the 'sound' energy partially transferred from the first tuning fork to the second. This is called *resonance*, and is one of the profound relationships between separate bodies, heavenly or human or, as with the temple, structural.

The word resonance comes from the Latin verb *resonare*, meaning to 'return to sound' - to sound and resound, like an echo. Usually we think of resonance in terms of objects like bells which when struck continue to ring or resonate the original sound. The example of the tuning forks is another type of resonance called *sympathetic* resonance - a merging created when energy moves back and forth between two or more bodies.

Throughout time people have used sound to heal, and all the current sound therapies are based upon this principle. The old Greek concept of *sympatheia* asserts that all forms and processes in the Universe are in sympathetic or *resonate* relationships. Sound therapy is a discipline that helps the body to both remember and reinstate its natural harmonics and vibrational patterns. The body, hearing the call of its natural harmonics, resonates with the frequency it is being given. And since the body will always return to its optimal functioning when given appropriate input, it moves towards balance.

During the process of resonance, energy transfers from one object to another. The second object then responds to this additional frequency energy by vibrating. 'Movement' is caused in the other object without touching it. So . . . one can *energize* 'other' through structure, thought, sound and light – or vibration. *The pure vibration of the geometry of the temple expressing itself through the physical medium of pure stillness/silence aligns and balances those experiencing it, pulling them into resonance with awakened consciousness.*

In quantum physics, matter is perceived as *resonant* particles, and currently understood simultaneously as a particle *and* a wave. Because it is a wave, each particle has also a specific frequency or vibration. Therefore, in the subatomic world, the *way* something vibrates determines what it is. Particles are created when electrons and protons are accelerated to near the speed of light. Collisions create a burst of energy in which a particle emerges. The energy of the collision keeps resonating or feeding back into itself, like a bell that is struck and continues to ring. Only the collisions that *resonate* create particles. (Physicists who explore the sub-atomic world for particles call what they are doing 'resonance hunting').

In more human terms, the state of merging with another is called *empathy*, meaning 'into feeling.' We allow ourselves to resonate with the vibratory field of the other until we *are* the other – or *empathetic* resonance. Resonant communication, or empathy, on a spiritual level is called *communion*, and it is this communion that the geometry of the temple seeks . . . and it is critical at this stage in our development that we learn and experience compassion for 'other.'

It is necessary here to present a caveat to all that has been said about the

228

importance of resonance and the temple's intention to create entrainment with all that is in harmony with Creation's purest and deepest tones. This, by itself, is misleading . . . and only a part of the whole, as the *compliment* to resonance is dissonance. Dissonance happens when energy moves back and forth between two or more bodies *without* merging into unified pulsation. The pulses beat *against* one another, as it would in the temple if the circular wall sections were not scalloped inwards. It is an aspect of 'that which is naturally so' . . . and it is important to acknowledge this.

As already mentioned, the Nobel Prize winning physicist, Dr. Ilya Prigogine, discovered the importance of dissonance while investigating chemical systems. He termed his discoveries 'order from chaos.' Prigogine proved that for a system to change and go into a different state of functioning it must first pass through a state of disruption or chaos. The sonic term for chaos is dissonance.

Prigogine points out the crucial role dissonance plays in living systems evolving into higher levels or devolving into lower levels of order, or resonance. He discovered that all living systems dissipate more and more energy over time caused by fluctuations or dissonances inherent within the system, before evolving to a higher or lower state. As time passes, these dissonances increase in intensity, causing the system to move further from equilibrium. At some point everything begins to wobble. The wobbling increases until all preexisting order within the system shatters, causing the system to leap into chaos. 'Chaos' is really a *stage* in the expression of evolving order.

This seems to describe the process of entropy, so inherent to *inert* or *inorganic* forms that science has mistakenly projected onto the entire dynamic of the Universe – but the entropic process is reversed by levels of consciousness inherent to advanced/complex living systems, or negative entropy. Many scientists (Arthur Young, Lee Smolin, Freeman Dyson, John Wheeler, Julian Huxley, Gregory Stock, Christian de Duve, Albert Einstein, Heinz Pagels, and Paul Davies, etc.) have suggested and are discovering that consciousness itself, reversing the law of entropy, is inseparable from and interwoven into the 'manifested' Universe . . . a Universe that is alive and expanding in another dimension – consciousness. Or, more appropriately, a Universe that *Is*.

As mentioned, Prigogine terms the precise moment a system goes from order to chaos a bifurcation point. As a system approaches bifurcation it only takes a very small and seemingly inconsequential event to create chaos. From chaos the system reorganizes itself into a new system functioning at another level of resonance and finds a new equilibrium (*see Critical Mass:, page 215*).

Further, the resonance of the new system, whether a higher or lower level, is determined by what mathematicians call 'strange attractors,' and these strange attractors can be visualized as seeds of the new order sown during the old order. (Similar to the Tibetan idea having to do with *imprints*, thought-forms in a person's subconscious that actually 'imprint' or *determine* one's life experiences, which itself is similar to Rupert Sheldrake's theory of morphic resonance). When the old order disintegrates, the new order re-forms around the vibration of these attractors, imprints, or *seeds*.

Resolution happens when dissonance becomes resonance – the most significant way that the temple seeks to transform the chaotic nature of our contemporary world into an alignment with 'that which is naturally so.' The following describes just how this is to be accomplished.

As already mentioned, there is a cavity between the surface of the Earth and the outer ionosphere. In 1954, a German physicist named Winfried Otto Schumann discovered a *standing wave*[237] resonance at a main frequency of 7.83 Hz. This phenomenon occurs because of electromagnetic waves interfering with each other – one generated from the conductive surface of the Earth and the other from the similarly conductive outer surface of the ionosphere - through the stationary, non-conductive medium of the air, creating a standing wave that remains fixed in a stationary or standing position. The Earth and the ionosphere are in *resonance* – and this resonance is known as the *Schumann Resonance*. It has been called the heartbeat of Earth.

A colleague of Dr. Schumann's, Herbert König, was able to demonstrate a very close correlation between the Schumann Resonance and the frequency of the mammalian brain, which includes humans, in the alpha/theta state – the frequency of consciousness most often associated with mystical states, meditation, the receipt of extrasensory information. There is also a frequency assigned to the Earth's magnetic field of around 2kHz, generating very weak signals. In Chinese teachings, the balance of these two frequencies – the stronger *Yang* frequency of the Schumann Resonance with the *Yin* frequency of the Earth – is ideal for mental, emotional and physical health.

Various scientific studies - in particular one done by Professor R. Wever from the Max Planck Institute for Behavioral Physiology, which completely screened out electromagnetic fields, including the Schumann Resonance – demonstrate the essential importance of the 7.83 Hz frequency for psychological and physical health. When the human brain wave resonates with this frequency, the result is *health*, including the increase of immune protection. When the frequency is blocked, the opposite happens. I recently saw a film entitled *The Grounded 2*,[238] about how walking without shoes, in direct contact with the Earth, is healing. At one point in the film, the late Edgar Mitchell, the astronaut who had a transcendent experience while in space and the 6th man to walk on the Moon, is being shown two sunflowers – one of them, the healthy one, is surrounded by wire mesh that is connected to a grounding rod to Earth - the primary idea behind the creation of the electromagnetic field surrounding the inner temple - connecting Earth to Sky, Sky to Earth.

Intimately connected to this phenomena, especially for today's world, is the fact that more and more of the surface of the planet is flooded with harmful, artificial electromagnetic frequencies – WiFi everywhere (6 *billion* mobile phone subscriptions worldwide),[239] mobile phones/ towers/transmitters, electrical wiring and appliances. These harmful frequencies tend to block the healthy, naturally occurring frequency of the Earth and the Schumann Resonance, causing stress, fatigue and illness. In June of 2016, Tom Wheeler, head if the Federal Communications Commission, gave a speech on the roll-out of 5G – the coming nationwide installation of the wireless network on steroids – completely untested.[240] It is only getting worse.

By enhancing this naturally occurring Earth/Sky resonance through the creation of a subtle, natural, protective electromagnetic field surrounding and interpenetrating the temple ... the marriage of Heaven and Earth in perfect balance . . . it is hoped that the ideal environment will be created for the experience, acceleration, transmission and reception of higher thought-forms and energy – and a realignment towards harmonic resonance with the frequency of Earth.

As mentioned earlier, the mechanism of this reception/transmission of these enhanced thought-forms will be the 64.8' sphere of a Vector Equilibrium crystal spherical matrix, above and below ground level, perfectly proportioned to the Stillpoint geometry of the temple. This is the third layer sphere of the Stillpoint/Flower of Life geometry upon which the temple is based, of which the Moon is the first layer, and is the layer that connects the Moon to the Earth within this geometry – the Sun being the 400th layer (*see Appendix A and www.stillpointdesign.org for a visual animation of this reality and the section below entitled The Quartz Crystal*).

The temple itself is nothing more than a technology for guiding consciousness towards a higher and higher frequency of awareness through an entrainment or harmony with 'that which is naturally so' . . . at the deepest levels. Certainly, resolution through entrainment and resonance is one of the temple's primary intentions.

Another aspect of its purpose is to assist one through the *intervals,* or dissonances mentioned, that naturally occur, because . . . as Gurdjieff observed in his Law of Octaves . . . it is in these intervals that one loses one's 'aim,' or goal. At the level of civilization, we are currently experiencing the most critical interval of our collective history. The danger lies in losing the larger aim of initiating purpose – *to awaken through the eons long process called the evolution of consciousness.* The threat is that this purpose will be seduced and co-opted – as it certainly now seems most likely – by the survival/fear based darkness that envelopes our world.

Or will we somehow find our way through this archetypal interval towards our healthy future? While these 'dissonances' are what we are normally running from, the awakened awareness that is being sought *includes* this aspect of creation as impartially as it includes the ladder of harmony used to reach it. We must keep our aim through this time of collective darkness. We must meet ourselves in the light of truth, acknowledge what we see of our darkness and find the courage and strength to do whatever it takes to connect to our higher selves. With the temple, we may have a way to keep our aim collectively. At times like this, it often takes a willingness to attempt what has never been done before.

Again . . . the temple's success is based entirely upon raising the frequency of human consciousness through resonance - through the sacred geometry of the temple - and most importantly through resonance with the higher consciousness responsible for creating this solar system . . . the bodhisattvic higher consciousness that has left us a profound message having only to do with fully enlightened awareness.

⊕ HARMONY:

The following is a slightly more detailed description of the principles of frequency and proportion addressed above as they relate to music . . . and ultimately, the temple.

Some objects, like the strings of a guitar, have a number of modes of vibrations that are all, as Pythagoras discovered, mathematically related. The frequencies of the vibrational modes in a string create standing wave patterns called harmonics: frequencies that are in some integral multiple (whole number relationship) of the fundamental frequency.

These vibrational modes are the basis of Western music and musical scale. Specifically, each vibrational mode of the string added a node, or *still, point* in the string that would divide the original string into one more 'pieces,' which would resonate at its fundamental frequency. In each case, the string would break into one or more shorter pieces, which vibrate at integer multiples (whole number proportions) of the fundamental frequency. Harmony happens this way. The temple's geometry seeks this harmony.

'One of the intentions of the historic temple was 'to surround people with harmony, and in doing so, they in turn would recreate harmony in their world. If geometry shapes the universe than surely, the ancients posited, the living geometry of a temple can help reshape people . . . . Geometry and strategic angle were used in temples – inside, outside, overtly or invisibly – to initiate a process of sensory manipulation that begins to open the body's electrical circuits, making it more receptive to finer, more penetrating frequencies and vibration.'[24]

The relationship between geometry and musical harmony is intrinsic: if the fundamental frequency for a string is 100 Hz, the first harmonic, the next vibrational mode would be at 200 Hz. This makes sense as the string is exactly 1/2 the length, and therefore would vibrate twice as rapidly. This second vibrational mode of the string is called the second harmonic, and is exactly twice the frequency of the fundamental. The Third Harmonic occurs when we break the string into three seemingly independent vibrational pieces, which happen to vibrate at three times the rate of the fundamental. The fourth harmonic, and those beyond, vibrate at an integer multiple of the fundamental frequency. The 2nd, 4th, 8th, 16th, 32nd, 64th, etc., are all perfect octaves of the fundamental.

Worthwhile to mention here is that octaves do not only consist of these whole notes, but also include half notes, or *intervals*. Similar to the discussion above regarding dissonance, one of the most basic principles within Gurdjieff's 'Law of Octaves' is the existence and necessity of these intervals. In each Western scale (based upon Pythagorean whole note proportions) there are two intervals . . . one near the beginning of an octave, and one at the end, just before the octave transforms into a higher level or frequency.

According to the Law of Octaves, as Gurdjieff defined it, everything that happens is subject to the rhythm of the interplay between the whole and half notes and can be seen as a series of octaves. The Law of Octaves states that it is within these intervals, or half-notes . . . especially the last one in the octave . . . that the *intention* of the octave is challenged. It is here that Prigogine's 'wobbling' occurs. It is here that harmony naturally breaks into chaos - and is guided into the next order, or higher octave, by one's *aim* and by *seeds of the new order sown during the old order*.

The harmony of the whole notes is balanced by the chaos of the intervals. They are necessary.

If one can see the entire span of the last 25,920 years - the Great Year, the Winter Solstice once again aligning with the center of the galaxy - as an octave, it is very easy to see that we, as civilization, are experiencing the last interval in this octave. It is now, more then ever before, that we need to 'keep our aim' – that is, we must do whatever we can to not succumb to the temptations inherent in the fear-based darkness that has been so seductive for so long. We must fill this interval with actions of pure intention, connecting us to our Source and our higher Self. *This* is what is being proposed. *This* is what this *action* is entirely about.

I would like to remind the reader of the images created by the Harmonograph (*see page 190*). These images, just as the whole number harmonious frequencies mentioned above, make clear the importance of whole number proportion – as *fundamental* as action itself. From Arthur Young, in *The Reflexive Universe*:

> 'It was Planck's epoch-making discovery that action *comes in wholes*, a discovery which in retrospect we can see to be true of human actions. We cannot have 1 ½ or 1.42 actions. We cannot decide to get up, vote, jump out the window, call a friend, speak, or *do* anything one-and-a-half times. Wholeness is inherent in the nature of action, of decision, of purposive activity. Planck's discovery about light touches home: it is true of our own actions.'

If one can imagine extending the principle of harmonics to infinity, the following quote from John Beaulieu, N.D., Ph.d., may be more clearly understood:

> 'Imagine that the whole Universe; everything we know, including cars, computers, airplanes, houses, buildings, lakes, oceans, continents, our bones, flesh, and nerves is a fountain of dream images generated and sustained by a submerged sound. Further imagine that everything we do and think, whether good or bad, moral or immoral, is an attempt to seek out and merge with that sound. Our goal is to return to the source of the fountain. Although we may identify with the object of value - i.e., a man or woman, a car, etc., the real attraction is the *resonance* we experience when in the presence of that person or thing. The experience vibrates us like a tuning fork and becomes a sonic homing buoy confirming our inner journey.'

Once again, this is reminiscent of Buckminster Fuller's observation:

> '. . . the physical is always the imperfect experience, but tantalizingly always ratio-equated with the innate eternal sense of perfection – thus the mind induces human consciousness of evolutionary participation to seek cosmic zero [the ultimate embedded sound].'

As the pure geometry of the temple is being used by human consciousness to recreate things as they are 'naturally so,' a conscious, profoundly silent 'sound' will be generated . . . mirroring the Original Sound. Resonance with this original 'sound' and the resulting harmony is one of the primary intentions of the temple.

⊕ THOUGHT / PRAYER / MEDITATION :

From Frederick Lehrman's book *The Sacred Landscape*:

> 'Consciousness, which is formless, gives birth to the form we call reality.   The Universe is coherent, alive and growing. Consciousness forms and inhabits body and cosmos. The landscape that we see around us is something which we both generate and contain. 'Reality' mirrors the mind. Like Alice, we might wish to pass through the Looking Glass to find the magical landscape beyond. But in truth, if we could only remember it, we are already standing in the world which the mirror displays. This *is* the magical landscape.'

Just as color and sound . . . as well as everything else . . . is vibration, so too is thought. Thought is subtle vibration expanding out from the center:

> 'Rigorous experiments using directed human thought have shown how our electromagnetic impulses are capable of altering the random movement of machines, even alter the beat of a computerized drumbeat, proving something revolutionary about human consciousness that was once believed limited to the field of mediums and magic.'[242]

Just as one vibration can entrain another (tuning forks), so too can one thought affect the vibration of the 'physical' world.   Much can be said about this, but nowhere has this been demonstrated so convincingly and beautifully as in the stunning photographic work of Japan's Masaru Emoto.

The first photo below is polluted water from the Fujiwara Dam, Japan, the second and forth are photos of water taken from springs known for their purity in Japan, the third of a water crystal from the fountain in Lourdes, France; the fifth of a water crystal exposed to the words, 'Thank you' in Japanese, and the sixth photo was taken of water crystals exposed to heavy metal music.   There is also a photo of a water crystal from water labeled with the following: 'You make me sick, I will kill you.' It made me feel unwell just looking at it and I chose not to include it here.

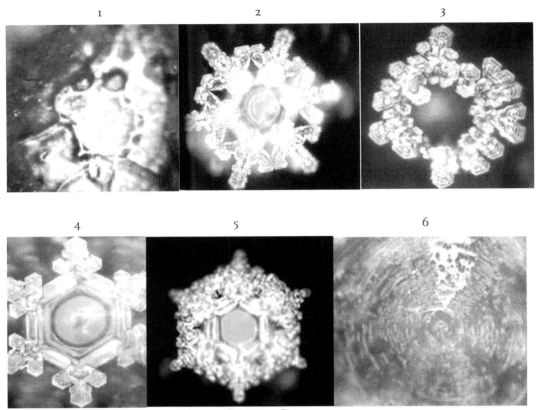

WATER CRYSTAL PHOTOGRAPHY
(Photographs Courtesy of Masaru Emoto)

There is also a beautiful photograph of a water crystal from a reservoir in Japan after a Buddhist prayer was offered. We see here, imprinted and held in the memory of water: The essential purity of untampered nature, the memory imprint of contamination, chaotic music, and kind and harmful thoughts.

Human beings, as well as the surface of the planet, are 70% water.

As the following will make clear (*see The Quartz Crystal & Water:, page 237*), the temple is intended to be an amplifier and transmitter of focused healing energy. Emoto's photographs demonstrate the principle behind this idea. One of the primary intentions of the temple is that the thoughts, meditations, and prayers of those entrained by the pure geometry of the temple, will be amplified and empowered to transform and heal on a scale not presently available.

For years, despite my passion for what architecture is capable of and my obsession with sacred geometry - in particular this structure - I could not convince myself that a building, nor any amount of 'loving-kindness' mediation, could effect the change, the paradigm shift so needed now. The work of Masuru Emoto and others demonstrates that the power of the mind - of consciously directed thought forms - *does* effect change in the environment.

It's also been scientifically verified that a sacred site by itself emits its own elevated frequency, similar to focused meditation – a critical aspect of what is being proposed. From *Common Wealth,* by F. Silva:

> 'Roger Nelson, a doctor of Psychology, used a REG (Random Event Generator) to measure electrical impulses relative to consciousness and place. After trips with meditation groups to Luxor, Karnak and several pyramids, the effects were six times that of ordinary . . . . the evidence infers that a 'consciousness field' exists and that intentions or emotional states which structure the field are conveyed as information that is absorbed into the distribution of output of the machine. In fact [the results] represented the largest effects ever seen. But what astonished Nelson was that the results from twenty seven sites were even higher whenever he walked around the sites in respectful silence, with a portable version of the machine sitting in his pocket. For him it proved that the *spirit of place* itself registered effects as high as the power emanating from a meditating group. While temples resonated a high degree of consciousness, the combination of focused group veneration and the temple seemed to create an expanded consciousness that had a marked affect on a machine.'[243]

Fundamental, evolving, consciousness is responsible for all that is. Consciousness precedes manifestation. Focused, aware, compassionate consciousness is capable of creating miracles. We are now at the stage of collective consciousness on Earth where a miracle is required if we are to continue.

⊕ THE QUARTZ CRYSTAL AND WATER:

As already mentioned, when purpose emerges from the Great Mystery, through the entry-point of the Stillpoint, it is the photon of light having no discernible mass, only a probability of location, and a huge amount of rotational energy and freedom. It 'descends' into matter and then, as the Universe cools, other particles form and organize themselves into atoms and finally into molecules - losing energy all the while, becoming more restricted and less free. The epitome of the end, or nadir, of this descent is the molecular world where everything is now organized in preparation for the 'turn' and the reversing of entropy and the awakening of life and consciousness (*see Arthur Young's explanation of the molecule chlorophyll, page 218*). At the heart of this most *organizational* stage of evolution is the quartz crystal, whose rigidly organized molecules have the ability to hold, store and focus energy (whether it be radio waves or thought-forms). From here the light begins to 'ascend' through plants, to animals, and to and through human beings and beyond. Seven steps – once again, the cosmic cycle so ingrained in our own daily lives through the cyclic presence of the Moon.

The quartz crystal is unique. From Jeremy Narby's *The Cosmic Serpent*:

> 'As early as 1923, Alexander Gurvich noticed that cells separated by a quartz screen mutually influenced each others' multiplication processes [and] deduced that cells emit electromagnetic waves with which they communicate. Quartz crystals have an extremely regular arrangement of atoms that vibrate at a very stable frequency. These characteristics make it an excellent receptor and emitter of electromagnetic waves, which is why quartz is abundantly used in radios, watches, and most electronic technologies. Quartz crystals are also used in shamanism around the world. Amazonian shamans, in particular, consider that spirits can materialize and become visible in quartz crystals. What if these spirits were none other than the biophotons emitted by all the cells of the world and were picked up, amplified, and transmitted by shamans' quartz crystals?'

Before his death some years ago, Marcel Vogel - an IBM scientist responsible for some of their breakthroughs - had set up a private laboratory in San Jose, California, and was working exclusively with quartz crystal research and assisted with the design of the meditation hall that served as a prototype for this temple . . . specifically with the design of an exact crystal layout.

Marcel was able to demonstrate in his work that the same principles of musical harmony, which have contributed to the forms of the temple itself, are also fundamental to crystal growth, and that plants respond to energy channeled through crystals - a subtle process, and one that applies to this work and to healing in general.

For the Mesa Temple, 12 crystals will be placed at the vertices of the Vector Equilibrium – 4 crystals just below the ceiling, 4 crystals at floor level, and 4 crystals below ground . . . 12 points around the Stillpoint at the center - defined by a 64'-8" diameter sphere. This is the 3rd layer sphere of the Flower of Life generated by the Moon's diameter – the first sphere - the sphere that connects the Earth, Moon and Sun to the Stillpoint geometry.

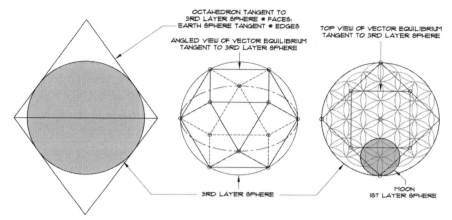

OCTAHEDRON TANGENT TO
3RD LAYER SPHERE ● FACES;
EARTH SPHERE TANGENT ● EDGES

TOP VIEW OF VECTOR EQUILIBRIUM
TANGENT TO 3RD LAYER SPHERE

ANGLED VIEW OF VECTOR EQUILIBRIUM
TANGENT TO 3RD LAYER SPHERE

3RD LAYER SPHERE

MOON
1ST LAYER SPHERE

Crystal Layout
(3rd Layer, 64.8' diameter sphere – 3 x 21.6,' the Moon sphere)

These crystals will surround the Stillpoint at the center of all the temple geometry. In plan view, or floor plan, this means the center of the circular wall of the inner temple at the entry floor level. *Zero . . . Stillpoint.* This crystal matrix is intended to be the 'brain' of the temple. It is the receiver, collector and transmitter of energy/thought/prayer from within the temple – those mediating, chanting, singing, toning, praying – as well as from the higher consciousness that placed this message in the heavens, and the Akashic Record.

This crystalline Stillpoint matrix will be grounded to Earth with a 20-ton monolithic black granite stone with a concavity in the center of its smooth, horizontal top surface that fixes the Stillpoint at the center of the crystal matrix . . . the symbolic/virtual/real center of everything. An anchor.

The Stillpoint itself does not 'exist' in our *four*-dimensional world and is implied. Originating in another 'dimension' it represents Spirit . . . and the *spiritual fire* of light. This 20-ton black granite monolith is the *Omphalos* stone. It is intended as an anchor for the Stillpoint and for the entire temple . . . grounding and controlling, or 'piercing,' the telluric/'dragon' energy moving through the structure.

Traditionally, this is called an *omphalos* – 'the navel, umbilicus, the central point.'[244] The omphalos designates the center, as well as the power center of the temple.

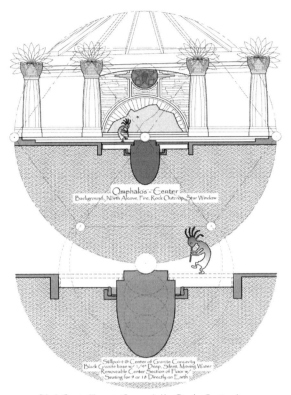

Black Granite Fountain Surrounded by Circular Seating Area

STILLPOINT / BLACK GRANITE FOUNTAIN / OMPHALOS (SECTION)

In many historical cases it rests atop a sacred mound . . . in this case the mesa rising out of the northern end of the deepest valley in North America:

'. . . the primeval mound, the central stone, and its encompassing temple then becomes a source of initiation and knowledge . . . the omphalos is the sacred pillar marking all at once the primeval mound, the umbilical cord with a creator god, and a place where the knowledge can be sourced . . . the place where the power of the land meets the descending power of a god serving as a focal point of energy.'[245]

There is another element to this particular Omphalos stone: It is also a fountain – and there are many things special about the element of water besides being one of the essential elements that permit life to exist. While 'water' in all of its forms is common in the Universe . . . *liquid* water is not. It is extremely rare.

Massimo Citro, in his book *The Basic Code of the Universe: The Science of the Invisible in Physics, Medicine, and Spirituality*, speaks about water's ability to hold *information*. Citro explains how water becomes medicine through *homeopathy*, not because of any chemistry involved, but because *information* is stored and transferred. Water is extremely *coherent*, or highly ordered. From *Basic Code*:

> 'If matter comes to interact with its own electromagnetic field, it can reach high levels of coherence . . . the system's state of minimum energy becomes a *coherent ground state* when molecules and fields oscillate in phase with each other. When matter and field oscillate in phase, they produce high levels of consistency and coherence' . . . and 'are called domains of coherence. These domains of coherence allow water to record information and 'imitate' anything . . . molecules that oscillate in phase can emit intense and penetrating waves . . . like a laser . . . in other words, water is able to receive, retain, and return information because it fluctuates between coherent and non coherent states. This enables water to be a medium for excellent communication.'

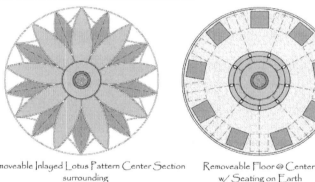

Removeable Inlayed Lotus Pattern Center Section
surrounding
Black Granite Fountain

Removeable Floor @ Center
w/ Seating on Earth

STILLPOINT / BLACK GRANITE FOUNTAIN / OMPHALOS (PLAN VIEW)

The ability of water to receive, mirror and hold thought-forms is powerfully demonstrated by the photographs of water crystals after being projected with such by Masaru Emoto, discussed above. Moving water surrounds the concavity in the middle of the fountain that implies/creates the Stillpoint, moving silently outwards towards a circular *infinity line*[246] - imagine a circular waterfall of a perfectly horizontal, 1/4" layer of moving water over polished black granite, falling over a precise, circular edge of this granite, disappearing almost silently beneath the floor – moving, glistening, shining, reflecting, absorbing, transmitting.

The most common arrangement of liquid water molecules is *tetrahedral* – similar to carbon and diamonds and synonymous with the Stillpoint geometry. It's own innate coherence will be focused and amplified by the pure geometry of the temple, as well as the elevated thought-forms, prayers and chanting of those experiencing the inner temple. It is hoped, similar to the healing waters of Lourdes, this energy will be reflected back into the environment of the inner temple, adding to the overall coherence. A self-reinforcing consciousness mechanism.

There is also another mechanical system within and surrounding the temple that supports this intention in a powerful way.

⊕ ELECTROMAGNETIC FIELD / VORTEX:

While the creation of a naturally occurring electromagnetic field/vortex as part of the temple's design has been introduced, it will be addressed in more detail here. In a tangible, physical manner, the temple will be a conduit for naturally occurring energy moving from Earth to Sky and Sky to Earth. It has been theorized that the stones of the megalithic period served, among other things, as acupuncture needles for the body of Earth. These stones, located on or at the intersections of *ley* lines - the paths along which energy, often in the form of underground waterways, flows within the Earth - served to receive and release energy at critical points in the Earth's energy grid. The temple is a conscious attempt to do just that . . . a structure designed to permit energy to flow naturally to and from the Earth and Sky – a constant movement of natural energy through the structure, imbued with loving, healing, conscious thought-forms.

From *Common Wealth*:

> ' . . . sacred power centers serve the same function as the endocrine glands of the human body. Just like human power centers such as the heart, liver and lungs, they serve to maintain the whole organism in balance. Thus, a temple built as a mirror of the cosmos and situated at an energy node, if correctly used, will serve the human body in the same way – first by influencing the connection of its spirit to the stars, followed by its denser organs in relationship to the environment' . . . and 'since number is an expression of universal concepts, it was considered very carefully when drawing up the designs of the temples. In this the architects sought a correspondence between their work in stone and benevolent cosmic forces. The net effect was to attract those forces, which would

enhance the environment for which the structure was built, thereby inviting a numinous state in the people attending the temple.'

A *vertical* copper rod will extend into Earth from the center of the temple at a depth equal to the distance from the center to the top of the spire atop the apex of the roof pyramid. Earth to Sky, Sky to Earth. This copper rod divides into four cables, each traveling to one of the four columns surrounding the inner temple at the northeast, northwest, southwest and southeast locations.

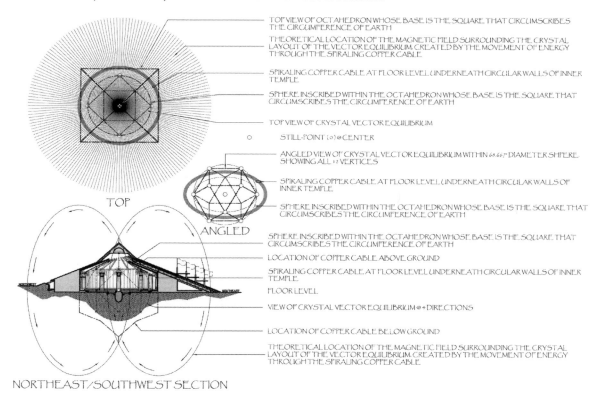

TOP VIEW OF OCTAHEDRON WHOSE BASE IS THE SQUARE THAT CIRCUMSCRIBES THE CIRCUMFERENCE OF EARTH

THEORETICAL LOCATION OF THE MAGNETIC FIELD SURROUNDING THE CRYSTAL LAYOUT OF THE VECTOR EQUILIBRIUM CREATED BY THE MOVEMENT OF ENERGY THROUGH THE SPIRALING COPPER CABLE

SPIRALING COPPER CABLE AT FLOOR LEVEL UNDERNEATH CIRCULAR WALLS OF INNER TEMPLE

SPHERE INSCRIBED WITHIN THE OCTAHEDRON WHOSE BASE IS THE SQUARE THAT CIRCUMSCRIBES THE CIRCUMFERENCE OF EARTH

TOP VIEW OF CRYSTAL VECTOR EQUILIBRIUM

STILL-POINT (○) @ CENTER

ANGLED VIEW OF CRYSTAL VECTOR EQUILIBRIUM WITHIN 68.667" DIAMETER SPHERE SHOWING ALL 12 VERTICES

SPIRALING COPPER CABLE AT FLOOR LEVEL UNDERNEATH CIRCULAR WALLS OF INNER TEMPLE

SPHERE INSCRIBED WITHIN THE OCTAHEDRON WHOSE BASE IS THE SQUARE THAT CIRCUMSCRIBES THE CIRCUMFERENCE OF EARTH

SPHERE INSCRIBED WITHIN THE OCTAHEDRON WHOSE BASE IS THE SQUARE THAT CIRCUMSCRIBES THE CIRCUMFERENCE OF EARTH

LOCATION OF COPPER CABLE ABOVE GROUND

SPIRALING COPPER CABLE AT FLOOR LEVEL UNDERNEATH CIRCULAR WALLS OF INNER TEMPLE

FLOOR LEVEL

VIEW OF CRYSTAL VECTOR EQUILIBRIUM @ + DIRECTIONS

LOCATION OF COPPER CABLE BELOW GROUND

THEORETICAL LOCATION OF THE MAGNETIC FIELD SURROUNDING THE CRYSTAL LAYOUT OF THE VECTOR EQUILIBRIUM CREATED BY THE MOVEMENT OF ENERGY THROUGH THE SPIRALING COPPER CABLE

TOP

ANGLED

NORTHEAST/SOUTHWEST SECTION

ELECTRO MAGNETIC FIELD / VORTEX

The four cables then circle in opposite directions *horizontally* around the base of the circular wall of the inner temple four times, the energy running up each of the four columns, continuing upward along the hips of the pyramid roof to connect to the top of the apex and golden spire. As this naturally flowing energy moves back and forth along this spiraling path, a toroidal electromagnetic energy vortex[247] forms that is perpendicular to the energy flow surrounding the inner temple. Torus energy . . . mirroring he toroidal electromagnetic energy vortex surrounding Earth.

GROUNDED TO EARTH SUMMER SOLSTICE 2013

The main idea here is to create a structure that has minimum resistance and permits natural energy to flow through it easily, creating the electromagnetic field/vortex. Besides creating an activated energy vortex within the inner temple, it may be a physical way to participate in the Earth's healing – releasing energy when

244

necessary from the Earth body and receiving energy from above. The field will permit the energy to move, while the vortex will help to amplify and focus this energy. The entire mechanism is yet another way the positive and negative elements interplay so harmoniously within the structure. This is also about enhancing the naturally occurring resonance that exists between the Earth's ionosphere and the surface of the Earth. As already mentioned, this is called the Schumann Resonance – the heartbeat of Earth. I refer the reader here once again to the film The *Grounded 2*, where two sunflowers are compared – one planted in a pot in the normal way, and the other surrounded by an electromagnetic field created by a wire mesh grounded to Earth[248] - the latter sunflower the healthy one.

'Temple legends describe how many of the power places of great importance – the Great Pyramid of Giza, the Oracle of Delphi, the site of Chartres Cathedral and Temple Mount, for instance – were chosen above underground fissures and streams leading to all parts of the land, by which a vital spirit was terrestrially dispersed. The chthonic mysteries were performed in galleries and chambers below these temples, into which ran metal rods connected to the tips of the buildings typically a Benben coated in a conducting material such as electrum. Just like a lightning rod, it allowed positively-charged electromagnetic energy to be conducted from the atmosphere and in the water veins below the temple and the metals in the rock, making the temple – and its practitioners – an instrument of fusion between the elements above and below: the sacred marriage in sky-ground dualism . . . the essence of this life force, or *mana, Baraka, prana or pneuma,* can be confusing but energy it certainly is. It has been described as *telluric current, earth energy, and geomagnetic force,* and it involves the intertwined forces of electricity and magnetism. As the anthropologist William Howells attempted to classify it, 'it was the basic force of nature through which everything was done . . . [its] comparison with electricity, or physical energy, is here inescapable. [It] was believed to be indestructible, although it might be dissipated by improper practices . . . It flowed continuously from one thing to another and, since the cosmos was built on a dualist idea, it flowed through heavenly things to earthbound things, just as though from a positive to a negative pole.'[249]   F. Silva, *Common Wealth*

⊕

⊕ Pyramid Shape / Energy:

This subject has been so trivialized over the years that including it in any way here is complicated. On the other hand, there have been serious people doing conscientious work - regarding the Great Pyramid specifically - for many years, and this is not accidental . . . this remarkable structure has intrigued people since it was created. It has been measured and re-measured, with theory after theory about why and how and when it was constructed and for what purpose. Like many of the other subjects addressed briefly here, there really is not time or space to give this the attention it deserves. Having said this, this particular shape plays such a critical function in the design of the temple that it is important to introduce at least a couple of areas of research that are pertinent (*see page 180 regarding its being a message*).

Remember too that this particular pyramid is an essential part of the Earth/Moon diagram, and includes both of the magical proportions Pi (π) and Phi (Ω) - the very special relationship between the two expressed only in this particular phi, or Great, pyramid geometry – and is an aspect of the Stillpoint geometry.

Regarding the interplay of energy within the temple (*see The Quartz Crystal & Water:, page 237: and Electromagnetic Field/Vortex: 242*), it appears that the geometry of the Phi Pyramid may generate *and* focus energy. Dr. Bill Schul was a staff member of the Department of Preventative Psychiatry at the Menninger Foundation, and had done extensive work and research with the Great Pyramid shape. Some years ago he had written a book called *The Secret Power of Pyramids* and I spoke with him, asking him to verify what he discovered about the location of the King's Chamber in the Great Pyramid.

He told me that he had built a glass pyramid to scale and shot a laser into it. At each point that the laser beam touched a wall he affixed a small mirror. He discovered that when he built this pyramid exactly as the Great Pyramid, *with each side slightly indented along each base similar the to original*,[250] all of the laser beams focused at the King's Chamber. It's likely that this room, rather than being a burial chamber as is commonly believed, was an initiation chamber – a specific space designed to create and receive amplifications of energy intended to stretch and magnify the initiate's experience . . . to grow consciousness.

What if this were so? . . . that yesterday's *community* built the Great Pyramid for archetypal *individual* awakening, while today's community – represented by a small group of spiritually motivated people - would build the temple for *communal* awakening. This is the irrepressible demi-urge evolving – having moved through more primitive forms of communal consciousness towards individuation, it is now moving beyond individual awakening towards an awakened collective.

There will be a central light sculpture hanging in the location of the King's Chamber, its design not yet considered. By positioning a light fixture in this auspicious place we are symbolizing our return to the light – in this sense, Spirit. It is through our awareness that the return to 'divine light' is made and placing the light sculpture at this particular focal point of the space seems an appropriate symbol and acknowledgment. This light sculpture is also the clapper of the 'pure silent bell.'

Physicist Jerry E. Bayles has done some theoretical mathematics which seem to support the idea that there is a scientific, physical connection between the geometry of the Phi-pyramid . . . in particular the Great Pyramid in Egypt . . . and the generation of energy itself. In a paper he wrote entitled *Ancient Power Generation*, which includes the complete theoretical mathematics to support his conclusions, he states:

> 'My conclusion is that the natural spiral (golden spiral), the Fibonacci ratio (golden ratio), the Golden rectangle, and the geometry of the great Pyramid are all related to the fine structure constant and the ratio of the electron's field energy to its rest mass energy at the Compton radius of the electron. The Great Pyramid therefore is capable of transforming energy from energy space to our real space via its static to dynamic geometry parameters of E (Epsilon, the fine structure or electron-photon coupling constant and the first of Rees's six numbers), Ø (phi) and α (ratio of the electron's field energy to its rest mass energy at the Compton radius of the electron). There also exists a geometry of the torus volume and area that transforms macroscopic space electron field energy field density to the correct geometry for the electron potential field energy at the Compton radius - the electron and the ratio of the two geometries is equal to the ratio of $4/\pi$. This suggests that around the Great Pyramid there may exist a torus shaped energy field with the pyramid apex poking out through the center of the torus.'

While the above quote is undoubtedly fairly incomprehensible to the average reader, it is included because of it relates the geometry of the pyramid and Golden ratio to the generation of energy, and because of the remarkable connection between the pyramid and torus shapes – the two major design elements of the temple being this pyramid shape protruding through the center of the swirling torus. The combination of these primal forms is essentially the basic design of the temple for a variety of reasons, and Dr. Bayles has connected these two forms purely through the mathematics of their 'physical' properties . . . a possible verification, from a completely unexpected and unrelated source, of the deep truth witnessed here.

⊕ Light, Symbol, and Life-force:

Light. The quality of light in the inner temple will be subdued and, above all, give the space a visual sense of subtlety and the sacred, with the colors, textures and materials being equally critical. The inner temple will be sensitive to external light and the path of the Sun in the sky as the Earth turns.

There are four, high, arched windows facing the four directions that penetrate the pyramid shape at the top of the temple. Light coming through these windows will be filtered through stained glass – adding to the subdued, quiet, sacred quality of the inner temple. Symbolically, these openings into the pyramid, with its masculine expression of straight lines and upward movement, are receptive openings for light from above – Earth and Air - another balancing of opposites. The serpentine clerestory window located in the space that is created where the spiraling roof circles around on top of itself *(see Northwest/Southeast Section below)*, will also feather southern, indirect, light through stained glass around the spiraling curve of the roof of the inner temple, and is designed to suggest the form and movement of the serpent or dragon, the ancient symbol for the life force.

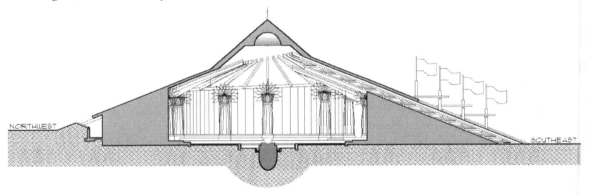

NORTHWEST / SOUTHEAST SECTION
(Showing serpentine clerestory window)

Also to be incorporated into the design of the temple will be a way to guide sunrise sunlight on Summer Solstice to the Stillpoint at the center of the fountain in the temple floor at the moment the Sun rises over the eastern mountains on that morning. The circular window facing southeast is provided for this reason *(see Temple Model, page 258)*.

Symbolically, the serpentine clerestory window seen above represents the snake . . . or dragon energy, the life force. Before the advent of the Hebrew Jehovah, and later Christianity, the snake, serpent, or dragon represented the life-force itself in ancient wisdom teachings ranging all the way from the Chinese to the Maya civilization in Central America. Our Sun also represents, and in fact *is*, this life-force for us. In the Yucatan area of Mexico at Chichen Itza, the Maya/Toltec civilization constructed an amazing astronomical observatory, a pyramid called Kukulcan, dedicated to their cosmology. Every spring equinox people gather there to watch a shadow-play celebrating and honoring the return of this life-force, or Sun/serpent energy. Around 4pm in the afternoon on the equinoxes only, the rays of the Sun cast a shadow that looks like an undulating snake onto the side of the north stairway of the pyramid.

This snake has *seven* coils, with an imaginary eighth forming at the top of the pyramid . . . *the beginning of the next octave.* A serpent's head carved in stone at the base of the stairway completes the picture of a deity-serpent descending from the sky, like the Sun's energy from above. This solar-snake of Yucatec Maya mythology is Kukulcan. The Yucatec rattlesnake (*Crotalus durissus durissus*) has a small round design resembling a 'solar' face near its rattle. It is believed that the Maya recognized these natural designs as symbols of the Sun deity, the rattle representing the Sun deity's crown.[251]

KUKULCAN, THE SERPINTINE LIFE FORCE:
SPRING EQUINOX, CHICHEN ITZA
(Image from Wikipedia)

The number of segments in a snake's rattle increases every year, and thus the rattle, for the ancient Maya, was a perfect symbol of cyclical time. In addition, snakes shed their skins at the height of the hot season, which symbolized the Sun's annual rebirth. As such, for the ancient Maya the snake was a compelling symbol of time and renewal.[252]

The group of stars known as the Pleiades, also called the Seven Sisters, is the source of many myths regarding extraterrestrial visitors. Also, in 'early Toltec mythology, the god Quetzalcoatl was originally identified with the Pleiades,'[253] and is Kukulcan's mythological counterpart. Quetzalcoatl was the god who, through self sacrifice, permitted the first Sun to rise . . . permitted manifestation to take place. The Yucatec Maya word for the Pleiades is *tzab*, which also means 'rattle.' In the Kukulcan Pyramid symbolism, the 'rattle' is the top of the pyramid – a rectangular, layered box with ornate embossed stone. More significant than any other date in the Mayan Calendar is the year 2012 . . . the 'last' year. As mentioned earlier, I believe that this date was chosen to symbolically represent the entire transit of the Winter Solstice alignment of Earth and Sun with the center of the galaxy – representing the completion of the Great Year and the return of the Light.

There were a number of remarkable celestial occurrences in that year. The first was May 20th, 2012. Just as our Winter Solstice alignment has been precessing towards and is now aligned with the center of the galaxy, in Sagittarius, so too has the Sun's 'zenith' alignment (when the Sun casts no shadow at solar noon) been approaching an alignment with the Pleiades, on the opposite side of galaxy, in Taurus (30°). On this day, in the spring of 2012, there was a conjunction with the Sun and the Pleiades, at midday when the Sun cast no shadow – also called the *Heart of Heaven*. The tail, or rattle of this seven coiled solar-snake, points to this rare precessional alignment in the zenith.[254] There was also an almost total eclipse of the Sun (94.39%) by the new moon in Taurus on this day, May 20th, 2012 – a perfect alignment with the Sun, the Moon and the Pleiades.

As mentioned, this symbol of the life-force, the solar snake, is represented in the temple by the undulating, snake-like form in the clerestory window located in the space created as the helical roof spirals back around upon itself. The serpent in the Kabbalistic teachings represents the wavelike, vibrating quality of light or energy itself – an aspect of the life-force and called the Serpent of Wisdom. The window is designed this way as symbol and reference to this primal energy, but it also is one of the actual paths that electromagnetic energy will flow as it travels through the copper cable described earlier (*see Electromagnetic Field/Vortex: page 243*). The serpent also represents esoteric, hidden knowledge: that which must be understood before our extended chakras can be activated – the Kundalini energy. Once activated, it allows us contact with higher energies, and signifies the birthing of a new cosmic vibration on Earth.[255]

The serpent has traditionally represented spiritual wisdom, life, healing, and the life-force in many ancient myths worldwide. In the ancient Sumerian texts, the half brothers Enki, Lord of Earth and Waters, and Enlil, Lord of the Air, competed for power and inheritance . . . eventually causing Enlil to twist the truth regarding the serpent, making it the evil Satan depicted in the Bible. Below is an image of ten Sephirot of the *Tree of Life* - 'the 10 attributes or emanations through which God reveals himself', with Kether, The Point of Creation . . . the Stillpoint . . . at its crown - intertwined with the serpentine life-force.

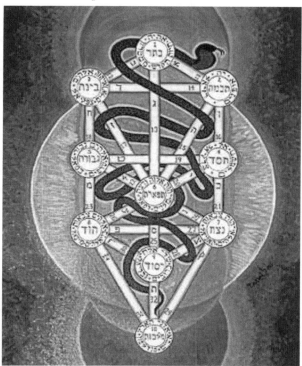

THE KABBALAH'S TREE OF LIFE

The ten Sephirot interconnected by the Serpent of Wisdom through twenty two paths.
Thirty two around one. The Vector Equilibrium/Flower of Life geometry.

⊕ WINTER SOLSTICE ... 2012:

There was yet another, even more significant date in the Mayan calendar in the year 2012 - December 21st . . . the Winter Solstice. *This* day - after thousands of years of orbiting our Sun while our solar system spirals its way around the center of the galaxy - was chosen, I believe, as the *symbolic* end of the Mayan calendar. On this day in 2012, the shortest day of the year, marking the return of the light, our Earth and our Sun on Winter Solstice are aligned with the center of this huge galaxy of the

100 billion swirling stars we live within, the Milky Way – demarcating this 'return of the light' on a much more profound level. It is *this* center, more even than our local Sun center, that is most deeply significant to us, as it signifies the center that our Sun and solar system circle. And it is this center . . . this Stillpoint . . . that the Mayans considered the source of their spiritual wisdom – the 'cosmic center' - and were trying to call attention to with the end date of their calendar.

While our Winter Solstices have been transiting the center of the galaxy since 1987 and will continue to do so through 2018, it is only in 2012 that all three alignments would occur – the Winter Solstice alignment with the center of the galaxy in Sagittarius, the zenith alignment with the solar Sun at the Heart of Heaven, with the Pleiades, in Taurus, and the alignment of the Earth, Moon and Sun in an almost total solar eclipse - and it is for this reason that I believe that the Mayans chose this date as the symbolic end-date for their calendar. This extraordinary alignment demarcates the end and beginning of the Great Year . . . the 25,920 year cycle of the Precession of the Equinoxes . . . and symbolically represents the nadir of and the return of the Light – a new Age, a new Great Year. *Now*.

The Mayans believed that these conjunctions with the center of the galaxy, as well as with its geometric compliment, the Pleiades, to be transformative, signaling a World Age shift, with the December solstice being the most important . . . symbolizing the darkest day - and the return of the light. Their entire calendar was based upon what they called the Great Cycle . . . or 1,872,000 Earth revolutions or days . . . or one World Age or growth cycle, the end of which signifies the next stage in its growth development. The symbolic end date of the Great Cycle, the end date of the Mayan calendar, is December 21st, 2012.

The cosmological calendar of the ancient Mayans is based entirely upon the importance of the return to the center . . . using our most profound *physical* center, the center of our galaxy, to demonstrate this. They considered this center as a fertile Creation Place or cosmic womb. This is, on another, even more symbolic level, Kether, the Kabbalistic Point of Creation at the crown of the Tree of Life . . . the entry point of the Great Mystery, symbolized by the veils of the Ayn, the Ayn Soph, and the Ayn Soph Aur – the Limitless - into our world of three dimensional form.

⊕ COINCIDENCE²:

As made clear earlier, and pertinent to the preceding discussion of our Winter Solstice, solar system and galaxy, there are a couple of remarkable facts regarding their relationships, similar in their improbability to the dimensions of the Earth, Moon, and Sun. The path followed by the Earth and Moon and other planets as they orbit the Sun is called the ecliptic. The band of stars we see in the night sky that we call the Milky Way form the plane of the galaxy of which our solar system is a part.

What is remarkable is that our ecliptic, amongst an *infinity* of possibilities, is aligned with the *center* of the galaxy. If this were not so, *solstice alignments with the center of the galaxy would not even be possible*. There is no explanation in all of science to explain this beyond 'coincidence.' To visualize this, hold two plates, one in each hand, and observe the infinite possibilities of their orientation to each other. There is only *one* possibility that permits the plane of the first plate to be aligned with the center of the second plate. Also, there is an infinite number of possible *angles* that the planes of the two systems could find themselves in relationship to each other - an infinite number of angles as the first plate rotates around its axis while still oriented towards the center of the second plate. The ecliptic of our solar system crosses the ecliptic of the Milky Way at almost precisely 60° - the angle of the Vector Equilibrium . . . the geometry of the Stillpoint.

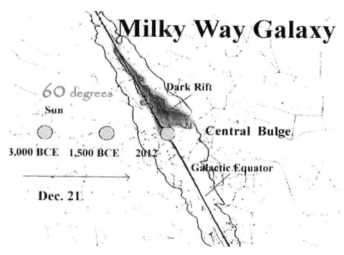

60° ANGLE BETWEEN ECLIPTIC OF SOLAR SYSTEM & GALAXY
(Courtesy John Major Jenkins)

This alignment allows for winter solstice alignments with this center every 25,920 years - the time it takes for the Earth's axis to complete one 360° 'wobble' . . . or full precession or circle - the distance from the Earth to the center of the galaxy also is close to 26,000 light years, meaning that the light emitted from the center reaches us precisely one precession . . . one Great Year later.

This suggests just how important the stillness at the center is. It is everything. Within it lies the key to understanding that we are One Being . . . One. And within this understanding, compassion for 'other' is born. Awareness of this absence of separation and the compassionate action that this awareness awakens is the temple's truest goal. The practical aim of the temple's technology – pulling a critical mass of global consciousness into the infinite awareness of Oneness by connecting to, *invoking*, the *evolved* consciousness responsible for embedding the Stillpoint geometry into the heart of the solar system - is pure alchemy, an alchemy of compassion.

The Mayans considered the center of our galaxy to be the source of their spiritual wisdom, and that the time that our Winter Solstice transits this galactic center would be a time of transformation and of increased influence from that center. It is for this reason - accompanied by the overwhelming evidence that our civilization is creating havoc as never before experienced on the planet due to human influence, and threatening all life on Earth - that I feel how critical it is that we act *now*.

SPIRAL GALAXY SIMILAR TO OURS

The 'coincidences' witnessed in our solar system, summarized in Appendices A, B, and C, are in fact aspects of an intentional *message* from the evolved higher consciousness that created this solar system for the acceleration of the evolution of consciousness. The center of our own galaxy was the source of the Mayan's spiritual wisdom – and their calendar pointed directly to this particular time, when our Winter Solstice is aligned with the center of the galaxy and when this particular information is appearing for the first time.

⊕ SPHERE'S WITHIN SPHERE'S ... THE STILLPOINT :

As mentioned above, one of the temple's intentions is the creation of a template for a new and powerful morphic field here on Earth – the intentional creation of a new form . . . the creation of a new paradigm – a field of compassionate, expanded consciousness.

This idea of an invisible field that provides the blueprint for creation/information is clearly reminiscent of the geometric, regenerating quality of the geometry of the Stillpoint . . . the Vector Equilibrium/Flower of Life (*see, Number:, page 195*). In Gregg Braden's book *Awakening to Zero Point*, he says:

> 'The term 'matrix' describes a collection of grids, appearing as stacked upon one another . . . that are hierarchical in nature, providing a structure for the gradual transition of energy/information from one zone of parameters to another. Energy/information/light originates within a source of consciousness and is propagated from this focal point through the multi-dimensional matrix – spheres within spheres. The planes of intersection within each of the spheres become the framework along which Matter is inclined to crystallize, as it descends into more dense levels of the creation matrix. It is within these intersection points that the true nature of creation may be glimpsed. For it is at the intersection planes of the spheres that the patterns of creation begin to coalesce. Everything is what it is because energy has congealed following one, or some combination of bonding patterns described by the Platonic solids. How does the energy 'know' where to form and how to bond? To understand the answer to this is to understand the morphogenetic fields, or blueprints, of creation.' [256]

The two-dimensional Flower of Life, representing the three-dimensional 32-sphere Vector Equilibrium matrix, spheres within spheres within spheres, is the geometry of this matrix. If one can accept this idea, than it is not a far leap to accept that this matrix of infinitely interconnecting Stillpoints underlies all of Creation, providing the beyond-quantum nuts and bolts of interconnectedness of all things. In the biology of organic organisms, this geometry very possibly provides the matrix that instantaneously transfers information through the process of morphic resonance across 'impossible' distances . . . the geometric grid of ant, bird and fish – and human - thought that short-circuits time/space as we understand it . . . what Jung called synchronicity.

The French philosopher/geologist/paleontologist/priest, Pere Pierre Teilhard de Chardin, a Jesuit most of his life, speaks about these ideas in the religious framework of his world . . . although his ideas were much larger and more expanded than those of the church he belonged to – a church that wouldn't permit him to publish his ideas. While he used the word 'God' in the language of his world, his experience went far beyond any common understanding of the word itself. He said:

> 'Because it contains and engenders consciousness, space-time is necessarily of a convergent nature [the idea that the accelerating expansion that we are experiencing now, will someday begin to collapse] and must somewhere in the future become involuted to a point which we might call Omega, which fuses and consumes them [space-time] integrally in itself. By its structure Omega, in its ultimate principle, can only be a distinct Center radiating at the core of a system of centers. In that final vision the Christian dogma culminates . . . *God is vibrant in the ether.* Through Him, all bodies come together, exert influence upon one another and sustain one another in the unity of the *all-embracing sphere.* And so exactly, so perfectly, does this coincide with the Omega Point that doubtless I should never have ventured to envisage the latter or formulate the hypothesis rationally if, in my consciousness as a believer, I had not found not only its speculative model but also its living reality.'

It seems clear that Teilhard not only knew deeply that it is, in fact, the very same Stillpoint that we have been addressing, the Stillpoint that is home to all of infinite space-time, outside of yet including time, the original and final vision of God itself . . . the Alpha and Omega . . . but that he experienced it directly – outside of all conjecture and labored ideas (*as found in these pages*).

Teilhard goes on to say:

> 'We are faced with a harmonized collectivity of consciousnesses to a sort of superconciousness. The earth not only becoming covered by myriads of grains of thought, but becoming enclosed in a single thinking envelope, a single unanimous reflection.'

This statement seems to mirror Buckminster Fuller's observation that 'the physical is . . . tantalizingly always ratio-equated with the innate eternal sense of perfection – thus the mind induces human consciousness of evolutionary participation to seek cosmic zero (*see Number:, page 195*).'

⊕ MESA TEMPLE:

'We are disconnected from some primal source with whom we once felt comfortable, and which served to restore our umbilical resonance with worlds that cannot be seen or touched. Even though we are experiencing the ills of disconnection, lying all around us in plain view are the means through which ancient cultures once maintained themselves in balance for thousands of years. They are called temples . . . . they all were built by faceless experts from forgotten ages to the same end: to be mirrors of the heavens so that ordinary men and women may be transformed into gods' . . . and 'through the knowledge and energy of the temple, the soul is transfigured, and suitably empowered it reaches into realities the normal senses cannot grasp.'[257]

It is here, in this time of transition and imminent chaos, that the tools necessary for the transformation . . . the seeds of the new order sown during the old order . . . will spring, seemingly spontaneously, into being. The Mesa Temple is *one* of these seeds . . . and there will be others. Speaking about the tendency of like ideas affecting each other across time and space, Sheldrake says:

'Synthetic chemists find that new compounds are generally very difficult to crystallize. But as time goes on, they generally get easier to crystallize all over the world. In the transmission of ideas or forms, art forms, by morphic resonance, there are two things. One is the number of people who do it or the number of times it's being done, and then there must be some variable of intensity. It must make a difference if someone is absolutely intensely involved with an idea and dwells on it with huge intensity compared with someone who just flicks through a magazine and sees a picture of your particular kind of thing for a few seconds and it's very superficial - there must be a big difference.

There must be an intensity factor and if somebody in solitude works away in an extremely intense way it may indeed set up a morphic field. In fact, we know that something like that does seem to happen, because it's very common in art, in fashion design, in science and technology for different people to have similar inventions, and very often they're trying to keep them secret.

I went to a symposium in London recently of fashion designers. We were discussing morphic resonance and the zeitgeist in relation to fashion. I [went] along with various fashion designers, and fashion retailers . . . . and also people in the stock market world, and the art world, some art critics. Everybody in all those worlds was very familiar with this phenomenon. You have something that's 'in the air' and these fashion designers are trying to keep their designs secret. They don't want other people to know about them, and they often find that other designers, even in other countries, come up with very similar designs.'

I can only hope that because this information has 'been in the air' for some time, that the world is pregnant with other, equally transformative attempts. It is difficult, of course, on this side of this 'temple experiment,' to know exactly how the technology will affect those experiencing it – or if it will truly fulfill the aim of its profound prayer . . . the direct connection to the higher consciousness/frequency that created this beautiful Earth for the acceleration of the evolution of consciousness. The conscious intention and hope is that the complete alignment and entrainment with the opening that the Stillpoint certainly is – essentially recreating that opening - and all that it means in this powerful, untried way, will provide a breakthrough in the growth of human consciousness.

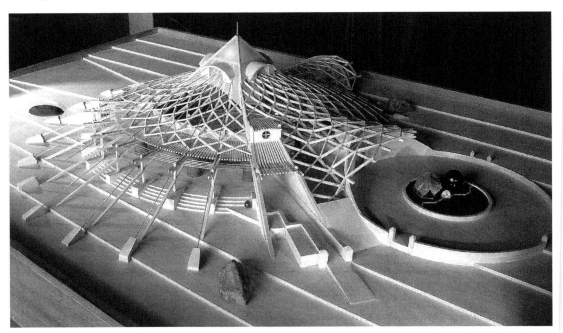

THE MESA TEMPLE FROM THE SOUTHEAST

The mechanism of this quantum leap in consciousness will be resonance with that higher consciousness. The structure is connected consciously to the very Earth from which it grows in proportional rhythm and surrounded by an electromagnetic field with the frequency of the Schumann Resonance, the heartbeat of Earth . . . and connected to and in rhythm with the Stillpoint, the soul, the center, of Gaia's consciousness – which is in turn is intimately connected to the even deeper rhythm and consciousness of the center of our solar system and the center of our galaxy and . . . the Center that is everywhere. After thousands of years of human evolution, we have an opportunity to physically manifest a perfect moment of awareness . . . . mirroring back to the higher consciousness that created this corner of the cosmos and to the Center . . . to ourSelf . . . that we understand, that we have stopped the world for the moment and are here, ready to be in the moment . . . and to receive in a way that perhaps we have not been able to before.

I've tried to imagine this moment . . . the moment of consecration.

Consecration: 'association with the sacred . . . to make or declare sacred or holy; sanctify.' The dedication of the temple. I can imagine the time when the temple's construction is complete. I can imagine a ceremony of consecration: chanting, toning, singing, prayer, meditation . . . a ceremony of *invocation*.

The seed is planted.

⊕ Floor Plan & Orientation, Structure & Materials, & Scale:

Each of these subjects is absolutely critical to the creation of a specific response to this message in the heavens, and relate to the precise construction of a portal to access this higher world in our time of crisis – the Mesa Temple described above.

How the building is laid out - its Floor Plan - as well as how it is oriented on and to the land, the materials used in its construction and the scale chosen to assure its whole number proportion to the dimensions of the Earth, Moon and Sun (and consequently to the Stillpoint geometry itself) become preeminent once that stage in this process is reached.

But this book has more to do with the stage preceding this and the startling discovery upon which all of this is based, and I have chosen to place each of these subjects in Appendix F for anyone interested in the specifics . . . the true nuts and bolts regarding the creation of the temple.

These are the truly practical considerations that are so important later on – and the subjects that matter so much to someone like myself who loves sacred architecture: what type and size the different structural elements are, as well as the specific details created to make them real, what sort of stone and wood used, how one enters the temple, which direction it faces, and the scale used to proportion everything precisely, and the details that truly bring life to a project such as this.

Having said that, I encourage the reader to go to Appendix F to see the graceful way the Floor Plan is laid out, the beauty of the Reflected Ceiling Plan showing all the radiating rafters (seen in the model), and the Elevations, accompanied by a somewhat complete description of all that is happening. When looking at the Elevations, envision a living roof – everything planted, and in a pattern subtly following the path of the spiraling purlins (the beams seen on the roof of the model) that are supported by the 28 radial rafters.

The section on Scale, to me, is fascinating . . . that is, just how did we come to use the term 'inch', 'foot' and especially 'mile', and the mind boggling fact that we somehow came to settle on a 'mile' that just happens to be intimately related to the dimensions of the Earth, Moon and Sun.

For now, though, the story continues regarding the importance of *where* any such sacred structure must go . . . and in particular the land that became the muse for the completion of the Mesa Temple's design.

⊕ Site & Placement:

Where the temple is located is almost equally as important as the design of the temple itself *(see also the quote from Common Wealth, by F. Silva, page 245)*. There are many places on the Earth's surface where energy collects – often called power spots. Rupert Sheldrake talks about the importance of determining the right site for a structure such at the Mesa Temple:

> 'If we are willing to consider morphic fields for minds and societies and animals and plants, it would also seem sensible to think of fields for ecosystems, or even for particular places. In fact, there is a 'science of places,' geomancy, which is concerned with just this kind of field. Geomancy is an ancient system for exploring the interrelationship of places and features of places; for locating power spots for building cathedrals and churches and temples, and for avoiding unsuitable places which have harmful influences.
>
> In ancient Europe there is no doubt that stone circles such as Stonehenge (and other places of ritual importance) were chosen

geomantically, in relation to the lay of the land, the flow of water, the direction of wind, the vegetation, and the orientation to the sun. There is also no doubt that cathedrals and churches in medieval Europe were built on sacred places which had been geomantically located; on places of power, sometimes sacred wells or sacred graves, sometimes places which manifested a powerful relationship to other features of their environment.'

Regarding landscape temples of history, as well as consciously designed temples, from *Common Wealth*:

'The ancients appreciated that human beings are, first and foremost, individuals, and that their journey towards spiritual enlightenment is an individual act where success is based on persistence, patience and perseverance. The hazards along this road are plenty and the distractions immense. Therefore, a little help on the journey has always been sought. And as far back as even the Aborigines can remember, we have sought places on the land *where the veil between worlds is thinnest*. Ancient traditions describe these as resident places of the spirits – what western scholars interpreted to be 'gods.' They are power places that help enlighten the individual and where the greater good of the community is served. And contrary to our modern perception of power as a monetary or political tool, they are repositories of energy, insofar as *they provide a more direct connection with an astral reference library and with the Great Spirit that flows through life.*'

While ancient cultures were aware of the importance of connecting with this spirit, we have lost this today – at a time when we need such a connection more than ever in our history.

I am now going to include a quote for the *third* time. I do this because of its profundity and because of its uncanny appropriateness given what this is all about, as well as my frequent criticism of orthodox science. I came across the scholarly occult book called *The Tarot* by Mouni Sadhu many decades ago. He was born the same year as my father, 1897, and named Mieczyslaw Demetriusz Sudowski - Polish through and through. Steeped in the Western occult tradition in his early life, he found his true teacher in 1950 . . . the Indian spiritual master Bhagavan Sri Ramana Maharshi. I was always fascinated that *after* he spent time with Ramana Maharshi, he wrote his remarkable book on *Western* occultism. As far as I'm able to determine, he used the name Mouni Sadhu as a writer . . . which means *silent monk* . . . because he felt that Ramana was speaking through him.

'Occult' according to Sadhu means the study of a 'deeper spiritual reality that extends beyond pure reason and the physical science' – or 'knowledge of the hidden . . . knowledge of the paranormal,' as opposed to 'knowledge of the measurable' - usually refereed to as science. He describes occultism as the explanation of 'phenomena which otherwise cannot be explained.' He is also clear about what he means by this idea in regards to spirituality:

> 'The first term, occultism, calls for the supremacy of the invisible to transcend the narrow framework of the physical manifestation of matter . . . . The second term, spirituality, transcends this world of illusion, and for a spiritual man the pronoun 'I' becomes identical with the consciousness of the Whole.'

Here again is his remarkable quote from *The Tarot* that I feel describes perfectly the intention and astral/physical nature of the temple, as well as the importance of finding the correct site for its location:

> 'Tourbillons, or vortexes, are astral creations of force which are the bases of all astro-mental realizations. Tradition ascribes the funnel like forms to them. Knowledge of the laws ruling over the tourbillons and their construction is one of the foremost principles of magic. The most guarded secrets of Hermetic magic are: *finding the point of support for the tourbillon on the physical plane, and the formula of transition from the astral to the physical world.*'

I do believe that what we would be attempting borders on the magical in the sense that it precisely invokes the spiritual world into our material world by using a precise 'formula of transition' . . . which the temple is intended to be, based as it is on the information contained in this most profound and sacred of messages – that found in the proportions of the Earth, Moon and Sun.

And this is not the only way to work with energy. The Chinese system of geomancy, Feng Shui, has been trivialized over the years, but its oldest concepts – especially those regarding site – go back thousands of years and have always been sound. The ideal site for a sacred structure, accordingly, is not on the mountaintop or valley floor, but on a hillside plateau with the view to the south (northern hemisphere) and protective mountains to the north, and bordering ridges running north/south to the east and west. The following is a description of the ideal site by Andrew L. March, in Frederick Lehrman's 'The Sacred Landscape:'

> 'The ideal 'site is ringed by a series of hills at various distances with water flowing among them, and is embraced by the two nearest of these as by a pair of arms, Green Dragon on the left and White

Tiger on the right. Behind the Site the ground should be high, and in front low, with water. Close by the front is a low eminence, a kind of repoussoir [a term used in painting layout meaning a figure or object in the extreme foreground: used as a contrast and to increase the illusion of depth]; at a distance beyond it is a view of the Facing Mountain. The highest mountain of a region, dominating the whole local Dragon system, is called taizu, 'Grand Progenitor,' typically a wilderness where 'human forces do not reach.' A high bump along a Marching Dragon is taizong, 'Grand Ancestor."

While the temple is not *necessarily* site specific and could be built in any appropriate location in the northern hemisphere if oriented as mentioned above (*see Floor Plan, page Error: Reference source not found & Floor Plan & Orientation, page 341*), and in the southern hemisphere if the present south elevation is oriented towards the north, it has been designed for a remarkably perfect site at the very northern end of the Owens Valley in eastern California.

In 2000, this land was owned by the U.S. Government (ahhh, the irony) with the intention of being publicly offered at auction. At the present time, 2018, this is still the government's 'intention,' but bureaucracy rules and nothing is being done. While I no longer believe that it will be possible to site the temple in this location, all that I've written would not have been possible had I not moved to the Owens Valley and been introduced to this site. My friend and I drove into Bishop in March of 2000 with the intention of moving here. I did something that I have never done before, and asked a man having dinner across the isle if I could join him and ask him questions about where might be the best place to move to in this vast valley.

The first and only land he mentioned was a mesa on land owned by the government – and this was the first place we visited. I was smitten. I returned maybe a hundred times in the next few years to walk the mesa and bask in the vastness it commanded. I have decided to include all of the magic surrounding this site, as it complies perfectly to the definition of such a site by many sacred descriptions and can serve here as an example of such.

So . . . what follows is a description of the site and placement as it applies specifically to this particular location. There are many incredible places in the world that would be appropriate, and perhaps even more ideal, locations for the construction of the temple, but, as the reader will come to see, there is a certain serendipity, perhaps even magic, surrounding this particular location. In every way, this land has served as my muse. It is for this reason that this particular site will be described in detail, and why the temple is called the *Mesa* Temple.

Perhaps more than anything else, there is a sense of overview and protection here. The Mesa Temple, as designed, sits towards the middle of a rocky mesa that rises out of an alluvial fan at 4,600 feet elevation flowing out of the over 14,000 foot White Mountains just three miles to the east. This mesa forms the 'low eminence' to the front, or south, of the temple site. In the near distance to the north is a another massive alluvial fan coming from Marble Canyon in the White Mountains, defining the northern geographical boundary of the 100-mile length of the deepest valley in North America, the Owens Valley. This great valley is defined by the north/south running White Mountains on the east the even taller Sierra Nevada on the west . . . both snow covered ranges filled with running streams year round. Farther north in Nevada, running east/west and protecting the 'rear' of the temple, are the almost 9,000 foot Excelsior Mountains, with Excelsior Mountain itself 40 miles almost due north of the exact site of the temple. About a quarter mile due north of the site is the largest rock outcropping anywhere in the area. There will be a pond of water in the walled garden area just to the south of the temple site.

ONE OF THE OLDEST LIVING BEINGS ON EARTH
(Photograph by Jörg Dauerer)

The White Mountains form the so-called embracing arm of the 'Green Dragon' ridge to the east of the temple site, and are populated with an almost endless array of 'Grand Ancestors.' They are home to some of the oldest living organisms on earth . . . the sometimes 5,000-year-old Bristlecone Pines. One of the 'Grand Ancestors,' 12,751' Mt. Hogue, defines the horizon 6 miles due east. The north/south running Benton Range, with an 8000' peak 8 miles due west, aligned with 11,600' San Joaquin Mountain at the eastern edge of the Sierras, forms the embracing 'White

Tiger' ridge to the west). 45 miles due south of the site, the prominent landmarks of the almost 14,000' Birch Mountain, just to the north of and aligned with 14,058' Split Mountain, rise up out of the steep eastern scarp of the Sierras . . . another range filled with 'Grand Ancestors.' In the classic Chinese geomantic model, these two mountains are the perfectly placed 'Facing Mountain' to the south.

One can see that the temple site is 'protected' all around, centered precisely within an array of eight major peaks at each of the eight directions. The site commands a 360-degree view, with the view to the south the 100-mile length of Owens Valley. The temple itself is located at the heart of the site, nestled between rock outcroppings and the gentle hill shapes of the rocky mesa.

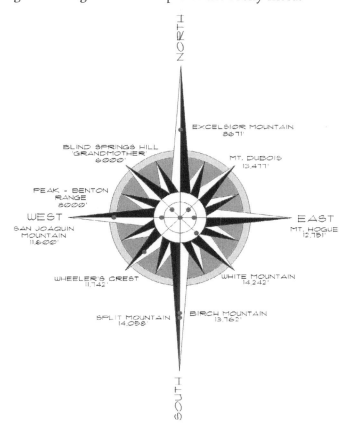

THE EIGHT DIRECTIONS
(Orientation of Stillpoint at center of temple to surrounding mountain peaks)

In the four cardinal directions just described, there are prominent mountains at the east, south, west, and north. In the quadrants between these directions, there is also a prominent mountain at each direction. Ten miles to the southeast is the 14,246' peak of White Mountain itself, the preeminent local Spirit . . . the 'Grand Progenitor,' and tallest in the range. This mountain catches Owens Valley's last light each day. Twenty-two miles to the southeast is the dramatic, carved granite summit of the northern end of Wheeler's Crest of the Sierras . . . defining the northern entrance to Owens Valley proper from Highway 395. Five miles to the northwest, standing by itself, is 6,000' Blind Springs Hill. Completing the circle, five miles to the northeast, is 13,477' Mt. Dubois in the White Mountains (*see The Eight Directions, above*). There is a feeling of power and protection . . . and of balance. The Mesa Temple grows from and is oriented to the Earth-body, and at a center of the local geography of the surrounding mountain ranges.

⊕ CHAKRAS:

As amazing as all this certainly is, there is much more. I hesitate to include the following information as I'm aware that it is far outside the credibility of the average reader. But I do so because what I describe exists – how one interprets it is perhaps the question. I myself don't really know how to hold all of this. I offer this only as an example of what may be possible when locating a building such as this.

Many sources attest that virtually all ancient temples and landscape temples are located atop telluric lines of force that move through the entire planet. From *Common Wealth*:

> '. . . there are places on the land where the geomagnetic field interacts with another force, and the effect intensifies. In physics it is called a telluric current; ancient people call it a spirit road . . . . the Sioux call this energy *skan*, and when concentrated at power places it is claimed to influence the mind, creativity, as well as elevate personal power in the form of spiritual attuning. In essence, the energy raises one's resonance, and contact with multiple power places builds up a kind of numinous state of mind . . . . and regardless of whether they visit sacred caves, mounds or mountains, devotees continue this practice to acquire the numinous energy of place, and in correctly harnessing this power they are able to receive visions.'

There is also a body of knowledge regarding how energy moves through the physical *human* body.   A seemingly endless amount of written information is available regarding the Sanskrit term 'chakra.' To simplify any lengthy discussion, the term has been defined in this way:

> 'Chakra is a concept referring to wheel-like vortices which, according to traditional Indian medicine, are believed to exist in the surface of the etheric double of man.   The Chakras are said to be 'force centers' or whorls of energy permeating, from a point on the physical body, the layers of the subtle bodies in an ever-increasing fan-shaped formation (the fans make the shape of a love heart). Rotating vortices of subtle matter, they are considered the focal points for the reception and transmission of energies. Seven major chakras or energy centers (also understood as wheels of light) are generally believed to exist, located within the subtle body. Adherents of Hindu and 'New Age' tradition believe the chakras interact with the body's ductless endocrine glands and lymphatic system by feeding in good bio-energies and disposing of unwanted bio-energies.'[258]

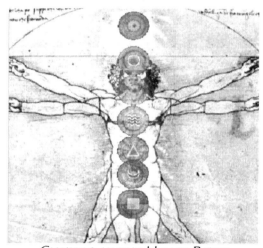

CHAKRAS IN THE HUMAN BODY
From top to bottom: 7:Crown, 6:Head, 5:Throat, 4:Heart, 3:Solar Plexus, 2:Sacral, 1: Root

After spending countless hours walking the area surrounding the temple site - this mesa arising from the alluvial fan pouring out of the White Mountains to the east - I decided to explore the entire area due south of the site – an area called Fish Slough. This is the land that the First People populated sometime after the glaciers had receded and is filled with remnants of old dwellings, petroglyphs and arrowheads. Lost in this vastness, I first visited the Scorpion rock outcropping that was covered with petroglyphs and well known locally from Paiute legend. It is a large outcropping, covering an area of perhaps 220 yards suare, in the shape of a huge scorpion – claws facing east, stinger facing west. It clearly had been an ancient gathering place, with a circular area in the center, perfect for ritual and ceremony.

I continued to explore, and stumbled upon an amazing phallic outcropping, seemingly rising from the Earth from out of nowhere *(see photos of sacral chakra below).* What?! I continued my exploration towards the south and came upon an ancient grouping of many circular stone foundations that had been used as the footings for living structures. These were the remnants of an old, permanent, camp, and it sat on a rise, surrounded and protected by rock outcroppings, above an old lake bed. There is nothing like this anywhere else in the entire Fish Slough area.

Then, driving north, I came upon another ancient gathering place – this time huge boulders filled with petroglyphs at the edge of an ancient river bed - the main drainage for the entire area. Unlike any of the areas already mentioned, this place exuded power.

I continued north, past the scorpion outcropping, and came upon what can only be considered a 'head' – a large rock outcropping next to the dirt road, staring directly at me. It was here that the light bulb lit up. A camp (survival), a phallus (sacral), a powerful place of gathering (solar plexus), a huge, prominent and raised area with a circular central area for ceremony (throat) . . . and now this head! No way . . . but there it was. At this point, I couldn't resist the almost impossible thought that perhaps I was witnessing actual chakras in the Earth, but was disoriented and had no idea where I was in relationship to the temple site.

I drove home and located each site on the three-dimensional topographic map of the entire area that I'd placed on a wall in my home – the first image of Owens Valley below. Oh my. Each was approximately the same distance from each other, all of them laid out in a north/south direction and in a spiraling, moving pattern.

The only chakra that didn't find *me* . . . the only chakra that I had to go looking for . . . was the heart chakra. What would be the chance that such phenomena would appear as physical manifestations in the *Earth*, arranged in a vertical, north/south direction directly south of the temple site – appropriately and obviously the location of the Crown chakra - and . . . *in correct order?* Below are 3d topography maps – the first of the Owens Valley from Boundary Peak in the White Mountains, on the border of California and Nevada at the north, to Mt. Whitney in the Sierra Nevada at the south, the second a close-up of the northern end of Owens Valley.

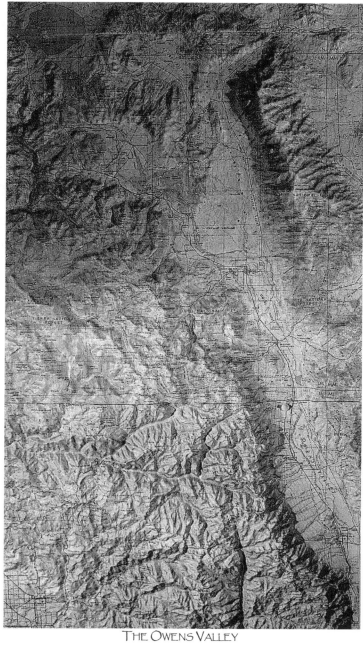

THE OWENS VALLEY

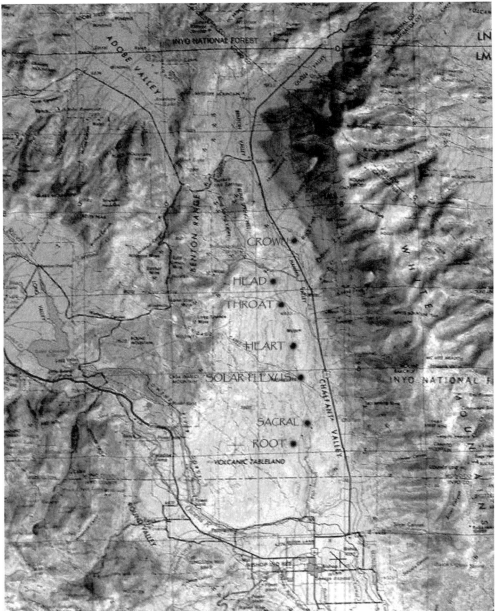

UPPER OWENS VALLEY: EARTH CHAKRAS

(Crown Chakra: Temple Site)

The second image above is a closer view of the same area at its northern end . . . the site of the Mesa Temple and the location of the seven chakras – the Crown chakra the site of the temple. I couldn't help but be reminded of the Lakota's Red Road . . . the spiritual path. The following was written by Wambli Sina Win, J.D., for Native Times:

> 'The Red Road or 'Chanku Luta,' as it is known by the Lakota, has been traveled by our ancestors long before us. Today, some may call it the road less traveled. According to Lakota belief, the Red Road begins even prior to conception and is a path which is available to those who are spiritually inclined. The Red Road which runs north and south, is a unique spiritual path, a way of life and enlightenment which has no end. During times of difficulty, the Lakota people could always rely upon the Red Road for strength and renewal.'

One can see it spiraling from South to North, North to South . . . vertical - the caduceus and the DNA molecule – all in *order*, with the temple site perfectly located at the Crown position.

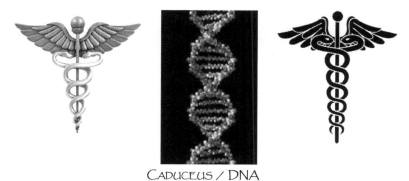

CADUCEUS / DNA

These chakras, or landscape temples, suggest powerful telluric energy lines running north/south in a spiraling motion, culminating at the crown chakra, under the 'dragon's head' in the north alcove of the temple.

Coming full circle now from the beginning section where it was mentioned that the hallucinogen 5-MeoDMT from the Alavarius toad was used by Mayan shamans to travel to the 'cosmic center,' there is a connection between the pineal gland in the center of our brains in the Head Chakra, to the Crown Chakra, or the temple site at the top of this Red Road. The pineal gland is the only gland in our bodies that produces natural DMT. From *Common Wealth*:

'Lastly, and possibly most importantly, is the effect that telluric energies in temples may have on the pineal gland, a pine cone-shaped protuberance located near the center of the brain. Fluctuations in the geomagnetic field affect the production of chemicals made by the pineal, such as *pinoline*, which interacts with another neurochemical in the brain, *serotonin*, the end result being the creation of DMT, a hallucinogen. It is believed that this is the neurochemical trigger for dream states – the hallucinogenic state of consciousness *that allows information to be received*. In an environment where geomagnetic field intensity is increased, people are known to experience psychic and shamanic states. It is one of the reasons why the temples were built, and why anyone would wish to attend.'

It is one thing to honor and recognize areas of the landscape where these earth energies travel and converge . . . and another altogether to enhance and focus this energy with a conscious structure based upon the laws of nature.

'It is well known that all temples and places of veneration are strategically sited upon the telluric lines of force that crisscross the face of the Earth, and whose fluctuation are sufficient to influence the body's electromagnetic field . . . but temples go well beyond this because they are a combination of layers of forces at work, creating specific environments that induce a pre-conceived effect on the body's receptive organs.'[259]

Below are photographs of the Earth chakras depicted in the topographic maps above. Much of my hesitation about including this controversial information, besides its being beyond the credibility of most of us, has to do with the importance of being there – that is, photographs can only relay so much information.

I am well aware of the human tendency to see patterns in images that are really projections – that is, patterns that aren't really there . . . it's called *pareidolia*, or *apophenia*. Perhaps this is what is going on here . . . but the 'light bulb' didn't turn on until I'd stumbled upon six of the seven chakras and returned home to discover how they were laid out on the land. I was as dumbfounded as anyone might be.

Keep in mind, as you see these photographs, that there is a infinite presence that cannot be captured . . . as well as the fact that they are found in order and are laid out north to south in a gently spiraling, caduceus-like pattern.

CROWN

Head of the Dragon . . . Temple site facing North

HEAD

Speaks for itself

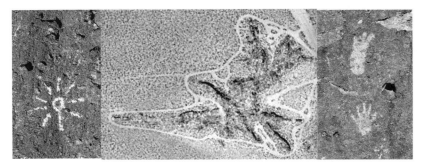

THROAT

About a ¼ mile x ½ mile in area and sitting on a raised plateau, seen from above it is a huge
Scorpion facing almost directly due east towards the White Mountains (aerial photo lower
middle) - head, arms, claws, with stinger pointing directly west. Covered with petroglyphs,
one of the earliest forms of symbolic communication, it was a sacred place for the First People
in the area and filled with legend. Behind the head is a circular sandy area that invites meeting
and ceremony – surrounded by infinity.

HEART

Of all the chakras, this site requires being there. More than all of the others, the Heart chakra is located in the center of a vast, open area. Like the heart itself, it is centered between the 1st and 7th chakra and more than any of the others, one has the sense of being at the center . . . and nowhere else along the series does is one so obviously surrounded by a huge circle of mountains . . . the Sierra and the Whites.

SOLAR PLEXUS

Just as the solar plexus are associated with power, this chakra exudes the most power of all seven of the chakra sites – and it located due East of the most powerful mountain in the area – White Mountain itself. Huge boulders adjacent to the ancient channel of the largest drainage flowing into the area are covered with petroglyphs. Like the Scorpion/Throat Chakra, this was obviously a place of power and gathering for the First People, when water poured through the adjacent canyon.

SACRAL

Completely by itself and approximately 75 yards from the base of a 300 foot ridge, it speaks for itself. This rock outcropping is approximately 16' tall.

ROOT

This site is slightly elevated from the ancient lake which is now Fish Slough, is surrounded by a rough circle of nine very large rock outcroppings, and consists of nineteen stone circle remnants, all opening to the east, which were the foundations for the homes of the First People. An eminently protected area, it served as shelter, home, a place for food preparation and the experience of daily life . . . a place to *survive*.

MESA TEMPLE SITE

Temple site in foreground . . . the Crown . . . looking East towards the heart-shaped valley
cut into the granite of the White Mountains by Falls Creek

HEAD OF THE DRAGON
Temple site looking North

⊕ DRAGON/SERPENT . . . LIFE-FORCE:

The serpent, or dragon, is an ancient symbol of the life force. The Owens Valley is filled - at all scales *(pun!)* - in fractal manifestations, with dragons. The Valley of Dragons.

Given the world we live in, this discussion has to begin with the association that most Westerners have with the serpent . . . the Christian image of the snake representing Satan himself in the Garden of Eden tempting Eve, or St. George fighting the evil dragon. This, of course, is another huge subject that can't be addressed in depth here, but suffice to say for now that this is a particularly Judaic/Christian viewpoint and is far from the experience of other traditions.

In megalithic times, the Mother, Goddess of Fertility, reigned. The Moon and the Earth were worshiped and venerated as sacred Goddesses, while pagan cultures flourished. Gradually, patriarchy began to supplant paganism and organized religion imposed itself on the world. The Judaic/Christian beliefs began their slow spread across the world, aided by the rapid spread of colonialism, eventually becoming the dominant paradigms in the Western world as they displaced the old world mythologies and pantheons with the monotheistic dogma of 'one *(male)* God.'

In the eighteenth and nineteenth century came the expansion of materialistic scientific rationalism and the advent of industrialization. Technology began to progress at an accelerating rate, and art and other right-brained values became submerged beneath this onslaught of practicality. Magic became esoteric, occult, and hidden. Ritual, the dance of the common people with the elements during matriarchal times, became the cloistered province of the patriarchal priesthood. Those few who continued the ancient ways did so in secret, fearing the wrath of the all-powerful church and the Inquisition. It was during this Judea/Christian epoch in our history that the serpent, symbol for the life-force so inherent to the domain of women and nature based cultures, became demonized.

The timeless, organic, Spirit-world that women are naturally more intimately connected to was a threat to the control oriented, fear-based patriarchal system. 'Witches' were burned at the stake by the tens of thousands. It was common to build churches on sacred pagan sites in order to suppress the dragon/serpent energy associated with the Earth Spirit (Glastonbury's Tor for example). Fear of the uncontrollable life-force represented so accurately by the serpent, transformed the snake into a symbol of Satan himself.

But this serpentine symbol for energy exists in many ancient cultures all around the world, and nowhere is it held with the same fear and aversion as with the entrance of the Hebrew Jehovah and with Christianity. In fact, it is often the snake that is the teacher in the ayahuasca visions of the Amazonian Indians.[260] The umbilical cord itself is the serpentine connection between mother and new life, and appears on the Goddess's images from the earliest times representing birth and regeneration. The Mayas and Aztecs, ancient Egyptians, and Druids all revered the serpent as a symbol of life-giving power. Today, this idea is honored in the Hindu texts, as well as by the Australian Aborigines and many Amazonian tribes. The *Caduceus* - two snakes coiling up the staff of the Greek Messenger of the Gods, Hermes (*see Caduceus/DNA, page 271*) - has been appropriated by Western medicine, and is the symbol of the life-giving physician. It has always been a symbol of magic and healing.

Both the snake, or serpent, and the dragon are traditionally symbols for the flow of life-force or energy. The ancient Chinese saw dragons in the landscape, representing the flow of terrestrial energy. I do too. The serpentine landscape temple at Avebury (the huge megalithic stone circle in England) was the site of seasonal invocations to fertility and the life-force.

> 'When NASA discovered magnetic energy spiraling inside tubes linking the Earth to the Sun – even employing the metaphysical word 'portals' – they essentially validated the ancient master geomancers and temple-builders who sourced these very same flux events on the land, because they were all too aware of their connection to territories far and beyond the confines of our terrestrial sphere. The image they chose to represent this elusive telluric force was the serpent or dragon, and in time this would become a culturally-shared archetype describing the energy's winding behavior along its earthly course as well as the skies. . .'[261]

To the Aboriginals of Australia, the great prologue to the beginning of time (the Ayn of the Kabbalistic tradition, the Nun of the Egyptian tradition) was called *alchera, or* the Dreamtime. It is the time of myth, when the Ancestor Spirits created the World. One of these ancient beings is the timeless Rainbow Serpent, *Almudj.* In what the Mayans call 'Zuvuya' - the circuit by which everything returns to its source - the great serpent of time, Tiamait/Orobororos, swallows his own tail . . . the torus shape, infinity . . . thus we re-turn to where we began. This is so reminiscent of the story from the Kabbalah where everything begins in the Limitless, or Ayn, and moves through the Stillpoint, beginning the evolutionary process defined by the ten Sepherot, and once again returns to the place it all began – the Alpha and Omega point.

In the ancient Mayan Creation mythology, the Milky Way (Mixcoatl) is this great serpent of time, whose gigantic mouth is the dark rift (the dark section of the Milky Way next to the galactic center in Sagittarius).

The Mayan word 'CannaC,' a *mirror* word, means 'serpent' - where Matter *and* Spirit are One - mirrored. The mouth of this great serpent is called the dark rift, and is the Path to the Underworld . . . the opening into the central bulge of the Milky Way, the pregnant Heart of Creation or the 'Womb of the Great Mother' . . . out of which the Universe was born.[262]

THE GREAT RIFT / SAGITTARIUS ... THE CENTER

According to the Mayans, this galactic core, *Hunab Ku,* is the source of transmissions to our species from Spirit . . . the source of *spiritual* energy . . . and is aligned with the idea that the galaxy itself, is a sentient, *conscious* being. As untenable as this may be in our modern world, this idea of an alive Universe holds a far deeper understanding of where we live and why we're here than that presented by materialistic science . . . or the world's major religions.

Through the alignment of our Winter Solstice and Sun with the center of the galaxy - an occurrence once every 25,920 years . . . *now* (which is only possible because the plane of the ecliptic, amongst an *infinity* of possible alignments, is aligned with the center of the galaxy, labeled 'quite a coincidence' by mainstream science), the Mayans believed that we would be experiencing an influx of spiritual wisdom. So much of the information proving the intentional creation of our solar system, as well as the message at the heart of it, has come to us during this Winter Solstice transit of the galactic center – the completion and beginning of the Great year. If, in fact, we *are* One, this means that *we* are trying to tell *us* something that is the very essence of *everything* - but 'the conduit is closing.'

As explained earlier (*see Light, Symbol & Life Force:, page 248*), the Winter Solstice in 2012 is the end date of the Mayan, cosmologically based calendar. They considered this point in time *'The Return of Kukulkan.'* Kukulkan of the Maya, or Quetzalcoatl, the mythological counterpart of of the Aztecs, is the 'feathered serpent' and is the major deity of the ancient Mayan civilization. Kukulkan has been understood to be a symbol for what Hindu Tantrics call *kundalini*, the serpent power that lies coiled in the root-center of the body, the 'serpent power' which resides in our nervous systems, rising up the spine to provide inspiration and illumination – or spiraling up the Red Road of the Earth-chakras due south of the mythical temple site - the feathered wings unfolding at the *crown* of the head (the same crown as that depicted as Kether in the Kabbalistic tradition, the *Point* of Creation). This energy travels up the spine and through the chakras, in a helical, spiraling, serpentine movement.

Even in the Christian tradition, the serpent in the Garden of Eden is the bestower of wisdom as well as the tempter, for wisdom is always a temptation to those who do not know how to use it. Thus Buddha conquers Mara, the Naga or serpent king, but Mara then protects Buddha. As such the serpent, Naga, is the keeper of the secret wisdom and the possessor of the life-force itself. The god of healing is a guardian of the underworld into which the initiate must enter in order to confront his or her unconscious. It takes a hero to encounter the dragon in its lair, free the damsel and find the treasure. We enter the depths to face our personal demons, monsters and dragons. When we have confronted them, we can be free of fear because we are no longer its prisoner. *This* is the treasure, and entering the underworld of the unconscious is the key to healing.

This kundalini, or 'life-force,' is the energy believed to animate us. There are many words for this force . . . bioenergy (Eastern European), ki (Japanese), mana (Hawaiian), orgone (Wilhelm Reich), prana (East Indian), skan (Lakota) and Qi (Chinese). The most important pathway of connection for kundalini runs from the base of the brain to the base of the spine . . . our own Axis Mundi . . . bringing in universal life-force. In *Hindu and Vedic* traditions, the male and female serpents *Ida and Pingala* entwine the central *'sushumna'* axis of our spine, weaving between the *'spinning wheels'* of our chakras. As mentioned above, chakra means 'wheel' in Sanskrit and has been described extensively in Hindu and Buddhist yogic literature. They are believed to be vortexes that penetrate the body, receiving, transforming and distributing Universal life-force.

It is important to note here that this energy, this life-force, is seen as *two*

282

intertwining snakes . . . again the perfect balance of polarity. In the temple, this is represented by the spiraling stainless steel cables moving in the opposite direction to the spiraling main rafters in the temple design, balancing this movement (*see Structure & Materials:, page 345*). The spiral, the symbolic abstraction of the dynamic, coiled serpent force, is a symbol for energy and, as mentioned earlier, the way energy travels in the Universe (*see Geometry, Form: the Earth . . ., page 174*).

In Jeremy Narby's book *The Cosmic Serpent*, he makes the extraordinary connection between the myriad of ancient creation myths from cultures around the world involving *twin creators* (the story of Quetzalcoatl and his twin Xolotl being but one of them) and the double helix of DNA, whose main purpose, as we are *presently* aware, is replication and *communication*. He points out that of the 80,000[263] or so plants existing in the Amazon rain forest, Amazonian natives are able to accurately discriminate between all of them and combine them to provide all kinds of practical, empirical drugs useful in their everyday life. Science's only answer for this, typically, is trial and error. Narby suggests something entirely different.

Amazonian shamans invariably say that they receive this information from the plants themselves . . . through accessing other dimensions by the ingestion of a herbal mixture known as *ayahuasca*. Essentially, this is one of many hallucinogens that transport normal, rational consciousness into realms where other, more esoteric, information is available. Ayahuasca consistently induces visions of entwined snakes that reveal information normally unavailable. The entwined snakes . . . or the double helix of the DNA molecule . . . is found in *all* life forms – bacteria, potatoes, giraffes, humans.

Equally remarkable, Narby relates, the Australian Aborigines believe that the creation of life was the work of a cosmic 'rainbow snake,' whose powers were symbolized by quartz crystals, [and] the *Desana* of the Colombian Amazon also associate the cosmic anaconda, creator of life, with a quartz crystal' (*see The Quartz Crystal & Water:, page 237*). The drawings of the Desana represent this with the halves of the human brain separated by an anaconda . . . and *hexagons*. The hexagon, or quartz crystal geometry, is a two-dimensional representation of the Stillpoint, creation, geometry.

Narby asks:

> 'How could it be that Australian Aborigines, separated from the rest of humanity for 40,000 years, tell the same story about the creation of life by a cosmic serpent associated with a quartz crystal as is told by ayahuasca-drinking Amazonians?'[264]

It is also important to mention that the four bases that comprise the DNA molecule are *hexagonal* . . . like quartz crystals (*see The Quartz Crystal & Water:, page 237*) and the Stillpoint geometry. Interestingly, the genetic code found in DNA that is the same for all living beings uses 64 three letter chemical 'words' in a myriad of combinations to create the vast variety of life forms on Earth – the same number as the 64-tetrahedron/sphere phase of the Stillpoint geometry which includes the Vector Equilibrium and is directly related to the Tree of Life of the Kabbalah and the 64-hexagrams of the Chinese I Ching and to the ancient Egyptian Flower of Life, and the same number of cells found in a human embryo (in the shape of the original cell-sphere) one cycle of seven days after inception.

Thus, the geometry of the DNA molecule, the originator of replication and communication – both in number and in pattern – relates directly to the Stillpoint geometry, the quartz crystal, the cyclical evolutions of the human body, and ancient bodies of wisdom all over the world. And, if Narby's hypothesis is correct, DNA is represented in the twin-creator myths from cultures all around the world, as well as being the source of the ancient plant wisdom of the Amazon cultures.

As mentioned at the beginning of this section, the serpent, or dragon, is an ancient symbol of the life force and it can now be seen that this is also a symbol or manifestation of the DNA, life generating, molecule itself. The Owens Valley is filled with these dragons . . . and the crown chakra of the remarkable chakra system found south of the temple site closely resembles the head of a dragon. This large rock outcropping, seen below, is enclosed by the temple in its northern alcove (*see Structure & Materials:, page 345*).

HEAD OF THE DRAGON

A quarter of mile or so due north is another, much smaller, rock outcropping which also appears to embody the head of a dragon – looking directly to the south towards the temple site. These rock outcroppings weren't, of course, *put* there by human beings – and perhaps this is only anthropological projection.

DRAGON'S HEAD

Seen from above, the mesa is . . . undeniably . . . a dolphin. Dolphins appear often in Greek myths, invariably as helpers of humankind, and the ancient meaning of the primordial mound is an essential part of understanding what this may mean.

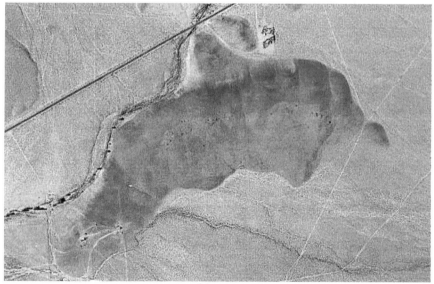

MESA

Aerial view of the site of the Mesa Temple aligned north/south,
with north to the right . . . the location of the Crown chakra . . . a dolphin.

285

Historical temples were often built upon sacred mounds, the hill of Giza that supports the Egyptian pyramids being one. Given the auspicious location of this mesa in the midst of this almost silent vastness, it is an ideal site for the temple (*see Site & Placement:, page 260*) . . . and the temple's design includes the larger outcropping above in its north alcove, with the smaller outcropping due north. It is, in fact, from the specific location of the larger dragon's head outcropping included in the north alcove that the temple is sited precisely on the land.

Very similar to Meher Baba's poetic words regarding the spark to awaken amidst the infinite oblivion of the beyond beyond state, in the ancient Egyptian creation myths dating to almost 5,000 years ago, the world appears first as a sacred mound, rising out of the lifeless waters of chaos. First to arise was the pyramid-shaped mound called the *benben* . . . from which the first sunrise emerged. *Atum* is the name given to the god considered to be the *first* god, having *created himself* (first cause) and is identified with this primeval mound. The name *Atum* means 'finisher of the world' as well as the 'underlying substance of the world.' Regarding this primeval mound as a sacred site, from *Common Wealth:*

> 'The essence of the god Atum first impregnates the site' and, 'once established, his energy physically manifests as a primeval mound . . . thereby helping to raise the level of awareness and in turn maintaining societies in balance for thousands of years. That was the intent from the moment the sacred mounds were created.'

These sacred mounds are also called landscape temples – perfect sites for a consciously created temple:

> 'A landscape temple is created from universal forces by a creator god for humans, and generates a wider spectrum of effect than a human-engineered site. A constructed temple, on the other hand, is built by humans for humans from the distillation of universal forces. It is an extension, a mirror of the undiluted original . . . by design.'[265]

The local Native Americans, the Paiute and western Shoshone, have lived in this valley for thousands of years, evidenced by petroglyphs that are many thousands years old. The Paiute word *Inyo* is used to describe the entire region and means 'a dwelling place of the Great Spirit.' The land is filled with Spirit.

These pages have necessarily focused on the Mesa Temple in an attempt to explain the many sacred principles that underlie its design . . . but, in truth, it is the land and its nature that is the underlying temple that the structure emulates and concentrates. The land is silent and grand . . . and the sky is vast. It is a cosmic natural cathedral.

The temple is oriented to the four directions and while it fits perfectly with this site, is is not necessarily site specific . . . it is adaptable.

The mesa is its natural home. It is my and the temple's Muse. Certainly it is where the temple was born . . . a structure based upon the first Word – the *most* sacred, Stillpoint geometry.

The Mesa Temple is designed to be a complete expression of the organic nature of the land that is its home, as well as the empirical laws of the Universe. Just as the Universe moves through a constant interaction of order and chaos, the mesa, shaped by billions of years of nature's unending movement, and impregnated by the god who created himself, is brought into full realization by the conscious order of the temple.

'Man models himself on the Earth,
Earth on Heaven,
Heaven on the Way,
And the Way on that which is naturally so.'

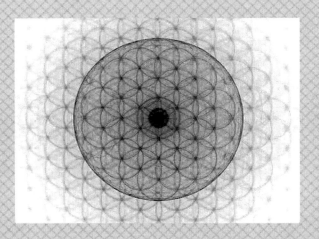

WHAT NOW?

New information has been presented that proves the existence . . . for the first time . . . of a higher, bodhisattvic consciousness responsible for the creation of our solar system. Other information has been presented that affirms this astounding reality. Further, the fact that all of this information is based upon discoverable, empirical, scientific data that has nothing to do with and is unnecessary for the evolution of *life* – yet is solely about *consciousness* - makes it clear that we are witnessing a grand *communication*. We are being told something very important.

This is clearly a *message*. At the heart of this communication is the geometry of consciousness itself, the Stillpoint or Flower of Life or Vector Equilibrium geometry that does not manifest anywhere in the known, observable Universe except here. It would seem obvious that *this*, the heart of the message, is the riddle we are being asked to unravel. *This* is the information we are being asked to understand.

The presentation of the evidence was followed by its implications as I understand them. Having experienced for myself the unequaled significance of the Stillpoint reality, I was certain that this message had everything to do with *consciousness*. I realized too that it was no accident that this communication regarding the geometry of consciousness was appearing *now* - for the first time ever in our collective experience. Also for the first time, the entire world is in crisis. We have *never* looked into an abyss so bottomless. Given this reality, it is clear that we are being given a message regarding a *global* shift in human consciousness . . . something never witnessed in the last 11,500 years and perhaps never.

People all over the world are attempting to find solutions to this global dilemma. But all such hopeful solutions – whether social, economic, technological, environmental, political, religious, or the switch from the suicidal energy we use to something that causes no harm - are born in the old paradigm of the last 11,500 years, and *none* of them are capable of effecting the change required now. The darkness that faces us is a crisis of *consciousness*. This was made clear by quotes from some of the people we respect the most. I repeat the following from M.K. Gandhi:

> 'An armed Conflict between nations horrifies us. But the economic war is not better than an armed conflict. This is like a surgical operation. An economic war is prolonged torture. And its ravages are no less terrible than those depicted in the literature on war properly so called. We think nothing of the other because we are used to its deadly effects. The movement against war is sound. I pray for its success. *But I cannot help the gnawing fear that the movement will fail if it does not touch the root of all evil – human greed.*'

The Dalai Lama, in his graceful way, called it a 'change of heart.'

'Consciousness' can be an ambiguous term, so it was made clear that the transformation of consciousness so desperately needed now is from our present survival-based awareness towards an awakened awareness of the Unity and connectedness of all things . . . a consciousness that ultimately expresses itself though enlightened *compassion*.

But how to accomplish this transformation of consciousness that, as far as we know, has never happened before? The Stillpoint geometry that is the heart of this message was embedded in the heart of our solar system such that its importance could not be denied. Our task now is not to endlessly discuss this incomparable phenomenon, but to *use* it - after all, we *are* running out of time.

I can come to no other conclusion other than we are being asked to recreate this opening . . . this portal . . . that the Stillpoint surely is, to access either the higher consciousness responsible for the creation of our corner of the cosmos, or the Akashic Record of all that evolving consciousness has learned in its eons-long journey – or both. I believe that if we were to be able to do this . . . use this information to create an opening to the Implicate world beyond the veil . . . that the transformation would not occur through our normal understanding of learning new information and acting upon it – asking questions, taking notes, intellectually coming to some kind of understanding of what is required, making decisions, and somehow figuring out a way to implement it. Rather, it would occur through some kind of deeper *experience*.

We are being asked to weave this information into a *response* whose very nature is one of *resonance*. *Resonance* with this higher consciousness will be the method used for the transmission and the transformation. Because of my life-long love of sacred geometry, it seemed obvious to me that this non-manifesting Stillpoint geometry needed to be recreated in our five-dimensional world (just as it was modeled for us) if possible . . . and I know of no other way to do this than through the creation of a technology – a physical *temple* - based entirely upon this invisible, non-manifesting geometry. A temple intended as portal.

After the evidence and its implications were presented, much of the following narrative concerned the sacredness of geometry and discussions of the eternal temple and its importance to us throughout history – and, in particular, the specifics of the Mesa Temple and the importance of its location. The timing of this action, this response, was made clear by John Michell, in his book *The Dimensions of Paradise*:

'One day, it is said, the Temple will be restored, the sacred world order will again be established and harmony between men and nature will once more prevail. That event, according to all prophecies, will take place at a period of extreme need and desperation.'

Yes, 'a period of extreme need and desperation.'

We are alive in the most critical moment in this latest phase of human history. I agree with those I've quoted and others who believe that we are headed into the abyss of our own creation unless there is a global transformation of consciousness in the short time we have. This transformation will not happen by itself. Unlike the preceding stages of evolution, through which our own existence was made possible, purpose will not mechanically pull from the future precisely what it needs in order to continue. We have evolved into *self-determining* beings and it is up to us to somehow figure out what is required. Arthur Young points out in *The Reflexive Universe* that there are off-shoots or 'whiskers' on the evolutionary tree that stopped evolving . . . 'experiments' that failed relative to the assent of awakening. We very well may be at that crossroads now.

These pages demonstrate, I believe – as nearly impossible as it is to *know* - that an *evolved* consciousness, infinite relative to our own, has intentionally placed within the heavens the most important information it has to share with us – discovered only when its import could be fully understood and was critically needed. The Stillpoint geometry, 'impossibly' embedded in the proportions of the Sun, Earth and Moon, defines the eternal present moment . . . the geometry of consciousness . . . what Buckminster Fuller considered:

'. . . the nearest approach we will ever know to eternity and God.'

This Stillpoint geometry is the Kabbalah's *Point of Creation*, science's 'point-of-singularity, one billionth the size of a proton,' where the Great Mystery exploded into existence with the purpose of awakening . . . 'that God should consciously know his own fullness of divinity.' It is the Rig Veda's 'core, central point' from which the Universe was born. Reductionist science recognizes it as the singularity from which the Universe was born. It is that important. There is no *reason* within known physics or cosmology, sacred geometry or mathematics, or any scientific discipline, or by any stretch of coincidence or chance for its inclusion in the dimensions of the Earth, Moon and Sun. This reality is not necessary for life to exist . . . it has *only* to do with the evolution of consciousness at the highest level – it is the opening though which we emerged, unaware of who and what we were, yet propelled into our future by the initial intention to awaken - and is the opening though which we are destined to return *if* we can use this key to access that same opening . . . consciously completing the cosmic cycle of awakening - fully aware. The Alpha and the Omega.

To repeat - this sacred information was not 'put' on this grand stage for information's sake. Its existence invites a response . . . a pure *action*. It is there to be *used*. 'Using' it means expressing it in *our* reality, the reality of our five-dimensional world . . . consciously bringing Spirit, or awareness, into Form – creating a direct connection – a portal - to higher consciousness.

As mentioned earlier, Amazonian shamans individually access empirical information available in other dimensions, or altered states:

> ' . . . they [shamans] talk of a ladder – or a vine, a rope, a spiral staircase, a twisted rope ladder – that connects heaven and earth and which they use to gain access to the world of spirits. They consider these spirits to have come from the sky and to have created life on Earth.'[75]

This is a description of the temple and its intention, as well as the consciousness that made all this possible – that is, created this Earth and its solar system with the perfect conditions for life. The most important aspect of the information presented is the empirical proof of the *evolved* consciousness ('spirits from the sky') that 'created life on earth.' But instead of journeying individually to 'gain access to the world of spirits,' the temple is a conscious attempt to access this world for *community*.

By using the five dimensions that define our reality and weaving the Stillpoint, creation/consciousness geometry, into every proportion of the temple's structure (including a double helix, 'twisted rope ladder,' spiraling roof), while empowering it with energetic mechanisms also birthed from this geometry, a purely *whole*-istic expression is created with the intention of opening a channel to the consciousness that made possible – 'created' the certainty of life on Earth, as well as the Earth itself and the rest of the solar system: that is, bringing spirit consciously into form by using the embedded code given to us in the 'impossible' dimensions of the Earth, Moon and Sun – the geometry of creation, the Kabbalah's *Kether, the Point of Creation* – the doorway to our origins in the Great Mystery, the Stillness; Alpha - and the doorway of our fully enlightened return - Omega.

Individual human beings have completed this journey to enlightenment – but in bodhisattvic manner return to this suffering realm to teach and to help the rest of humanity to awaken. That an almost infinitely evolved higher consciousness than our own put this message in the heavens for our benefit is the ultimate bodhisattvic gesture. Our future is not a matter of individual awakening. It must be a shared, *communal* effort with the intention of raising the general level of global human consciousness – triggering its inevitable result . . . compassion.

For all these reasons, the expression of pure non-physical reality in physical terms . . . the temple . . . with the intention of creating an opening to this higher consciousness, is the most appropriate response I personally can envision. This has been the aim of temple builders for thousands of years. And again . . . if there is a better idea, than that is what I'm for – what is most critical now is discovering the best way to use the information so majestically presented.

On another note, speaking of the often necessarily secret cults that protected the wisdom that temples embodied throughout time, and about the bodhisattvic responsibility they had to oppressed people everywhere, a last quote from *Common Wealth*.

> 'In essence, these organizations served these spiritual forces of Light, and through initiation into the cult of knowledge they sought to free humans from their enslavement to ignorance, not by physical force but through a wisdom that ultimately exposes the world rulers of this darkness so that the souls of men and women are no longer prostitutes to their will.'

Ahhh . . . 'the world rulers of this darkness' . . . what to do about *them?* They will not let go of that gun in their 'cold dead hands.' These are the psychopaths and sociopaths, and those who envy and emulate them, born hardwired to personal survival or made that way by their world, without any hope of experiencing or expressing empathy. It has often been said that 'they' need 'us' for this system to work. They have created a system that demands our participation. We are the fuel that keeps it going – until we are no longer needed. It is almost impossible for me to imagine any kind of change affecting the consciousness of those that are in control other than a transformation of *our* consciousness. These are the same people who, again in *this* country, 'legally' changed the status of corporations to *persons,* created the Federal Reserve and the Patriot Act and the War on Drugs and the endless list of such that make their wars and oppression and greed possible.

After all this was completed, I felt that I had to address this question of 'the darkness' straight on. It's been alluded to frequently here – but never completely displayed in its enormity. I know that most of us don't want to hear it, don't want really to know. But now is the only time we have and it is a time to not look away. I began to write about it in earnest, intending to include what I'd compiled in this book. It was not easy, after so many years of being assaulted by it, to have to sit down and return to the scene of all the crimes and write about it. In the midst of doing this, I fell into a depression about all that is happening. It is immense - and darker than one's imagination is capable of imagining. It brought on a feeling of deep hopelessness. But rather than disrupt these pages, which are essentially about the Light and its expression, I've listed many subjects that are a part of this vast, dark agenda and have chosen not to include the full story – after all, each of us is aware in our own way that the world has never been at this crossroads before, aware of the daily atrocities. All of this information is available by book, video, and the Internet. It's all there for those who wish to verify for themselves. Some of you already know and many of you don't – but feel it. It's real, and worse than you think.

A friend who was helping with the editing of this book told me that I had to end it on a hopeful note . . . that I had to offer people something to do that could help. I knew what he was talking about . . . but I had nothing to offer in the normal way. I clearly remembering watching Al Gore's *An Inconvenient Truth* where he recounted a litany of damning problems with little hope of their being solved . . . yet at the end he reminded us to recycle and to buy a hybrid car. Really? Virtually any book I've read that addresses the problems in the world, tries to end on a positive note. Still, I read Jane Goodall's *Reason for Hope*, trying to find *something* that had escaped my research. I don't know of anyone I respect more in this world more than this person. There is no 'denial' in her. Her book was filled with the many many problems that the world was cursed with in 1999 . . . problems that have only accelerated in the time between then and now *(if you need reminding, see this list on page 163)*. And she didn't put her head in the sand . . . ever. Imagine someone as caring and compassionate as she, someone whose love of animals is unparalleled, visiting a lab where scientists hooked up electrodes to the brains of animals and performed all sorts of inhumane experiments on them.

But she did this. She did not turn away. And yet her book was called *Reason for Hope* . . . and sure enough, towards the end, she did her best to come up with such reasons. One long paragraph included a list of 'ifs' . . . '*if* everyone biked or walked when it was practical, shared a car, or took public transportation – the reduction in air pollution would be dramatic,' 'imagine the difference *if* everyone stopped eating meat – even for a couple of days a week,' and so on. There were ten such 'ifs,' none of which are going to happen given the world as we know it today. And then a list of companies ravishing the planet for profit, with the comment: 'And unless you and I support those companies, by purchasing their products, they will never survive in the competitive marketplace.' But, you see . . . we will. And this exposes the weakness in hoping. Along with this came a whole lot of examples of amazing work towards the good that is happening all over the world. In fact, she reminded me of the innate goodness in people. I do believe that this is who we are . . . if given the chance. Jane also made it clear that she wasn't naive about any of this, and there is courage in her hope.

Bottom line:

> 'I do have hope for the future – for our future. But only *if* changes are made in the way we live – and made quickly . . . I had no doubt that, given time, we humans were capable of creating a moral society. The trouble was, as I knew only too well, time was running out.'[266]

In another part of her book, she acknowledges that what lies ahead requires more than hope. With the following quote, I added her to my growing list of the people in the world that we respect the most who have each said, in their own way, that without a global shift in human consciousness, we will not be able to alter the direction we are headed. From *Reason to Hope*:

> 'Our task, then, if we would hasten our moral evolution, progress a little more quickly toward our human destiny, is obvious – formidable, but in the long run not impossible. We will have to evolve, all of us, from ordinary, everyday human beings – into saints! Ordinary people, like you and me, will have to become saints, or a least mini-saints. The great saints and the Masters were not supernatural beings; they were mortals like us, made of flesh and blood.'

Yes, formidable is the word . . . but it is here that I have to totally rely on the magnitude of what the Stillpoint information implies. It is all I know of that brings genuine hope. It is the only possibility I know of that has the potential to meet the steamroller of darkness enveloping our world. I cannot foresee the future . . . but I *can* imagine a world where people are waking up in mass in some way that I can't totally understand. I can only offer this, once again, from *Passion of the Western Mind*:

> 'A threshold must now be crossed, a threshold demanding a courageous act of faith, of imagination, of trust in a larger and more complex reality; a threshold, moreover, demanding an act of unflinching self-discernment . . . [the masculine and feminine] synthesis leads to something beyond itself: It brings an unexpected opening to a larger reality that cannot be grasped before it arrives, because this new reality is itself a creative act.'

And for those of us who fancy ourselves as 'spiritual,' I offer this pertinent quote from Catherin Austin Fitts, who served as Assistant Secretary of Housing and Federal Housing Commissioner at the United States Department of Housing and Urban Development in the first Bush Administration, and was vilified and made bankrupt by the banking institutions she threatened with her progressive policies:[267]

'In the summer of 2000, I was giving a speech to a group of people who have a conference once a year on how to evolve our society spiritually, and I was giving them a speech called 'How the Money Works in Organized Crime' . . . and explaining how the U.S. economy launders 500 billion to a trillion dollars . . . of all illegal monies - it's much more than that now - and I asked them what would happen if we stopped being the leader in global money laundering. And they said 'Ah, you know, we'd have troubles refinancing the deficit, and the stock market would go down, so our IRA's and 401K's would go down.' So I said, OK, let's pretend that there's a big red button up here on the lectern and if you push that button, you could stop all narcotics trafficking in your neighborhood, your county, your state, thus offending the people who control 500 hundred billion or a trillion a year, and the accumulated capitol thereon. And, out of a hundred people dedicated to evolving our society spiritually, only *one* would push the button. So I asked the other 99 why didn't they push the button, and they said that they didn't want their IRA's and 401K's to go down, we don't want our government checks to stop, and we don't want our taxes to go up.'

The quote goes on, lambasting the hypocrisy of people in general, and the 'spiritually' smug in particular - a unique form of denial. Modern Man for the most part experiences himself as separate from whatever is outside of the skin that contains him. This is the great lie and the lie that the current and destructive scientific mechanistic view of the Universe is based upon. The truth is that *all* of this is One Being . . . one intricately, ingeniously, infinitely connected, alive Being. It is in the *process* of becoming aware of itSelf. This process happens within the asymmetrical flow of *time* . . . but the reality is that it is all one eternal present moment containing everything that ever was, is, or will be - *way* beyond the ken of our present level of consciousness. The all-important practical reality of all this is that a consciousness aware of this ultimate, instantaneous, interconnectedness, experiences the knowing that *I Am That*.[268] You are me. I am you. We are One. This instantly translates into the manifestation of compassion in our world or . . . love.

Humanity is lost in the lie . . . a lie that organized religion, with its God sitting high above, separate from His creation, as well as the study of parts, rather than the whole, and the purposeless, random reality of materialistic science, both encourage. And because of this we are looking into the abyss. We now know that a *evolved* higher consciousness of bodhisattvic intention not only exists but is communicating with us in the most unimaginable of ways. It appears that we are being offered a way out of this nightmare. May it be so.

'This Earth is one of the rare spots in the cosmos where mind has flowered. Man is a product of [over] three billion years of evolution, in whose person the evolutionary process has at last become conscious of itself and its possibilities. Whether he likes it or not, he is responsible for the whole further evolution of our planet.'

Julian Huxley

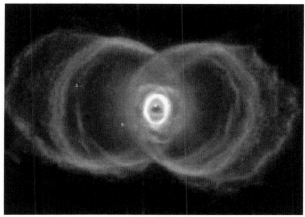

THE EYE OF INFINITY

The temple, in its absolute essence is: 'ratio-equated with the innate eternal sense of perfection . . . inducing the mind to seek cosmic zero.' Will it work? There is no way to know on this side of an experiment such as this. It is, in reality, the self-determined attempt of *community/civilization* to reach into the future, to reach into a new paradigm, accessible perhaps because of the alignment with, and open to, the energy emerging from the center of our *local* cosmos . . . the Stillpoint at the center of the galaxy of which we are a part. It is our responsibility to consciously reach into our future for the seed upon which the evolution of our consciousness and the survival of humanity depends.

You now have this information and I invite anyone interested to make what they may of it. The evidence is irrefutable. The implications I've shared are as I understand them . . . you may have other understandings. Because of the nature of the task, I have had to use words and ideas in an attempt to communicate what this is . . . but this is not expressible in words and now, I hope, I am done with them. My focus must now return to creating the response I feel is being asked for and can only hope there are others who feel as I do and that these people will contact me . . . I am looking for Arthurs to help me pull the sword out of the stone - it is time for action.

The temple is ultimately a sacred space - sacred in its geometric expression, alignment, harmony, and connection to earth and sacred in its purpose of creating a space for silence . . . with the Stillpoint, the eye of infinity, at its center . . . and sacred in its intention of creating an opening to the higher consciousness that made this all possible.

A Beginning.

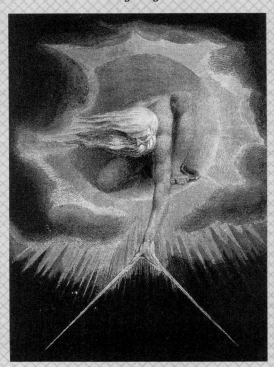

THE ANCIENT OF DAYS

LET THE SLAVE

From William Blake's
America: A Prophecy
. . . as interpreted and sung by Van Morrison in the song
Let the Slave

Let the slave grinding at the mill run out into the field.
Let him look up into the heavens and laugh in the bright air.
Let the enchained soul shut up in darkness and in sighing
Whose face has never seen a smile in thirty weary years
Rise and look out, his chains are loose, his dungeon doors are open.
And let his wife and children return from the oppressor's scourge.

They look behind at every step and believe it is a dream,
Singing, the Sun has left his blackness and has found a fresher morning,
And the fair Moon rejoices in the clear and cloudless night.
For empire is no more
And now the lion and wolf shall cease

For everything that lives is holy
For everything that lives is holy
For everything that lives is holy
For everything that lives is holy

What is the price of experience? Do men buy it for a song?
Or wisdom for a dance in the street? No, it is bought with the price
Of all that a man hath, his house, his wife, his children.

Wisdom is sold in the desolate market where none come to buy,
And in the withered field where the farmer plows for bread in vain.
It is an easy thing to triumph in the summer's Sun,
And in the vintage and to sing on the wagon loaded with corn.
It is an easy thing to talk of patience to the afflicted,
To speak the laws of prudence to the homeless wanderer,
To listen to the hungry raven's cry in wintry season,
When the red blood is filled with wine and with the marrow of lambs.

It is an easy thing to laugh at wrathful elements,
To hear the dog howl at the wintry door,
The ox in the slaughter house moan.
To see a God on every wind and a blessing on every blast,
To hear sounds of love in the thunder storm
That destroys our enemies' house.
To rejoice in the blight that covers his field
And the sickness that cuts off his children,
While our olive and vine sing and laugh 'round our door
And our children bring fruits and flowers.

Then the groan and the dolor are quite forgotten,
And the slave grinding at the mill.
And the captive in chains and the poor in the prison,
And the soldier in the field,
When the shattered bone hath laid him groaning
Among the happier dead.

It is an easy thing to rejoice in the tents of prosperity.
Thus, could I sing and thus rejoice . . .

But it is not so with me.

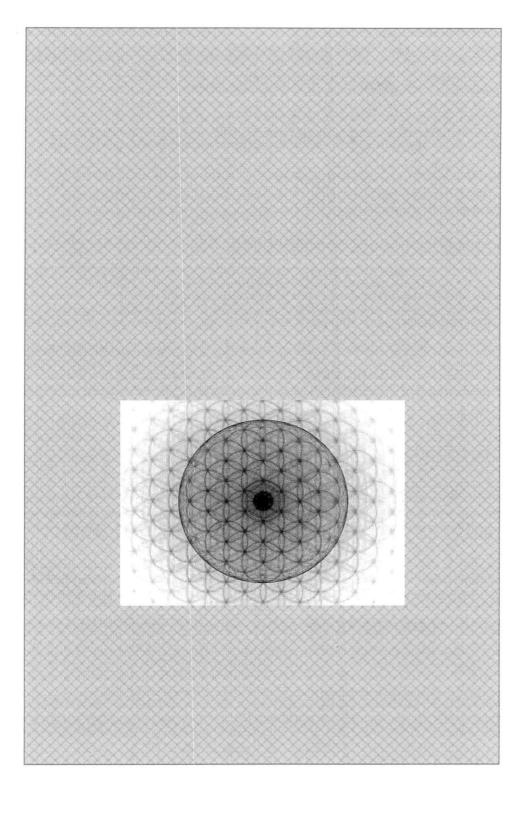

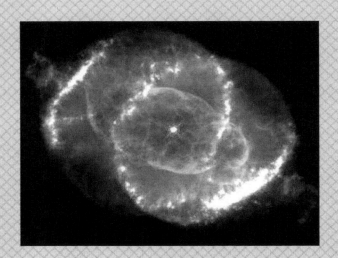

EPILOGUE

It ultimately comes down to action in our world . . . not the endless conversation. If you've read to the end of all this, your experience has been one of a lot of words and the thoughts/feelings that accompany them . . . but no real *action*. But thoughts precede action, just as consciousness precedes manifestation. I recently had an experience that fused all of these words into one, pure act.

The heart of the message from the bodhisattvic creator of our solar system is the Stillpoint phenomenon embedded in the Earth, Moon and Sun, and regards a possible global shift in consciousness – a rather significant act. The easiest way to open to this worldview-shattering information is its expression in the total eclipse of the Sun by the Moon. This is the interface where the implicate world most powerfully penetrates our day-to-day, physical reality. Until only very recently, my entire 'experience' of total eclipse has been contained within the solely intellectual understanding expressed in these pages and the photograph, and many others similar to it, included at the beginning of the book. I was aware of my limited understanding . . . but really had *no* idea how incomplete it was. Annie Dillard, in her 1982 book *Teaching a Stone to Talk*, captures this gap in understanding perfectly. She will help me tell this last story.

In February of 1979, Annie and her husband were living on the Pacific coast in Washington state and decided to drive east into the Yakima Valley to watch the upcoming eclipse. The story begins with observations of the trivia that filled their lives as they checked into a motel, and then:

> 'We would drive out of town, find a hilltop, watch the eclipse, and then drive back over the mountains and home to the coast. How familiar things are here; how adept we are; how smoothly and professionally we check out! . . . off we went, as off we have gone to a hundred other adventures.'

An event without parallel was experienced across the United States on August 21st, 2017 . . . a total eclipse of the Sun by the Moon traversed the entire country from Salem, Oregon to Charleston, South Carolina, taking a mere hour and a half from coast to coast. Perhaps as many as 20 million people witnessed the event . . . more human beings than any other time in history. And I am grateful that I could be one of them.

Annie had no idea. I had no idea.

As the days before the upcoming moment slipped away, each morning I watched the Moon before the Sun came up as she gradually moved east towards her rendezvous. Two weeks ahead of this meeting she was huge and full just before slipping beneath the Sierra on the western horizon first thing in the morning . . . the Sun immediately replacing her in the eastern sky. Each morning, as she grew daily smaller on her eastern journey, I found her higher in the dark early morning sky. Soon enough she was straight up and just above Orion's red shoulder Betelgeuse, and had waned to half of her fullness, at quarter Moon.

She now began her decent towards the eastern horizon, and before I knew it, she was at crescent phase and rising with a shining Venus above the White Mountains. I left my home on Friday morning, the 18th - my mother's birthday - on my way to Idaho, and camped near the Nevada/Oregon border. When I opened my eyes Saturday morning, there was Venus, bright and just-risen . . . the Moon, now just a sliver of her former self below, having passed her in her orbit round the Earth. Venus and the Moon bright in the eastern morning sky, now exactly a full hand's width apart with my arm fully extended . . . the Sun still below the horizon, but coming on.

VENUS-MOONRISE AUGUST 19TH, 2017

The farther I'd gotten away from my computer and phone and daily responsibilities, the more I was able to align with the sacred dance happening in the sky . . . a dance that happens so slowly relative to our frenetic lives that it is almost invisible. But now I was able to see and feel the full grandeur of the relentless, powerful, almost imperceptibly slow movement – the Earth on her axis, the Moon in her orbit and even . . . so very far away and only a feeling . . . the movement of Venus around the Sun. As I lay there, watching all this happen as the Earth slowly turned, the thought occurred to me that the movement could only be perceived if we stopped our own world to take notice. And not only that. It seemed to me that this gradual movement was at the very edge of human comprehension. One can only *just* perceive the Earth spinning on her axis towards the east – with the very gradual movement of the Moon only experienced at the daily morning checkpoints. This last was about to change . . . dramatically.

OUR SOLAR SYSTEM

But moving she was . . . also towards the east in her orbit around the Earth. In fact, all of the planets move in this direction – counter-clockwise as seen from above. As the Sun and planets condensed out of the hot cloud of gases that preceded their origin, this huge cloud mass began to rotate . . . as does everything in the manifested Universe. This is why the Sun, the Earth, and the Moon on their axes, as well as all the planets in their orbits – in *our* solar system - move in this easterly, counter-clockwise direction.

That day, like Annie and her husband, I drove towards the upcoming event to find a hilltop. Millions of people were gathering all along the route and signs of this began to appear the closer I got. Finally in the path, humanity was evident in all its glory . . . campers, trailers, sunglasses, shorts, lawn chairs, ribboned-off parking lots and rows of portable outhouses. The lady at the check-out counter in the grocery store had her 'Eclipse! August 21st, 2017' shirt on. The ambitious, anticipating the event, had tables set up with every imaginable eclipse accoutrement . . . hats, shirts, bracelets, belts with eclipse belt buckles, eclipse glasses and 'art'. There was a Burning Man type stainless steel sculpture of 'Eclipse Man' luring the curious into what was unabashedly and light-heartedly labeled 'Official Tourist Trap!' . . . a billboard across the street advertising the 'Family Eye Center' in nearby Payette, Idaho, with a girl wearing eclipse glasses staring, we assume, at the Sun, letting people know that 'we have viewing glasses & safety tips'. America was gettn' ready!

'ECLIPSE MAN!'

As much as I appreciated the excitement and anticipation of those who'd chosen to make whatever effort they did so as not to miss maybe the event of their lives, I knew that my own experience had to be one of complete isolation. Later that day I found an obscure, deeply rutted dirt road covered in sunflowers that led up to to a high ridge surrounded by a huge bend in the majestic Snake River . . . with a view that seemed to encompass the entire world and dead-center in the path of the approaching eclipse. I would camp on a shoulder of this ridge, and tomorrow - Sunday, the day preceding the eclipse - meet my son and his new wife of exactly one year ago today. I would bring them up to join me and to find their own quiet place to experience the fast approaching event with each other.

It was a joy to be high above the world, with 360° of surrounding views into the the seeming infinity of the beautiful, soft, rolling hills of late summer Idaho and Oregon . . . the deep silence of nature interrupted only by the occasional buzzing insect or zipping grasshopper. Far below, the ancient Snake wound its way through bright green fields short months from harvest; people driving here and there, going about their lives. I rolled out my bed for the night . . . going to sleep just after the Sun had set below the distant hills on the western horizon.

The day preceding the eclipse I awoke to the same view as the morning before . . . but this time, only Venus was visible in the eastern sky. Already, I was about to see something that I'd never witnessed – the rising of the last tiny sliver of crescent Moon, the day before she disappeared altogether, and became 'new' once again. Now, although days away from the pace of my normal, hamster wheel life, I was still impatient to see the first sign of her . . . and kept my eyes fixed so as not to miss it. It seemed like an eternity, but there she came . . . the very first glimpse I'd ever had of the tiny point at the end of one of her horns, arising ever so slowly now above the horizon. A spark, a point of light, a profound moment, her grand entrance ushered in by revolving Mother Earth. She'd covered a lot of ground in her orbit eastwards in the day intervening, distancing herself from Venus now high above her, and had almost disappeared into her renewal.

Sometime after this, the Sun made his appearance, and the day had begun.

I didn't want to leave, but needed to drive back to town to find Mike and Dylan and bring them back with me. We found each other, and each of us was feeling the growing excitement. We arrived back the next morning to the shoulder of what is called Indian Head Mountain with just minutes to spare before we needed to begin our ascent to the top. They were dumbfounded by the expanse. Up we went.

And so here it begins.

Annie had compared seeing a partial eclipse to a total eclipse as riding in an airplane to jumping out of an airplane. I'd seen an annular, or partial eclipse before . . . and was ready to jump. Still, words will never describe what jumping actually feels like, nor what happened.

We'd positioned ourselves far apart on the ridge that defined the top of the mountain, but still close enough so that we could yell to each other if need be. I took my shoes off, spread some tobacco as an offering (a friend taught me this long ago), and connected to Mother Earth. A long prayer, sometimes silent and sometimes in conversation, hands on earth, eyed closed. I was ready . . . I thought.

I put my eclipse glasses on and looked up. The event had just begun . . . the first black sliver disappearing the upper right area of that blinding disk we forever take for granted in the sky above. I yelled to them that it was beginning. I returned to my own private, silent space and was instantly mesmerized by what was happening. I couldn't turn my eyes away.

Once again, my daily pace was immediately challenged by the almost painfully slow advance of the Moon . . . but I kept my gaze constant – and I began to merge with this ancient, cosmic, ultimately dignified cyclic pavan of Sun and Moon and Earth. But I am ever so human and, as in meditation, my mind chattered away, interrupting my intention to become one with the almost imperceptible movement. In time, my chatter dropped away, leaving me in total, grand, infinite silence with our ancestors.

And this is where my jump into the unknown truly began.

I began to see the blinding circle of Sun, and the emerging black circle of the Moon sliding across it, not as two-dimensional disks, but as spheres . . . as three-dimensional *orbs* far far far away in the endless space. I could *feel* the distance now separating all three of us . . . me sitting on the surface of Earth, the Moon hundreds of thousands of miles away, the Sun many millions of miles beyond that . . . each coming into impossibly *precise, proportional, intentional* alignment with each other. I could *feel* the distance and the 'see' the spheres now . . . and felt at one with a kind of unspeakably profound, stately, ritual - an intentionally choreographed performance like nothing I'd ever experienced in my life, still painfully slow to this impatient human, but so very clearly, and eternally, real - and happening right *now*. I was being attuned to an entirely different frequency or vibration . . . I was being brought into resonance, entrained, with something way beyond human. I surrendered.

The almost inconceivable distance had transformed the speed of the Moon's orbit around the Sun . . . which is almost precisely *three times the speed of sound*, that now familiar whole number proportion . . . to a movement almost imperceptible to discern. My God . . . the scale of all this! It was impossible at this point not to feel the presence of something that was infinitely greater than myself.

I continued watching this grand drama unfold and then, precisely half way to totality, the vesica pisces appeared . . . the circumference of each circle precisely intersecting the center of the other . . . the most basic pattern in the Flower of Life, Stillpoint, geometry of consciousness. A teaching on many levels was happening before our eyes. But this aspect, the intellectual, was not available to those who lived so long ago.[269] I was soon to find out that humanity had been prepared at the instinctive/emotional level beyond my imagining for thousands of years.

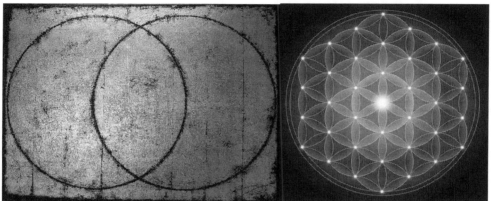

VESICA PISCES . . . SUN / MOON   FLOWER OF LIFE / STILLPOINT

In the back of my mind, I knew that behind me, to the west, the shadow of totality would be fast approaching . . . the Moon's orbital speed of 2,288 miles per hour translating to around 1,880 miles per hour (!) on the ground (in Idaho)[270] because the Earth is also spinning in the same direction. This was the main reason that Mike had wanted to find a high hill from which to view the event so that we could see the shadow approaching. I'd tried hard to imagine what this would look like. I'd seen many well-drawn animations of this clearly defined, polite circle moving across the landscape. I tried to imagine how 1,880 miles per hour would appear in real time.

The sky had been ever so slowly becoming darker and darker over the last hour, and in the last few minutes this darkness began to take on an other-worldly quality. But I'd become totally lost in the eclipse itself and had forgotten to look behind me. Suddenly Mike called out and I tore myself away. He was pointing towards the west. He'd caught the very beginning of the shadow in the distant western hills. By the time I turned around that distance had been halved. Before I was totally aware of what was happening, it was upon us.

Oh my God. *This* I hadn't expected in *any* way. The word 'shadow', while accurate, does not begin to describe the experience of what was happening. There was no polite circle rapidly approaching at some unclear speed. The diameter of this 'circle' (the way I'd always imagined it) was 60 to 70 miles across – but I'd seen it animated from a perspective high in the clouds. What it meant to us on the ground was that the entire western horizon was filled with an ominous, lightning-speed approaching, indescribable, incomprehensible . . . Darkness. What we were experiencing was in fact instantaneous immersion in a darkness that was tangible . . . a darkness that seemed to have mass and impregnated *everything* . . . almost as though being consumed by some black, Stygian presence.

There is nothing that could possibly have prepared me for this experience.

This was visceral . . and radically different than *anything* I'd ever imagined or experienced. What had been a warm 90° instantly turned cold with the disappearance of the Sun. No wind. No movement. The grasshoppers still. Silence. How fragile are our lives.

> 'For behold, darkness shall cover the earth, and thick darkness
> the peoples; but the Lord will arise upon you, and his glory will be
> seen upon you.'[271]

It is so easy now to understand these words from Isaiah almost three millennia ago - this darkness was that huge. *My* world was *completely* enveloped in a darkness that had no familiar memory with which to dismiss it into convenient denial. If I'd lived back then, I'd have no assurance that this wasn't the end. Now, as Modern Man, I knew of course that this would pass and knew exactly what was happening. But not then. Still, the unparalleled uniqueness of this moment had me sobbing, breaking down in unexpected emotion . . . not of sorrow, not of joy, but of holy awe. The showmanship!

The land itself became iridescent and magical and surreal beyond description. The world now glowed in what Annie had described as 'platinum . . . a color never seen on Earth.' I'd been transported to another world. I slowly spun around in a circle with my arms spread wide, not understanding what I was seeing or experiencing, but reveling in absolute awe.

In the back of my mind was the knowing that this phenomenon could never be happening were the Sun and Moon not expressions of the Stillpoint geometry - the geometry of consciousness . . . the primal geometric blueprint upon which all of manifestation is based and the very heart of sacred geometry and, most significantly, a geometry that does not manifest anywhere in the observed Universe, except here in our solar system. The fact that the Sun's diameter just happens to be precisely 400 times larger than the Moon's makes this so.

Regardless, what I and millions of other human beings were experiencing . . . and billions of other human beings over many thousands of years in the distant past had experienced . . . would *still* not be possible were the Moon – moving ever so gradually away from the Earth for the last 4.6 billion years - not now within the window where it was also precisely 400 times closer to the Earth than the Sun at totality. For the majority of this vast span of time, *perfect*[272] total eclipse on the face of the Earth was not possible . . . but humans are alive within the small window when witnessing – and understanding – this rarest of events *is* possible. Only the combination of these two phenomena happening simultaneously made this awe-inspiring moment a reality . . . and only on the surface of Mother Earth when and where we live.

This clearly was a communication on the grandest scale . . . and it was penetrating my bones.

Yes . . . I was reveling in absolute awe.

'In the last sane moment I remember,'[273] I somehow thought to grab my camera and looked up and took this photograph:

PERFECT TOTAL ECLIPSE: AUGUST 21ST, 2017

Extraordinary! . . . but how can I possibly tell you how totally inadequate this image is, relative to all that was happening? As Annie said so clearly:

'We had all died in our boots on the hilltops in Yakima, and were alone in Eternity.'

After taking the photo, I lost myself once again in the eclipse . . . now total. The sphere of the Sun . . . its diameter exactly 400 times larger than the Moon . . . *perfectly* covered by the sphere of the Moon . . . exactly 400 times closer to my spot on the surface of the Earth than the Sun during a moment encompassing many millions of miles, the precision of this cosmic design screaming to be understood. All of the all-but-invisible movement stopped for a full 2 minutes while this alignment froze time *in the particular place that I stood.* The world *stopped* for these two minutes. The tiny bullseye moving over the land at blinding speed, visiting the entire populated world at one time or another every eighteen or so months over vast amounts of time, never visiting the same place in less than *fifteen generations* – froze time in place. The world as we've always known it, regularly transformed into a magical, mystical, mythical, iridescent landscape of another world. A constant reminder that we cannot make assumptions about our world, that something else is going on. All at once. All new. All now. Nothing to compare it to.

There was no doubt that this was a powerful communication without parallel . . . a purely symbolic act performed in empirical, scientific accuracy, mirroring the Stillpoint/Movement aspects of reality on the grandest of stages, constantly and periodically flooding the populated world with a wake-up call for thousands and thousands of years. This was no 'happy coincidence'.

In my focus on the eclipse itself, I'd forgotten to watch for the night sky filled with stars and planets that I'd been told would appear. Surrounding this pure alignment of Earth/Moon/Sun, Mercury and Mars were very close by, and Regulus, Leo's brightest star, close enough to almost be within the Sun's corona, with Venus and Jupiter not far. The eclipse, happening just before noon, was high in the sky . . . Venus was at mid-heaven, just as it was at the time of my birth . . . cycles upon cycles upon cycles. I saw none of this . . . but they were all there none the less, adding their influence in graceful astrological precision.

And *then!* . . . out of this darkness . . . a pinpoint explosion of light appearing on the upper-right side of the Moon . . . a pinpoint of exploding light. And this explosion continued to expand at the precise timing of the almost imperceptibly slow movement . . . the movement itself, as well as the growing light, a seamless, slow-motion continuity of perfection. The wave-like connectedness of time and light . . . the similarity of the initial explosion of the Limitless Mind of God through the Point of Creation into manifestation inescapable . . . all seamless, expanding, growing . . . the other-worldly platinum ever so gradually replaced by golden light - '. . . but the Lord will arise upon you, and his glory will be seen upon you'. Indeed.

The world was not ending . . . but returning. And how grateful I was.

And how grateful billions of human beings have been over the last many many thousands of years. In olden times, there was no way to understand what had just happened, no way in the world to know that the Moon had just slid across the

Sun . . . only that our familiar world had been consumed in a Darkness we'd never known – a dragon perhaps? . . . never to see the Light again, never to return to the lives that we'd somehow managed to arrange in ways that made sense – except that the 'Lord' made it so.   Annie traveled back to that time when we had no way to know. From *Teaching a Stone to Talk*:

> 'The grass at our feet was wild barley. It was the wild einkorn wheat which grew on the hilly flanks of the Zagros Mountains, above the Euphrates valley, above the valley of the river that we called *River*, we harvested the grass with stone sickles, I remember. We found the grasses on the hillsides; we built our shelter beside them and cut them down. That is how we used to look at them, that one, moving and living and catching my eye, with the sky so dark behind him, and the wind blowing. God save our life.'

'Grandfather have mercy, have mercy have mercy.' In hindsight, and with the eye of our scientifically informed world, it is easy to see how even *this* unequaled phenomenon has become humdrum as we rush back to cling to our familiar world or, as Annie said:

> 'The mind wants to live forever, or learn a very good reason why not.   The mind wants the world to return its love, or its awareness; the mind wants to know all the world, and all eternity, and God.   The mind's sidekick, however, will settle for two eggs over easy.'

I realize now that I have been living side-by-side and surrounded by those whose yearning goes not much further than deciding what to have for breakfast or endless variations of such. I have blamed myself mercilessly for my complete failure to communicate the literally incomparable import of the Stillpoint phenomenon. But I now realize that my meager abilities perhaps had no chance to begin with . . . given the fact that humanity has been bathed in steady bombardment of such belief-shattering reality for its entire existence.

It has made no difference. The subtlety and elegance of this communication is sublime beyond our present ability . . . beyond our ken.

Still . . . I dare to hope.

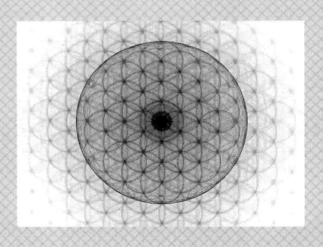

Appendix A

A Review of the Stillpoint Geometry embedded
within the Sun, Earth and Moon

For a 3d animation of the following 2d geometric proof, upon which all of these pages are based, see www.stillpointdesign.org.

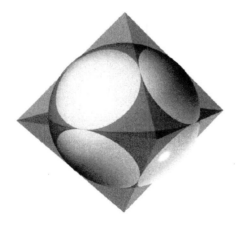

1. Sphere / Octahedron

The Sphere/Octahedron interface, creating or being created by the Vector Equilibrium through the intersection of the sphere with the midpoints of the *edges* of the Octahedron (twelve points around one) . . . the purest expression of the perfect balance of polarities that describe the Universe. Spirit/Form.

Seen in 3 dimensions: twelve interpenetrating spheres around one, twelve points around one . . . all points equally distant from each other - the Vector Equilibrium, the geometric expression of the perfect balance that science describes the universe to be - the geometry of the Stillpoint.

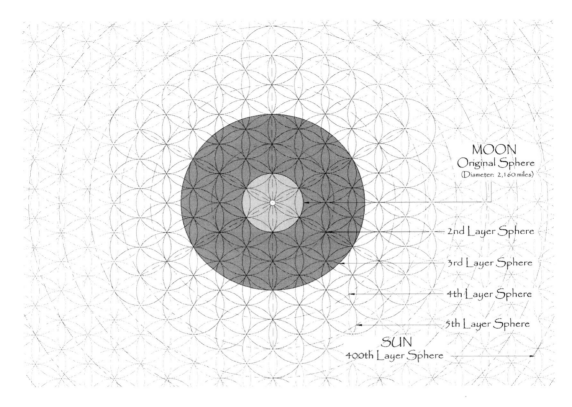

MOON
Original Sphere
(Diameter: 2,160 miles)

2nd Layer Sphere

3rd Layer Sphere

4th Layer Sphere

5th Layer Sphere

SUN
400th Layer Sphere

2. Flower of Life . . . Infinity

Seen in 2 dimensions, the original sphere, the 1st layer, is dimensioned to the diameter of the Moon. The 3rd layer sphere is accented. The center of this matrix is the Stillpoint . . . and is everywhere. This original sphere is multiplied in all directions through the Flower of Life/Stillpoint geometry . . . 12 point/spheres around the central point – infinity. If the original sphere is considered to be the first layer, all of the related layers of interpentetrating spheres are multiples of the original: first layer, second layer, third layer, etc. The 400th layer sphere is the diameter of the Sun.

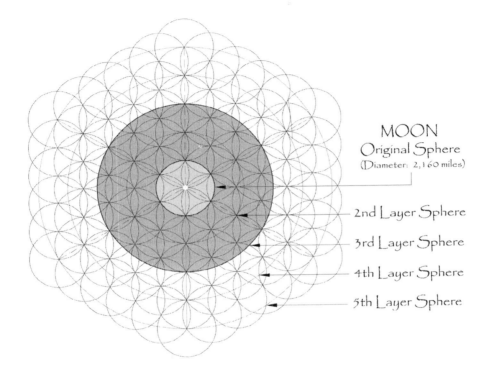

MOON
Original Sphere
(Diameter: 2,160 miles)

2nd Layer Sphere

3rd Layer Sphere

4th Layer Sphere

5th Layer Sphere

3. 64-SPHERE/TETRAHEDRON MATRIX

The original 64 interpenetrating spheres around the original point, the stillpoint, similar to the 64-cell stage of our own experience 7 days after conception. Each of the spheres circumscribes a tetrahedron (see #3 next page). The tetrahedrons embedded within each sphere orient the geometry . . . culminating in the double-penetrating 'star tetrahedron,' represented in two dimensions by the 'Star of David' geometry, depicted in 3A and 3B below. The Kabbalah's *Tree of Life* and the Chinese *I Ching* are derived from this geometry.

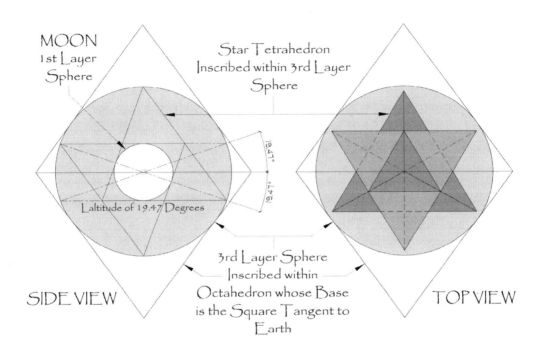

MOON
1st Layer
Sphere

Star Tetrahedron
Inscribed within 3rd Layer
Sphere

19.47°
19.47°

Latitude of 19.47 Degrees

3rd Layer Sphere
Inscribed within
Octahedron whose Base
is the Square Tangent to
Earth

SIDE VIEW

TOP VIEW

4. STAR TETRAHEDRON

4A. SIDE VIEW   4B. TOP VIEW

Moon Sphere tangent to faces of tetrahedrons . . . points
of tetrahedrons define 3rd Layer Sphere

Here is the introduction of an octahedron tangent to the 3rd layer
sphere at its faces. Within this 3rd layer sphere, two tetrahedrons are also
introduced with the axis of each tetrahedron aligned with the vertical
axis. If a sphere is created that is tangent to the two horizontal planes of
the tetrahedrons – which intersect the 3rd layer sphere at 19.47 degrees –
the diameter is that of the 1st layer sphere – in this case, the Moon.

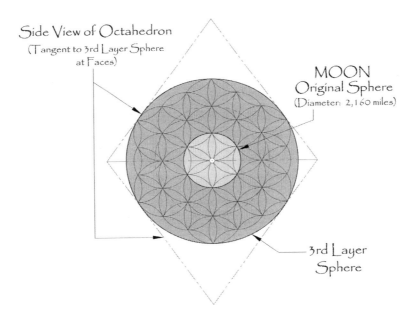

Side View of Octahedron
(Tangent to 3rd Layer Sphere
at Faces)

MOON
Original Sphere
(Diameter: 2,160 miles)

3rd Layer
Sphere

5. FLOWER OF LIFE

Here, the Flower of Life geometry is added, showing also how the 3rd layer sphere
is derived.

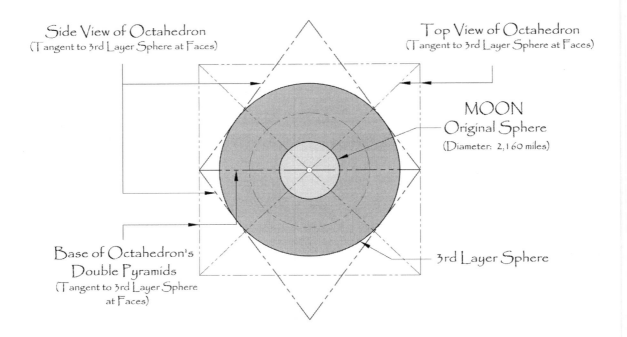

Side View of Octahedron
(Tangent to 3rd Layer Sphere at Faces)

Top View of Octahedron
(Tangent to 3rd Layer Sphere at Faces)

MOON
Original Sphere
(Diameter: 2,160 miles)

3rd Layer Sphere

Base of Octahedron's
Double Pyramids
(Tangent to 3rd Layer Sphere
at Faces)

6. SIDE VIEW OF OCTAHEDRON
An Octahedron circumscribes the 3rd layer sphere, the *faces* of the
Octahedron tangent to the sphere.

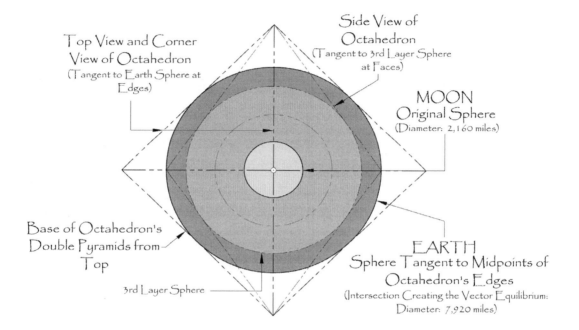

Top View and Corner
View of Octahedron
(Tangent to Earth Sphere at
Edges)

Side View of
Octahedron
(Tangent to 3rd Layer Sphere
at Faces)

MOON
Original Sphere
(Diameter: 2,160 miles)

Base of Octahedron's
Double Pyramids from
Top

3rd Layer Sphere

EARTH
Sphere Tangent to Midpoints of
Octahedron's Edges
(Intersection Creating the Vector Equilibrium:
Diameter: 7,920 miles)

7. Top & Corner View of Octahedron

The Octahedron from different angles – from corner and top -
with the addition of another, larger sphere, tangent to the *midpoints*
of the *edges* of the Octahedron.

The diameter of this sphere is the diameter of Earth.

The 12 points of the intersection between this Earth sphere and
the Octahedron is the Vector Equilibrium (see Sphere-Octahedron
graphic above). Both the Flower of Life, and the Vector Equilibrium
generated by the intersection of the Octahedron and Earth sphere,
are expressions of the Stillpoint geometry.

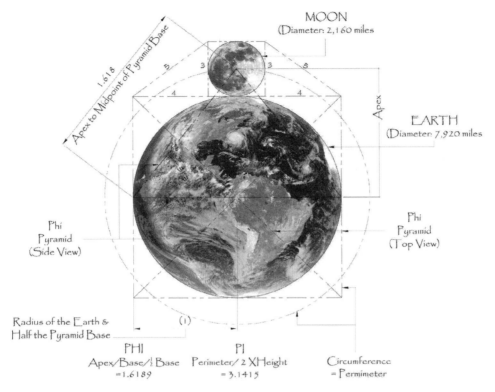

8. EARTH / MOON RELATIONSHIP

Beginning with the Earth circle (sphere), a square (cube) circumscribes it. A larger circle (sphere) is created that has the same circumference as the perimeter of the square (called 'squaring the circle' or 'marrying Heaven and Earth' in sacred geometric terms).

A point is created on this larger circle directly above the center of the Earth circle, and this point creates a triangle whose base is the diameter of the Earth. The triangle is a section of Egypt's Great Pyramid. It incorporates the proportions of Phi (the golden ratio, the proportion of life and growth) and Pi (the proportion that transforms a line into a circle).

A smaller circle is now drawn whose center is the apex of the pyramid, tangent to the circle of Earth. The square that circumscribes this circle creates the Pythagorean 'Builder's Triangle' - 3, 4 and 5 - and the 90° angle.

The diameter of this smaller sphere is the diameter of the Moon . . . and 'we arrive where we started . . .'

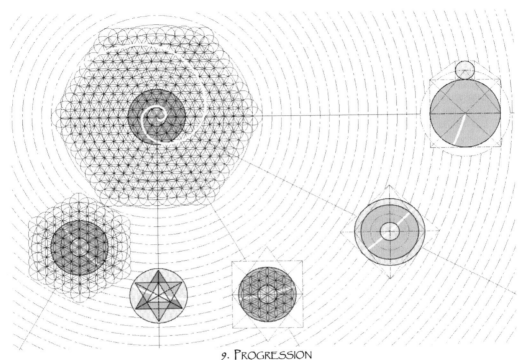

9. PROGRESSION

Summary demonstrating the progression from the infinity of the Flower of Life/Stillpoint geometry - the original sphere dimensioned to the diameter of the Moon, through the initial 64-sphere/tetra*hedron matrix of the expanded, three dimensional Flower of Life, through the creation of an Octahedron tangent to the 3rd layer sphere at the Octahedron's *faces* (generating the eight points of the Star Tetrahedron), through the creation of a larger sphere, tangent to the midpoints of the *edges* of the Octahedron – whose diameter is that of Earth, to the creation of the Earth/Moon Diagram.

The 1st layer sphere is the Moon. The 3rd layer sphere connects the Moon and Earth through being tangent to the octahedron at its faces (3rd layer sphere) and edges (Earth). The 400th layer sphere is the Sun. *There is no reason in physics or in all of science to explain this phenomenon. It is outside the laws of sacred geometry. It is far outside any law of chance or accident. It is not necessary for life . . . it is the stillpoint geometry describing the eternal present moment of awakened consciousness . . . the matrix underlying all that is and isn't.*

Appendix B:

The Message in Detail:
(From *Who Built the Moon?*, by Christopher Knight and Alan Butler)

366
The number of rotations in an Earth year
366
The number of Megalithic Yards in 1Mg second of arc of the Earth
366%
The percentage size Earth to Moon
400
The ratio of the size [diameter] of the Moon to that of the Sun
1/400th
The number of times the Moon is closer to the Earth than the Sun
40,000
The number of Megalithic Yards in 1 Mg second of arc of the Sun
40,000
The number of kilometers the Earth turns on its axis in a day
400
The number of kilometers the Moon turns on its axis in a day
10,000
The number of days in 366 lunar orbits
100
The number of Megalithic Yards in 1 Mg second of arc of the Moon
400
The number of times the Earth rotates faster than the Moon
109.28
The ratio of the size [diameter] of the Earth to that of the Sun
109.25
The number of Earth diameters across the diameter of the Sun
109.26
The number of solar diameters across the Earth's orbit at aphelion
27.322
The sidereal days in 1 lunar orbit
27.322 x 4 = 109.2
27.322%
The percentage size Moon to Earth
10,920.8
The size [circumference] of the Moon in kilometers

Appendix C:

Following are more complete descriptions of the six numbers written about by Sir Martin Rees (there may be as many as 36 of these constants) which, if each was not *precisely* as it is, we would not be here to discuss them. Also explained in more detail are some of the many 'coincidences' unexplained by science, sacred geometry, chance, and not required by life that are found in our own solar system within this Universe. The last list included in this Appendix is the full list of the 20 factors listed in *The Privileged Planet* that limit the estimated amount of habitable planets in the Milky Way Galaxy.

Six Numbers upon which our Universe Depends for its Existence:

These six numbers are found in the Universe's smallest and largest structures. Two relate to basic forces, two determine the size and large-scale texture of the Universe, and two fix the properties of space itself. They are:

> 1. E (Epsilon) - the nucleus of a helium atom weighs 99.3 percent as much as the two protons and the two neutrons that fuse to make it. The remaining .7 percent is released mainly as heat. So the fuel that powers the Sun - the hydrogen gas at its core - converts .007 of its mass into energy when it fuses into helium. That number is a function of the strength of the force that binds atomic nuclei together and determines how all atoms on Earth are made.
>
> If the number were only a mite smaller - .006 instead of .007 - a proton could not bond to a neutron, and the Universe would consist only of hydrogen. No chemistry, no life. And if it were slightly larger, just .008, fusion would be so ready and rapid that no hydrogen would have survived from the Big Bang (for the purpose of this preface and paper, science's best guess regarding the origin of our Universe, the Big Bang Theory, will be referred to . . . but, in fact, it is the idea of the stillpoint at the center of the torus, through which energy eternally cycles that is a more accurate description reality). No solar systems, no life. The requisite number perches, precariously, preciously, between .006 and .008.
>
> And this is just the first of Rees's six numbers. If you add to the equation the other five, life and the structure of the Universe as we know it become unlikely to an absurd degree. Astronomer Hugh Ross has compared the state of affairs to 'the possibility of a Boeing 747 aircraft being completely assembled as a result of a tornado striking a junkyard.'

2. N - equal to 1,000,000,000,000,000,000,000,000,000,000,000,000. The number measures the strength of the forces that hold atoms together divided by the force of gravity between them. It means that gravity is vastly weaker than intra-atomic attraction. If the number were smaller than this vast amount, 'only a short-lived, miniature universe could exist,' says Rees.

3. Ω (Omega) - which measures the density of material in the Universe - including galaxies, diffuse gas, and dark matter. The number reveals the relative importance of gravity in an expanding Universe. If gravity were too strong, the Universe would have collapsed long before life could have evolved. Had it been too weak, no galaxies or stars could have formed.

4. Λ (Lambda) - the newest addition to the list, discovered in 1998. It describes the strength of a previously unsuspected force, a kind of cosmic anti-gravity that controls the expansion of the Universe. Fortunately, it is very small, with no discernible effect on cosmic structures that are smaller than a billion light-years across. If the force were stronger, it would have stopped stars and galaxies - and life - from forming.

5. Q - which represents the amplitude of complex irregularities or ripple in the expanding Universe that seed the growth of such structures as planets and galaxies. It is a ratio equal to 1/100,000. If the ratio were smaller, the Universe would be a lifeless cloud of cold gas. If it were larger, 'great gobs of matter would have condensed into huge black holes,' say Rees. Such a universe would be so violent that no stars or solar systems could survive.

D - the number of spatial dimensions in our Universe – that is, three. 'Life could not exist if it were two or four,' contends Rees.

PHENOMENA WITHIN OUR OWN SOLAR SYSTEM:

1a. The *precise* dimensions of the Earth and Moon and Sun fit into some of the most sacred geometry known (*see Earth/Moon Relationship, page 7 & 179*). There is no law in physics that determines the size of celestial bodies orbiting other celestial bodies. There are sixty-five moons orbiting seven of the nine planets in our solar system (two do not have moons), and *there is nothing even remotely close to the proportions found in this relationship.* This is 'impossible' within the 'laws' of our Universe - whether they be physics or mathematics or geometry – or by any conceivable law of chance. This implies that this phenomenon is intentional on some level. This graphic relates the dimensions of the Earth and Moon to the ancient sacred geometry of marrying Heaven and Earth, Spirit and Matter . . . squaring the circle. In three dimensions, this relationship, the joining of the octahedron generated by the square with the sphere generated by the circle, leads directly to the geometry of the Vector Equilibrium (*see Number:, page 195*), which is the geometry of the stillpoint at the center of all things (*see Sphere/Octahedron, page 11*). The graphic relates these dimensions as well to the cosmic, mystical geometric proportions of π (pi – the mystical proportion relating the straight line to the circle) and φ (phi – the proportion found everywhere in *life* - the proportion of life itself) found in the phi-pyramid . . . as well as to the mystical, virginal number seven, also found in the phi-pyramid (it is almost precise in dividing the circle into sevenths).

1b. The *precise* dimensions of the Earth and Moon and Sun are encoded within the Flower of Life – an ancient two dimensional symbol (possibly 13,000 years old, from a time before the Great Flood, laser burned into the granite walls of an Egyptian temple dedicated to resurrection by a technology only 'discovered' in the last century) that, when expressed in three dimensions, again refers to and includes the Vector Equilibrium - the geometry of the stillpoint at the center of all things. Here again, the fact that the Earth and Moon dimensions are found yet again in another example of the most sacred of all geometry is 'impossible' . . . not explained by mathematics, physics, geometry - or chance . . . leading us to the stillpoint.

1c. The tetrahedron is the first and most primal three-dimensional expression. It is the first form . . . where Creation first expresses itself, or becomes. Unlike *any* of the other polyhedrons, the proportional relationship between the sphere that circumscribes it, touching at the apexes, and the sphere that is inscribed within it, touching at the surfaces, is 3:1 . . . a pure whole number proportion. There is something very special about this relationship, and it is the basis upon which physicist Nassim Haramein was able to directly connect in precise geometric expression, the 64-Tetrahedron/Sphere Matrix (*see Barbary Castle/Osirian Temple/64-Tetrahedron/Sphere Matrix, Appendix E, page 336*) with the Chinese I Ching (the short and long lines of the 64 tetrahedrons equals the exact number of short and long lines of the 64 Hexagrams of the I Ching), the Egyptian 'Flower of Life,' and the Hebrew 'Tree of Life' (this geometric two-dimensional figure can be laid directly over a section of the two-dimensional representation of the 64-Tetrahedron/Sphere Matrix, as well as the fact that the 32 paths of the 'Tree' are equal to the 32-Tetrahedron/Spheres that form the inner layer of the 64-Tetrahedron/Sphere Matrix in the shape of the Vector Equilibrium . . . the geometry of the stillpoint). It turns out that if the smaller sphere, the sphere inscribed *within* the tetrahedron is scaled to the dimensions of the Moon, the larger outer sphere, or 3rd layer sphere, that touches the tetrahedron's apexes is equal to the large circle/sphere of the 'Flower of Life' which in turn is equal to the sphere which is *inscribed* within the octagon whose base is the square that circumscribes the circle of Earth *(see Barbary Castle/Osirian Temple/64-Tetrahedron/Sphere Matrix, Appendix E, page 336)*.

All this really says is that, once again, the improbable dimensions of the Earth and Moon are synonymous with the proportions of the geometry of the stillpoint . . . this time including the primal three-dimensional form, the tetrahedron. *And* . . . the 400th layer sphere equals the diameter of the Sun, 'impossibly' including it also within this most sacred of all geometries.

Note: These three initial examples are listed as 1a, 1b and 1c because they are all related through, and can be generated through, shared geometry. Yet, taken as a whole, their improbability is unrelated to the following six examples, as is each following example unrelated to any other.

2. The Sun and the Moon are almost *precisely* the same size in the sky at the time of total eclipse. This phenomenon exists because the Sun's diameter is almost exactly 400 times larger than the Moon's (an astonishing fact in itself, given that it is a remarkable whole number proportion, proportions which have inherently sacred properties themselves *(see Full Solar Eclipse, page 8 and Proportion, Vibration & Balance:, page 189, and The Harmonograph, page 190)* and is almost exactly (within the small difference in the parameters of the shortest and longest orbits of the Moon around the Earth) 400 times the distance of the Moon to Earth away. This means that the Moon can *precisely* eclipse the Sun, exposing the Sun's corona, permitting some essential scientific discoveries. The amount of play in this proportion (inside of which the Moon completely covers the Sun and outside of which the Sun is visible around the Moon's edges) is approximately 1:5000. This fact again is 'impossible' . . . i.e. not explained by the laws of mathematics, geometry, physics . . . or chance. As mentioned earlier, there are 65 moons in our solar system orbiting seven of the nine planets and nothing even comes close to this proportional relationship. That is 'how' it happens to happens to happen, but not 'why.' More than any other significance, the geometry of the stillpoint, the Vector Equilibrium, *is* Unity. Expressed in our three-dimensional world, it is the sphere. Here again is the almost impossible likelihood of this relationship occurring as it does, while actually physically *entraining* *(see Frequency/Vibration/Sound/Resonance:, page 226)* our reality to Unity, in this instance through the repetition of cycles.

It is interesting to note here, regarding the properties of whole number proportions mentioned above, that part of the 'how' that makes the example in #1 above work geometrically – not the 'why' – is that, again outside of any laws of mathematics, geometry, physics, etc., the proportional relationship between the Earth and the Moon is *precisely* 11/3 . . . it doesn't *have* to be this way. The diameter of the Earth, 7,920 miles/the diameter of the Moon, 2,160 miles equals 3.6666667 etc., or 11/3.

Also, noted in the 2004 book *The Privileged Planet*, the eclipse of the Sun made two incomparably important scientific discoveries possible: the discovery of the makeup of the Sun's atmosphere, permitting science to extrapolate the makeup of distant stars, advancing cosmological research about our origins and the nature of the Universe, and the verification of Einstein's Theory of Relativity by observing that the gravitational pull of the Moon bent light rays from stars behind the Sun that wouldn't have been visible otherwise. *The Privileged Planet's* major premise is research verifying that the conditions in space permitting habitability also were the very conditions that permitted scientific discovery.

Regarding other observations about the Moon from *The Privileged Planet*, it was noted that not for the relatively large *size* of the Moon relative to the Earth, we also would not be here to experience it. The strong gravitational pull of the Moon on the Earth regulates the Earth's atmosphere by stabilizing the tilt of the axis, permitting the seasons and controlling the flow of warm and cold water in the oceans – everything so remarkably fine-tuned for life to exist and flourish.

Also from this same source of information, is the observation that of the vast spectrum of electromagnetic radiation emitted throughout the Universe (from very high frequency gamma rays, through x-ray, ultraviolet, visible, infrared, microwave, and finally radio waves at the lower end of the frequency scale) only the very tiny sliver of visible light familiar to us (one trillionth of a trillionth of the entire spectrum!) proves essential to plants, animals and human beings . . . almost the entire rest of the spectrum is useless to organic life. This is the *same* spectrum that is most informative to science regarding its discoveries of various structures in the Universe around us.

The Privileged Planet's point is that none of this discovery of the Universe is necessary for Darwinian survival . . . for life. All this implies that a purpose required that the Universe contain 'observers who can discover.' The basic premise upon which all these pages is based, is that the purpose generating all we know and see is the evolution of consciousness (not simply the attainment of knowledge through discovery).

It must also be mentioned that when the Moon was created it was next to the Earth and has been gradually moving away from the Earth for 4 ½ billion-years. The relatively small window where the total eclipse can happen began some 50 million years ago and ends in about 50 million years . . . about 2% of the total time since its creation and . . . 'fortunately' . . . this is perfectly timed to Man's existence on this planet.

3. The Ecliptic, the more or less flat plane described by the planets' orbits around the Sun, is *precisely* at a 60-degree angle from the plane of the Milky Way

Galaxy . . . it could be *any angle*. I'd seen in many books I'd read and websites I'd visited that this angle is 60-degrees, but I wasn't able to discover how anyone came to this conclusion.

I asked a friend who is a mathematician/astronomer if he was aware whether of not this was the exact angle. Not finding any references that gave a value for the angle between the Galactic Equator and the Ecliptic, he calculated it himself in the following manner.

He began with a value of 62.6 degrees as the angle between the Galactic Equator and the Celestial Equator (*not* between the Galactic equator and the Ecliptic) having found this value on several web sites and also in the book, *Celestial Mechanics* by Laurence Taff. Using the spherical triangle formed by the intersection of the Galactic Equator, the Celestial Equator and the Ecliptic and using the fact that the Celestial Equator and the Ecliptic meet at a 23.5 degree angle (the angle of the Earth's axis) and the fact the Celestial Equator and the Galactic Equator meet at a 62.6 degree angle, plus the arc lengths of the sides of the spherical triangle read from maps in the *Uranometria 2000.0* (which he had to physically scale), he was able to solve the remaining angle, which is the angle between the Galactic Equator and the Ecliptic.

Using the Law of Sines for a spherical triangle while using two different pairs of data gave the following values: 59.97 degrees and 60.63 degrees. Using the Law of Cosines gives the following value: 60.23 degrees. He said that the small difference in these values probably reflects the errors in estimating the lengths from the 'Uranimetria').

I mention all this to make it very clear that this number has factual basis . . . because *this* number reflects above all other numbers and relates more than any other number to the geometry of the Vector Equilibrium embedded within the geometry of the Flower of Life (*see Page 51, of John Martineau's A Little Book of Coincidence in the Solar System*).

It is my understanding that over 200 other 'planetary' systems have been detected by the observation of 'wobbles' in the star they orbit. Because of the importance of determining, if possible, whether this remarkable phenomenon (as well as related phenomena in #'s 4 and 5 that follow) is the rule or the exception, I contacted Dr. Brad Carter, an eminent astronomer/physicist specializing in the discovery of other solar systems who lives and teaches in Australia, and asked him if he found any alignments or orientations at all close to ours. In no uncertain terms, he told me that all observed phenomena relating to the orientation and relationship to the center of the galaxy was *random*. This means that, given the angle's uncanny *preciseness* in relation again to the stillpoint geometry, and the unlikeliness that the angle is *precisely* 60-degrees – and not 49 or 26 or 7.2 or 53.7 or . . . whatever – puts it comfortably into the company of the previous examples.

4. The fact that the Ecliptic is in alignment with the center of the galaxy in the first place. Here it seemed possible to me that the huge amount of gravity at the center of our galaxy could entrain this alignment to be as it is . . . but Dr. Carter, again in no uncertain terms said no . . . way too far away to make any difference on that scale.

What *is* happening, regardless, is an *entrainment* of another kind, again relating directly to the stillpoint. Because of the slant of the Earth's axis, seasonal time is demarcated as we orbit the Sun. The four cardinal points of this orbit are spring and autumn, when the Sun crossed the equator, and summer and winter when the Sun reaches its yearly limit above and below the equator at the Tropics of Cancer and Capricorn.

The Winter Solstice, the moment the Sun touches the Tropic of Capricorn, is traditionally the beginning of the year . . . the shortest day and the longest night in the northern hemisphere, and the moment of the literal and symbolic return of the light. Because of the slant of the Earth's axis and the 26,920 year cycle of its wobble, the Great Year, the Earth's continual movement around of the Sun permits the alignment of the rising of the Winter Solstice Sun to be marked against the background of the fixed stars.

Because, and *only* because our ecliptic happens to be aligned with the *precise* center of the galaxy out of an infinity of possibilities, a conjunction of the alignment of the Earth's axis with the Sun (the Winter Solstice) and the center of the galaxy is even possible. The ancient Mayans (*see Winter Solstice . . . 2012:, page 251, and Dragon/Serpent . . . Life-Force:, page 279*) believed that, while the Sun is our physical source of life, the center of the galaxy is the spiritual source. Certainly it represents *our* center of the highest concentration of energy of *any* kind this close to us, other than the Sun.

The nature of the magnetic field/vortex around the Earth pulls energy in at the poles. The Mayans created a remarkable calendar based solely upon the event of the conjunction of the Winter Solstice with the galactic center . . . when the Earth's axis is in alignment with both the energy of the Sun *and* the energy of the Galactic Center. The end-date of that calendar is December 21st, 2012 . . . the year in which there is both a Galactic conjunction *and* a 'zenith' conjunction with the Pleiades . . . on the opposite side of the Milky Way.

For all of human history, the sky has been the source of spirit . . . and knowledge . . . with a search for the center always being dominant. Before aboriginal people moved east and south over the land bridge between Asia and North America, they revered the pole region, with its stars whirling around the center. As they moved south, they were able to mark and map this center as it now rose and set on the horizon and, finding it to move, looked for a new, more eternal one. For a period, and only within certain southerly latitudes, they replaced this center with that of the Zenith – the two days each year the Sun is directly above in the sky and casts no shadow.

But as time went on and they continued to watch and to mark the rising and falling of the stars - and most importantly, the rising of the Winter Solstice sun - they located the true center. Finally, in relation to us and to our place in this infinite cosmos, they located that which would always stay fixed and would 'never' change . . . the stillpoint at the center of our galaxy.

They watched the skies for thousands of years, looking for and being entrained through the great cycle of time by the center itself. And *now*, through the years from 1987 - 2012, the rising of the Sun on the Winter Solstice gradually moves through, or transits, the alignment with this vast source of energy, the Galactic Center, for perhaps the first time in the Earth's history (if the theory that our axis tilted 11,000 years ago is correct), and certainly for the first time in human history as we know it.

And *this* is why the improbable fact that our solar system is in alignment with the *precise* center of the galaxy is so remarkable. Alignment with the *center*. Again . . . an undeniable similarity with the preceding examples.

5. The fact that it takes the wobble in the 23-½ degree slant in the Earth's axis *almost exactly* 26,000 years to make one rotation (called the Great Year) . . . *one* cycle . . . and it takes light *almost exactly* 26,000 years to travel from the center of the galaxy to the Earth. This is as bizarre as everything that has preceded it.

These figures come from the latest research (*close* to 26,000 light years is the number given by NASA for the distance to the center of the galaxy and 25,920 years is the most accurate number I've been able to find for the cycle of the Great Year) and . . . as you might imagine at this point . . . there is nothing in known physics that could possibly connect the two phenomena (Webster's: Phenomenon: Any event, circumstance, or experience that is apparent to the senses and that can be scientifically described or appraised).

This means that light leaving the center of the galaxy at any point in the circle of that Great Year (say Winter Solstice of any year) will be the light we see on that solstice exactly one Great Year, 25,920 years, later. This proportional relationship of *one to one* relates distance to time, on a profound scale. It is also a deeply subtle entrainment, through the rhythms of the Universe, to Unity . . . and to the center, the stillpoint. Once again, an absurdly unlikely phenomenon connected to the geometry of the stillpoint.

6. The fact that 1x2x3x4x5x6x7 = 7x8x9x10 = 5040 = the sum of the radii of the Earth (3690 miles) and the Moon (1080 miles) (*see Scale:, page 350 and page 31 of John Martineau's A Little Book of Coincidence in the Solar System*) Here, just as *precisely* as with the geometric models listed above, yet completely unrelated to any of them, the system of number that has been used for thousands of years and that we use and understand *today,* is *precisely* and seemingly *impossibly* related to the dimensions of the Earth and the Moon, as well as to the current measurement of a mile . . . as well as specifically to the number seven – the one number mainly associated with the phi pyramid that is part of the graphic mentioned in the first example; that is, it is this pyramid that comes so very close to dividing a circle into sevenths (one seventh of a circle = 51.43 degrees and the apex angle of the phi pyramid is 51.85 degrees).

There is magic in number and, pertinent to the theme of this preface, there is perhaps none more so than the number seven. The following is an excerpt from Michael Schneider's book, *A Beginner's Guide to Constructing the Universe*: 'Seven is known as the only number of the original Dekad (one through ten) that is not *born,* that is, its polygon, the heptagon, cannot be constructed using only the three tools of the geometer . . . the compass, the straightedge and pencil, the tools that mirror the methods of the cosmic creating process . . . and 'born' like the other shapes through the 'womb' of the vesica piscis.

It is 'virginal' in that it is untouched by other numbers in the sense that no number less than seven divides or enters into it, as two divided four, six, eight, and ten, three divides six and nine, four divides eight, and five divides ten. It is also considered childless since it produces no other number (by multiplication within the ten, as two produces four, six, and so forth).'

So, here we have yet another uncanny and completely improbable phenomenon, expressed in terms we understand *today,* that relates the dimensions of the Earth and Moon . . . the dimensions of the geometry of the stillpoint . . . to a profound expression of sacred geometry – that mentioned in phenomena #1 (*see Earth/Moon Relationship, page 7 and 179*). Yet here the emphasis is on the 'unborn' quality of the stillpoint expressed through the virginal number seven. As much as we take the 'point' for granted in our reality . . . it doesn't exist. It has no dimension. It is not *born,* inferring a reality outside of the mind.

Again . . . an 'impossible' improbability leading to the stillness at the center - in yet another way.

7. The fact that in *seven* days, a quarter of the cycle of the Moon's orbit around Earth, the original *single* cell that eventually becomes who *you* are divides and re-divides until it once again achieves a spherical shape, consisting of 64 spherical cells (*see When We Were Small: . . . page 188*) . . . the same number as the 64-tetrahedron/sphere configuration which is generated by the Vector Equilibrium and is geometrically connected to the Chinese 'I Ching,' the Egyptian 'Flower of Life,' laser burned into the Temple of Osiris (a temple *older* than the great cataclysm recorded in geologic records that occurred almost 11,000 years ago), and the Hebrew 'Tree of Life' (*see Number:, page 195, and Winter Solstice . . . 2012:, page 251*). In other words, in *one* octave, *one* cycle, of growth relating to the cycle of the Moon, our own experience began *in* and moved *through* a geometry that includes the stillpoint and is the geometric geneses of some of the oldest and deepest sources of spiritual knowledge on Earth.

8. Kepler's discovery regarding the mean orbits of the planets and their relationships to the Platonic solids. The following quote is from Lee Smolin's 'The Trouble with Physics,' speaking about an amazing discovery by Johannes Kepler in the early 1600's:

> 'The cube is a perfect kind of solid, for each side is the same as every other side, and each edge is the same length as all the other edges. Such solids are called Platonic solids. How many are there? Exactly five: besides the cube, there is the tetrahedron, the octahedron, the dodecahedron, and the icosahedron. It didn't take Kepler long to make an amazing discovery. Embed the orbit of Earth in a sphere. Fit a dodecahedron around the sphere. Put a sphere over that. The orbit of Mars fits on that sphere. Put the tetrahedron around that sphere, and another sphere around the tetrahedron. Jupiter fits on that sphere. Around Jupiter's orbit is the cube, with Saturn beyond. Inside Earth's orbit, Kepler placed the icosahedron, about which Venus orbited, and with Venus's orbit was the octahedron, for Mercury.'

9. The entire panoply of geometric relationships demonstrated by the orbits of the planets in our solar system displayed so beautifully in John Martineau's *A Little Book of Coincidence in Our Solar System*. The diagrams of the planets motions relative to each other found in this book are purely geometric . . . for no known reason. One of many examples is the relationship of the orbits of Mercury and Venus . . . 'one Mercury *day* is exactly two Mercury *years*, during which time the planet has revolved on its own axis exactly three times. *The reason for this harmony is still unknown . . . there must be a reason for this beautiful fit between the ideal and the manifest, but none is yet known.*'

THE 20 CONDITIONS FOR ADVANCED LIFE

Frank Drake originally put together a formula based upon the requirements for complex life that was capable of transmitting radio waves into outer space . . . i.e. a 'technological civilization.' His original formula was incomplete and contained many variables that could not be known. At the time, Carl Sagan estimated that there were perhaps a million such planets in our solar system. In Donald Brownlee and Peter Ward's book *Rare Earth: Why Complex Life is Uncommon in the Universe*, this formula was added to and made more exact as scientific discovery progressed. The authors of *The Privileged Planet*, Guillermo Gonzalez and Jay W. Richards, further expanded and refined this formula, making the likelihood of complex life extremely rare. Here are the 20 factors that are independent of each other. If a conservative 10% likelihood exists for each, and only the first 13 are factored in (due to the reliability of the variables) the probability of a planet like our existing . . . simply from the universal physical conditions required – that is, without even considering the previous list of 'impossible' coincidences – is $1/100^{th}$ . . . or, significantly less than 1. If any of the other factors are considered, it's obvious that it is *very* rare.

The idea is to take the total number of stars in the Milky Way Galaxy and multiply this number by the 10% probability represented by each of these factors (I will add clarification as best I can when needed, but please refer to the book or look up whatever is unclear on the Internet):

- fraction of stars that are early G dwarfs (required *type* of Sun) and at least a few billion years old
- fraction of such stars in the Galactic Habitable Zone (where life is most likely to develop – liquid water a must)
- fraction of remaining stars near the corotation circles and with low eccentricity galactic orbits (this has to do with being in an area that is affected by, but clear of the spiral arms of the galaxy)
- fraction of remaining stars outside spiral arms (being within a spiral arm will not permit life to evolve)
- fraction of remaining stars with at least one terrestrial planet in the circumstellar habitable zone (the Goldilocks zone that permits life)
- average number of terrestrial planets in the CHZs of such systems
- fraction of remaining systems with no more than a few giant planets comparable in mass to Jupiter in large, circular orbits
- fraction of remaining systems with terrestrial planets in CHZ with low eccentricities and outside dangerous spin-orbit and giant planet resonances

335

- fraction of remaining planets near enough inner edge of CHZ to allow high oxygen and low carbon dioxide concentration in their atmospheres
- fraction of remaining planets in the right mass range
- fraction of remaining planets with proper concentration of sulfur in their cores
- fraction of remaining planets with a large moon and the right planetary rotation period to avoid chaotic variation in its obliquity
- fraction of remaining planets with right amount of water in crust
- fraction of remaining planets with steady plate tectonic cycling
- fraction of remaining planets where life appears
- fraction of remaining planets with critically low number of large impacts
- fraction of remaining planets exposed to critically low number of transient radiation events
- fraction of remaining planets where complex life appears
- fraction of remaining planets where technological life appears
- average lifetime of a technological civilization

Appendix D:

This appendix addresses the representation of number in the temple:

Two is represented by the myriad of merging innate polarities already mentioned, and most importantly by the DNA-like double helix represented by the outer spiraling roof form in one direction and the spiraling stainless steel cables in the opposite direction. **Three** by the triangular sides of the Phi-pyramids and octahedron and by its integral role within the numbers six and nine, mentioned later. **Four** is expressed mainly by the all-important four directions, by the four apexes of each of the principle abstract tetrahedrons (of the 64-tetrahedron matrix), and by the four elements represented symbolically within the forms of the temple. **Five:** 'We now arrive at the mathematical middle of our journey to ten to dwell with the principles symbolized by the number five, which the Greek philosophers called Pentad. Beyond the Monad's point, the Dyad's line, the Triad's surface and the Tetrad's three-dimensional volume, what remains? The Pentad represents a new level of cosmic design: the introduction of life itself.'[274] It is the principal number of life, of nature and the main physical expression of the Golden Proportion. The four elements cannot create life on their own . . . they have to combine. The number five represents 'Quintessence,' a term used by the ancient Greeks to describe a mysterious fifth element - in addition to air, earth, fire and water, which produces life. The Golden Proportion is the expression of regeneration . . . of life (*see Golden or* Fibonacci Proportion, page 193). Pentagonal symmetry is the symmetry of life. Cut open an apple, look at a sand dollar or the leaf pattern of most flowers and you will observe the number five.   The fifth platonic solid, the dodecahedron, represents this 'Quintessence of Nature' with its 12 pentagonal faces. For this reason, this number used as the pitch for the spiraling helix roof - 5:12, or 5 vertical units for every 12 horizontal units. This is a way of recognizing (again, in whole number proportion) the remarkable relationship of the number 5 (the number found everywhere in regenerative nature, representing organic life) and the number 12 (the number which man uses to divide the heavens and represents man's rational understanding of the Spiritual Universe). This number is inherent to the life process represented by the swirling helical roof. **Six** is represented by the Flower of Life inlaid on the floor of the inner temple. It is the principal number for structure, function and order. The Flower of Life . . . a two-dimensional pattern representing the three-dimensional 64-tetrahedron matrix, with its repetitive flower pattern of six interlocking, interpenetrating circles upon the circumference of a central circle, all sharing a common central point (12 around 1 in 3 dimensions) . . . 'provides a finite model of what is essentially an infinite matrix, underlying the fabric of Creation.'[275] The Egyptian Emerald Tablets of Thoth, attributed to Hermes Trismegistus, mention this flower many times:

> 'Deep in Earth's heart lies the flower, the source of the Spirit that binds all in its form. For know ye that the Earth is living in body as thou art alive in thine own formed form. The Flower of Life is as thine own place of Spirit, and streams through the Earth as thine flows through thy form; giving life to the Earth and its children, renewing the Spirit from form unto form. This is the Spirit that is form of thy body, shaping and molding into its form.'
> Tablet 13

Seven is significant everywhere in the geometry of the temple and is found in both plan and section dimensions. This number occupies a critical place within the Dekad – the primary numbers one through ten. As mentioned earlier, it is the balance or pivot number: $1 \times 2 \times 3 \times 4 \times 5 \times 6 \times 7 = 7 \times 8 \times 9 \times 10 = 5,040$ . . . or the sum of the radii of the Earth and Moon in miles (3,960 plus 1,080) – giving the number seven above all others a very special part in the unique relationship between the Earth and the Moon.   Another of its unique properties is that it is 'untouched by other numbers in the sense that no number less than seven divides or enters into it, as two divides four, six, eight, and ten; three divides six and nine; four divides eight; and five divides ten.' It is called the 'virgin' number. It is the number **Seven**, through the establishment of the pyramid, that brings the temple into the third dimension, into form, from the unborn to the manifest. From Michael Schneider's *A Beginner's Guide to Constructing the Universe*:

'It's well known that the regular heptagon is the smallest polygon that cannot be constructed using only the three tools of the geometer; the compass, straightedge and pencil, the tools that mirror the methods of the cosmic creating process. In other words, an exact heptagon is not (and cannot be) 'born' like the other shapes through the womb of the *vesica piscis*. This explains why the virgin seven cannot be entered (divided), cannot produce numbers within the Decad, and cannot be captured: simply because it is not born.'

Exactly like the stillpoint that does not 'exist,' does not manifest . . . is not born.

Significantly, Arthur Young's theory of the evolution of consciousness, his process theory established in *The Reflexive Universe*, has seven stages, each with seven sub-stages – the number seven being critical to the process of evolution itself. Seven also represents the marriage of Heaven and Earth – 3 (Spirit) and 4 (Matter) – and the pyramid has a square base (4) and four triangular sides (3). The 51.85-degree pyramid, unique in its incorporation of the Golden Proportion Phi and Pi, and in its relationship to the dimensions of the Earth and Moon, also is very close to equaling one seventh of a circle (51.43 degrees). This very unique number is also the basis for the fractal of the Pythagorean octave of music. This is the octave we are so familiar with in the Western world, based upon ratios of simple whole numbers consisting of seven notes, returning to the original note an octave higher or lower.

'For the true knowledge of music is nothing other than this:
to know the ordering of all separate things and how the
Divine Reason has distributed them; for this ordering
of all separate things into one,
achieved by skillful reason,
makes the sweetest and truest harmony
with the Divine Song.'

Hermes Trismegistus

Eight is represented by the eight triangular faces of the original octahedron (*see Sphere/Octahedron, page 11*), by the eight apexes of the interpenetrating tetrahedrons of the 64-tetrahedron/sphere matrix (8 x 8), by the eight mountain peaks at each of the eight directions (*see The Eight Directions, page 265*) and by the stained glass 'Star' window (Lakota) above the fire pit in the north alcove of the temple (*see West/East Section, page 343, and North Elevation, page 348*).

And now the enigmatic number **Nine**. It is the last among Pythagorean number and is the limit of numbers' generative principles. It represents completion and the last horizon before transforming into a new level of expression. It represents transition . . . the last of the prime numbers, which – with the addition of one – appears to transform itself into ten – yet rests there only the length of a point/moment, immediately regenerating itself back into **One** (one plus zero). The removable Flower at the center of the inner temple's floor, surrounding the black granite Omphalos fountain, has 18 sections or seats on Mother Earth surrounding the fountain. There are fourteen main columns (the number of faces of the Vector Equilibrium), each 5.4' in diameter (a whole proportion of the Moon's diameter). **Ten** of these are laid out through the number nine (here again the transformative, mystical transmutation of **Nine** into **One**), while the four remaining columns, supporting the cantilevered pyramid roof, are laid out through the numbers Four and Seven . . . Four: one at each corner of the pyramid and four divisions of the 28 segments of the spiraling, radiating roof rafters – Four divisions of Seven. The columns alone reflect an intricate weaving of **Zero** and **One** and **Four** and **Seven** and **Nine** . . . and back to One after touching **Zero** – creating a tone of transition and transformation within the inner temple.

Appendix E:

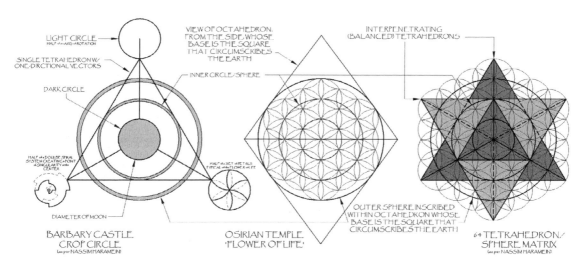

BARBURY CASTLE CROP CIRCLE / OSIRIAN TEMPLE 'FLOWER OF LIFE' /
64-TETRAHEDRON – SPHERE MATRIX

In the illustration above is the side view of the same octahedron that, when joined with the sphere of Earth (tangent to the midpoints of its *edges* (see *Earth-Sphere/Octahedron, page 11*)) creates the Vector Equilibrium. Here the octahedron is tangent to the third sphere of the Flower of Life (with the first sphere scaled to the

Moon) at its *faces*. This demonstrates the 'impossible' fact that the Earth and the Moon are not only a part of the 'Earth/Moon Relationship' shown on page 7, but also are an integral part of the *still-point* geometry. *(See Appendix A, page 303, for a complete explanation).*

That is, if the smaller circles found in the Flower are scaled to the dimension of the Moon, the third outer layer/sphere is tangent to the octahedron at its *faces* . . . the Earth sphere tangent to the same octahedron at its *edges*.

The precise dimensions of the Earth and the Moon are a part of the geometry of the still-point. Incredibly and 'impossibly,' the Sun's dimension is also included in the Flower of Life, or still-point, geometry - the 400th layer sphere being the diameter of the Sun – being exactly 400 times larger than the Moon's *(see Appendix A, page 303)*.

One can see, regarding the crop formation, that it expresses one half of the equation that is completed in the three dimensional 64-tetrahedron/sphere matrix: a spiral is shown going in just one direction and only one half of a flower is shown (the light, dashed lines in the illustration represent the completed half not included in the crop formation), one half of an axis of rotation, the two smaller spheres are dark and light – depicting two halves of something whole, a yin and yang, and the single tetrahedron (the first three-dimensional form, consisting of four equilateral triangles) which can only point in one direction, with no counterbalancing vectors in the opposite direction, is balanced and completed in the 'star' tetrahedrons of the 64-tetrahedron matrix (these observations made by Nassim Haramein). One interpretation of this information is that it is a code pointing in the direction of wholeness . . . of Unity. The geometry of the interpenetrating tetrahedrons of the complete matrix represents this balance and unity . . . and within this 64-sphere/tetrahedron matrix – the layer of sphere/tetrahedrons represented by the larger, outer circle – are 32-sphere/tetrahedrons in the shape of the Vector Equilibrium . . . the geometry of the still-point.

Appendix F:

⊕ Floor Plan & Orientation:

The orientation of the temple is as a medicine wheel - the never-ending cycle of life bound to the four directions. The entrance, as in Native American traditions, faces east because, in Black Elk's words, 'it is from this direction that the light of wisdom comes.' It is towards the east that the Earth rotates and for this reason it also feels correct that the entrance opens to face that direction and align with that movement – and to face the Sun at first light. From F. Silva's *Common Wealth:*

> 'Orienting the temple to the energy of specific creative processes enhances the action taking place inside the sacred space. The orientation is typically defined by the direction in which the entrance faces, for it welcomes the energy into the temple's inner sanctum . . . . Entrances in the east symbolically face the light of the reborn Sun. Such temples benefit from the purified morning air, and thus represent that element. Dawn is the awakening from darkness, the Sun emanating out of the underworld, the light returning to the land, hence these are temples associated with enlightenment.'

One enters from the east into an entrance foyer through a pair of tall, heavy, stained glass, thick wooden doors. The outer gate. They are each 42" wide x 10' high. This foyer is circular, 18' in diameter, with an arched wooden ceiling. After entering, to the right and left, are openings to practical concerns – shoes, bathrooms, hallways and coats – a recess in the circular wall provides access, the recessed wall a place for art. A thangka perhaps.

It is here, in this circular foyer, that the silence of the inner temple is sensed for the first time. One knows to soften the voice, center oneself . . . remove the shoes.

To the front or west upon entering this foyer, are two more doors opening to the inner temple that mirror those facing east, to the outside.

Once one is ready to enter the inner temple, this last pair of doors are opened. Quietly. Softly. Respectfully. They open into a smaller, round, secondary foyer - twelve feet in diameter, with a domed ceiling - which is adjacent to and open to the circular inner temple. The entry is designed this way with the intention of guiding one into the inner temple gradually, leading one though stages from the often chaotic and noisy world outside to the silence within. The immediate feeling upon entering will be one of overpowering, sacred, protected, silence.

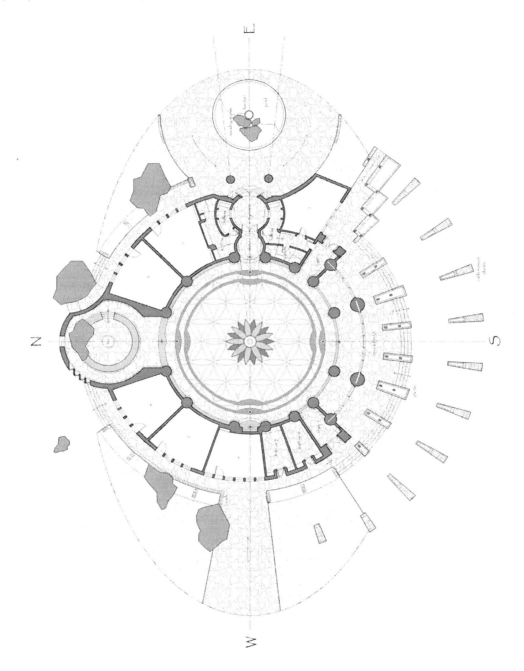

FLOOR PLAN

Once one enters the inner temple and is within the second, smaller, circular, domed foyer, two 7" steps lead down into the circle of the inner temple itself. Once on this entrance level, there is a 6'-8" wide sitting/walking area to the right and left, around the circumference of the circular room. Adjacent to this level, towards the center, and one 7" step lower, is a concentric step or sitting area 3'-4' wide . . . room enough for a zaftan (square cushion). Finally there is one more 7" step to reach the floor level of the circular central wooden floor, inlaid with the Flower of Life pattern – four steps in all. At the center, in the middle of the inlaid lotus blossom, is the black granite fountain whose concave upper surface establishes the Stillpoint from which all forms are generated.

To the right, or north, upon entering the inner temple, is an alcove with a wooden arched roof that surrounds a dragon shaped rock outcropping. In front of this outcropping is a recessed fire pit surrounded by circular, recessed seating (*see Floor Plan above*). There are also four tall, vertical windows in this alcove facing the west, intended to bathe the rock outcropping in the light of the setting Sun . . . especially during the Summer Solstice sunset.

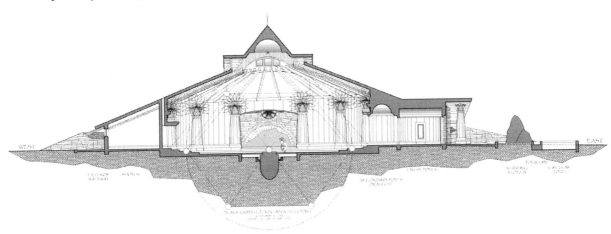

WEST/EAST SECTION
(Facing North)

As in Lakota and other Native American traditions, the place of honor is in the west. Directly across from the entrance to the inner temple, to the west, is an alcove for this purpose - for a teacher, speaker, or presenter. There is a small skylight, invisible from the interior, in the Earth covered spiraling roof of the inner temple just above this alcove that will feather indirect light down this granite surfaced wall, subtly lighting the area behind the speaker.

Depending upon the tradition or inclination of those using the temple, the north can also be used as a natural place for the speaker or teacher. In the north is found the constancy of the North Star and, in the northern hemisphere, this direction holds the quality of snow, or purity. For these reasons, a speaker may wish to align him or herself with this direction, and the layout of the temple supports this.

To the left, or south, upon entering, the inner temple is almost completely glass. Three large openings along the circular south facing wall will have 60' of curved sliding glass doors that disappear into large columns, permitting the inner temple to be completely open to the outside and the vastness of the view to the South. The center opening, facing due south, is almost 28 feet, with five, 5'-6" x 10' sliding curved sections. Flanking this opening to either side are openings of 15' each, with three, 5'-0" x 10'-0" sliding curved sections. Flanking each of these openings to either side is another opening of 6,' with a pair of swinging glass entrance/exit doors, each side entrance having its own foyer. The curved south wall has a total of seventy-two feet of 10' high opening glass doors facing the south and the view of the entire 100-mile length of Owens Valley (*see Floor Plan & West-East Section above, and South Elevation, Page 348*).

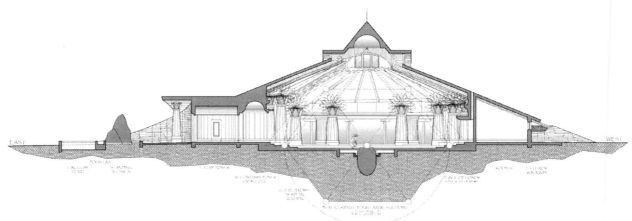

EAST / WEST SECTION

(Facing South)

The inner temple is accessible only by the main entrance to the east, and by the southern glass doors that open to the enclosed garden area to the south . . . there are no doors that open into this area from any of the ancillary rooms, contributing to the sense of protection and sacred space. The expanse of curved southern doors opens onto the covered porch and out into what is essentially a very large walled garden. While the desert will remain untouched in the larger general area around the temple, the immediate area to the south will be enclosed by a somewhat invisible wall (stone columns at the corners, with 6"x6" wooden posts, 6' on center, with vines growing over the wire mesh) permitting this outside space to become an extension of the interior space of the temple. It will be a protected outside area that includes many rock outcroppings, a pond, sweat lodge, and a garden of trees and plants. Around this silent inner temple are six ancillary rooms for private meetings and smaller groups. The outer, spiraling wall of each of these rooms is filled with continuous 4' high windows that are 5' from the floor, providing privacy from the exterior walkway as well as abundant light and views of the sky. Each room has a 9' tall door with a similar window at the same height as the others. The inclusion of these ancillary rooms is both practical and symbolic. Insight obtained from the inner temple is shared and discussed here with the intention of coming to understanding, this understanding disseminated to the world - insight from the Stillpoint, processed into understanding through the helical process of life represented by the rooms themselves.

Outside of the spiraling outer walls of these ancillary rooms is a 5' wide access walkway, partially roofed and bordered on its outside by a 4'-6" high stone covered retaining wall (*see Floor Plan above*) that permits the slope of the spiraling roof to extend all the way to the desert floor – permitting the temple to appear as a small, unassuming, hill in the desert. As mentioned, all of these outer rooms are accessed either from the outside or from interior halls . . . the inner temple remains free of any interaction from these more utilitarian spaces.

⊕ STRUCTURE & MATERIALS::

The inner temple is approximately 46' high at the center measured from the granite fountain to the top of the dome. The circular wall defining this room begins approximately 20' high at the southeast diagonal and is approximately 25' high at the clockwise completion of one revolution, as seen from above. This circular, scalloped wall (*see Frequency/Vibration/Sound/Resonance:, page 226*) will be surfaced with vertical tongue and groove wood. The ten large 5.4' diameter columns that are on the nine-fold layout, as well as the four similar columns that are on the twenty-eight-fold layout – fourteen in all reflecting the number of sides of the Vector Equilibrium - will have planters at their capitols just below small, 12" diameter skylights – *a circle of trees in the desert surrounding water . . . an oasis.* There are 28 main supporting rafters, the extensions of their spiraling ends forming a circle in plan-view, surrounding the vertical axis of the center of the temple – the Axis Mundi.

These spiraling, helical rafters are the support for the feminine, nautilus, spiraling lower roof, as well as the pyramid roof with its interior dome discussed below.

REFLECTED CEILING PLAN

Atop these rafters run two other layers of 6"x8" rafters, or purlins, spiraling out From the center in opposite directions (*see Reflected Ceiling Plan above*), supporting the 3x12 tongue and groove decking running down the slope. The slope of this roof extends past the outer walls of the outer rooms, all the way to the desert floor and is covered with rigid insulation and 12" of living/planted earth – a hill in the desert. Inside, the entire circling wall and spiraling ceiling of the inner temple are wooden – there is nothing high-tech about this, everything natural.

Modeling a rainbow with muted colors – lower frequency light/color towards the bottom graduating to higher frequency light/color towards the top - the main rafters are red, the first layer of spiraling rafters orange, the second layer, traveling in the opposite direction, yellow, and the decking green . . . the colors reminiscent of a Buddhist temple.

While it is in the nature of a spiral/helix to travel *in and out* at the same time, balanced in this way, expanding or contracting depending upon the direction it is going, the spiraling rafters themselves will be counterbalanced - as are the two snakes coiling up the staff of the *Caduceus* (*see Dragon/Serpent . . . Life Force: page 279*): that is, while the spiraling rafters gradually *rise* as they move in a clockwise direction seen from above, there is still an aspect of one-directional movement. In order to balance this movement, stainless steel cables, in place of rafters spiraling in the opposite direction, gradually become higher, as seen from above, as they move in the *counter-clockwise* direction. At the south side of the temple these cables are exposed and anchored to the ground with stone foundations. As they become lower while spiraling in this clockwise direction, they eventually disappear into the desert roof of the temple, reappearing inside the temple in the northeast. At this point these cables are not intended to be structural . . . but are primarily energetic in nature, to balance the counter-clockwise movement of the spiraling roof.

The flat ceiling that is the base of this pyramid roof, just above the arched windows and the spiraling rafters, is dark blue (*see Reflected Ceiling Plan above*). The ceiling just under this pyramid roof will be dome-shaped – its diameter 10.8,' the half-radius of the Moon (*see East-West Section, page 344*). It will be surfaced with plaster, colored with a dark violet-blue pigment embedded with very subtle, tiny, interspersed pieces of mother-of-pearl tile – diamonds or stars in the night sky. At the center of the dome is a diamond lit with a laser, with the Stillpoint at its center . . . the sacred point at the apex of the smaller Phi-pyramid whose base is the square of the circle of the Earth (*see Temple Layout, page 194*).

The floor outside the inner temple, including the perimeter walkway, foyer, entrance and patios, as well as the outer walls of the temple, will be surfaced with stone. Ideally this will be granite, which is basically silica, the main component in the Earth's crust and mantle:

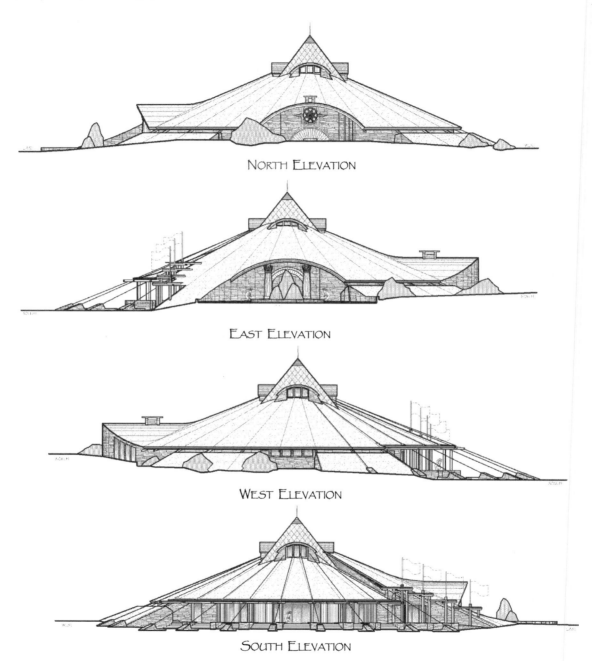

NORTH ELEVATION

EAST ELEVATION

WEST ELEVATION

SOUTH ELEVATION

'Granite gives off a significant electrostatic charge, and the type used in the Great Pyramid contains 25% quartz, iron, and magnetite . . . the correct use of stone is important. In addition to the magnetite, the stones used for sacred sites contain large amounts of quartz, a piezoelectric substance that not only can hold a charge, but like the silica used for computers, it can be instructed to do so.'[69]

The floor of the inner temple will also will be wooden, as are the walls and ceiling, with the large inner circular section of this floor (almost 60' in diameter) inlaid with contrasting type of wood in the Flower of Life pattern, each circle in the pattern being 21.6,' the diameter of the Moon. The central 21.6' diameter circle of inlaid floor will be removable (*see Stillpoint/Black Granite Fountain/Omphalos, page 239*). 18" below the level of this inner floor will be a circular section of Earth for special ceremony when it is necessary to be directly connected to the Earth. 36" above the ground, centered within the concavity of the *Omphalos* fountain, is the location of the second sacred 'zero' point, the center of the temple itself (*see Number:, page 195*).

The temple will be constructed with non-toxic, non-allergenic materials. It will be a protected place in every way, and will smell and sound and feel and *be* natural in all aspects. It will be heated by spiraling, radiant heating under the floor which will give heat silently, free of dust, and move naturally from the heated floor upward (a radiant heating system). The water that is used in this heating system will be heated by solar panels in a discrete location. The temple will be cooled naturally by convection, pulling air from large conduits under the Earth (where the air is cooled to the Earth's 55° temperature naturally), which are open to the outside air at their outer ends through screened vents. The natural convection will be helped when necessary by a silent fan in the pyramid roof, pulling the warm air up through the temple and out the vent on the top of the pyramid, while pulling in Earth-cooled air below. In this way, although the interior will be cool due to being covered with earth, any extra heat will be released to the outside when needed. This completely natural system will be augmented by a swamp-cooler system designed to be as silent as possible, but ready to use whenever necessary – even though the temple is designed with extensive roof and wall insulation, as well as being covered by 12" minimum of desert earth, as the site spoken about, in the high desert of California, can get very hot in the summer. Electricity will be supplied by the same solar collectors that are separate from the structure.

Many details need still to be worked out regarding the nuts and bolts of building this structure . . . . always following the pure intention of the temple's principles – to consciously bring Spirit into Form.

⊕ SCALE :

So much of what is being contemplated and hoped for regarding the design of the temple is based upon measurement, but above all it is a matter of whole number proportion relative to the dimensions of the Earth, Moon and Sun . . . and the Stillpoint. It feels important to give some attention here to the units of measurement being used . . . in particular, the *foot* and the *mile*. The temple itself will be measured out in terms of feet and inches . . . but what *is* a foot? And the term 'mile' has been used when alluding to the interesting *coincidence* that the sum of the radii of the Earth and Moon is 5,040 *miles* . . . the same as 1x2x3x4x5x6x7 = 7x8x9x10 = 5,040.

At first glance, one has to admit that this is at least an interesting coincidence . . . and perhaps it's just that. But the question is raised . . . just what *is* a *mile*? As will be seen, there is a fascinating and apparently arbitrary quality that has emerged through history to be agreed upon as a 'mile.' Without getting into an ocean of comparative numbers, the term 'mile' relates to a certain number of feet . . . but the Greek 'foot' is different in actual length from the Roman 'foot' is different from the Egyptian 'foot' (cubit) is different from the modern 'foot,' and then there are *nautical* feet. So . . . the term 'mile' could mean many things, but it is, in fact, based upon the modern foot. Hmmmm.

When I first began researching all this in 2000, all I could find regarding the historical origins of today's measurement systems were *arbitrary*. This all changed with the work done by Alexander Thom, Christopher Knight and Alan Butler. As mentioned in the general summary at the beginning, in Knight and Butler's *Civilization One*, they were able to unravel the mystery surrounding the earlier discovery by Thom of a common unit of measurement used in the construction of ancient megalithic stone circles – the Megalithic Yard. While Thom established that this was the unit of measurement used, Knight and Butler were able to establish that this was a *geodetic* unit of measurement - that is, a *whole number unit* of the Earth's circumference – 366 Megalithic Yards equal one polar second of arc. In their later book, *Who Built the Moon?*, they were dumbfounded to discover that the Moon and Sun also were measurable in these same units – 100 Megalithic yards to each second of arc for the Moon, and 40,000 MY for the Sun. Here is their moment of discovery in their own words, from *Who Built the Moon*:

> 'We had been very surprised at the way the Megalithic Yard bisected the circumference of the Earth, but what we didn't expect to discover was any direct connection between the Megalithic Yard and other bodies within our solar system. And there are none – apart from the Moon and the Sun.
>
> The Moon has a beautifully neat 100 Megalithic yards to each

second of arc, which could be a very odd coincidence if it were not for all of the other facts we discovered which point to a whole range of round numbers [see Appendix B, page 324]. And of course the Sun has an incredibly round 40,000 Megalithic Yards to each Megalithic second of arc.

It had struck us as quite amazing that anyone more than 5,000 years ago could have created a unit of measure that worked as a perfect integer of the planet within such an elegant system of geometry – starting and finishing with the Earth's number of 366. Whilst this was impressive, we were perplexed at the apparent impossibility of creating a unit and a geometry that produced beautifully round integers on the Earth, Moon and Sun. *To do so should be as close to impossible as anything can get.*

Units that are integer, within the same geometry, for two heavenly bodies would be very difficult – but three? That's ridiculous! And yet the sums spoke for themselves. The fact that the approach did not work for any other body in the solar system pointed to a very special relationship for the Earth, Moon and Sun.'

Yes . . . a very special relationship. The reader can see that they were still in their process of discovery – while at first glance it was 'impressive,' it also seems impossible . . . growing awareness that this phenomenon had nothing to do with ancient people somehow coming up with an integer common to *all three celestial bodies* - but with the inherent proportions of the bodies themselves . . . their 4 ½ billion year relationship had not changed and, because of the proportions – the proportions of the Stillpoint geometry - one size (unit) fits all – the Megalithic Yard.

Yes . . . a very special relationship far beyond coincidence.

Getting back now to the subject of scale . . . how in the world could a modern *mile* (which is made of of modern feet and inches) fit into all of this? How could the sum of the radii of the Earth and Moon be 5,040 *miles* . . . and relate so impossibly (coincidentally?) to the number 7 (1x2x3x4x5x6x7 = 7x8x9x10 = 5,040) – the *unborn* number which most relates to the unborn nature of the Stillpoint and to the magical geometry of the Earth/Moon diagram?

It turns out that the modern mile is also made up of whole number segments of the Megalithic Yard . . . 1,920 MY equal one *modern* mile! Without going into the whole story, it turns out that all of our modern units of measurement of *weight, distance and time* can be traced to this remarkable unit of ancient measurement . . . and, oh yes, weight and distance were also related to the size of the *barley seed*. Oddly, Edward the 1st decreed this to be so too – see below. What is pertinent here as regards the scale of the temple is that the use of feet and inches relates to this unit of measurement that is common to the Earth, Moon and Sun.

Interesting too is that the world still is not generally aware of this and remains stuck in the past as it understands the origins of today's' systems of measurement. Following was my own understanding of this until some years ago . . . and which is still accepted as what's so by the world at large.

In ancient Greece, 6000 Greek feet equaled ten Greek *Stadia*, their 'mile,' which was theoretically 1/21,600[th] of the circumference of the Earth, or one minute of arc. (This number 2160 comes up a lot – the diameter of the Moon, the number of years of each astrological phase qualifying the heavens – each a month equaling 2,160 years of the 25,920 years precession of the equinoxes – the Great Year). But this was not to last. As the story goes, Edward I, sometime in the 13th century, decreed that three grains of dry, round barley make an inch, and twelve if these 'inches' made a foot. In the early 1100's, King Henry I decreed that the yard should be the distance from the tip of his *nose* to the end of his *thumb*, the foot then being one-third of this royal length. This led Queen Elizabeth I to declare, in the 16th century, that henceforth the traditional Roman mile of 5,000 Roman feet would be replaced by one of 5,280 of Henry's feet – or 440 'yards' – because three feet made a yard and 220 yards was one 'furrow long' (furlong) and eight furlongs made a mile. Go figure. And the number 5,280 was imprinted in history.

Still . . . the sum of the radii of the Earth and the Moon is 5,040 'modern' miles. This number has everything to do with the number seven and the number seven has everything to do both with the 'unborn' quality of the Stillpoint - the number seven is also 'unborn' - and with the pyramid that so precisely relates the proportions of the dimensions of the Earth and Moon to each other - it's apex angle almost perfectly divides a circle into sevenths . . . which eventually leads one to the geometry of the Stillpoint. For me, this also is no coincidence . . . it was designed to be this way as yet another cosmic marker or message, letting us know 'they' are there and 'they' are communicating to us.

But before being made aware of the significance of the Megalithic Yard, I was totally confused. How could this seeming arbitrariness about how we calculate a mile *today*, ostensibly based on King Henry's nose a thousand years ago, lead so precisely to the Stillpoint?

Thanks to the work of Thom, Knight and Butler, this question is now answered.

This central principle requires that the dimensions of the temple be a whole number proportion of the Earth and Moon's dimensions (*see Proportion, Vibration & Balance:, page 189*) *and* to be scaled to the human proportion. In human terms, the circular inner temple dimension of 79.2' diameter would provide enough floor area - including the north alcove and the area to the south by the curved glass doors of 7000 sq. ft. – for 400 to 500 people . . . homeopathy for cosmic global transformation.

The scale used thus far in determining the exact size of the temple is 1:528,000 or 1'-0" = 100 miles . . . thus 79.2' (diameter of inner temple) = 7,920 miles (diameter of Earth) and 21.6' (diameter of module for outer spiral) = 2,160 miles (diameter of Moon).

The diameter of 7,920 miles was chosen as the mean diameter of the Earth by averaging the circumference around the equator (24,901.55 miles), with the circumference around the poles (24,859.32). This difference of 42.23 miles was divided by 2 (21.115 miles) and added to 24,859.32 miles to equal the mean diameter of 24,880.435 miles for circumference of Earth/3.141592 (π), equaling 7,919.69 miles diameter. This was rounded to 7,920 miles and is the commonly accepted figure.

To extend this idea of the importance of scale, as well as to emphasize the now indisputable fact that a message of unprecedented importance is being communicated to us by the consciousness that created not only the Earth, Moon and Sun, but the entire solar system, I include now an amazing list of the information encoded within the precise, *scaled (1/43,200)* proportions of the Great Pyramid. From Ekhardt Shmitz's book *The Great Pyramid of Giza: Decoding the Measurment of a Monument:*

The following is a list of conclusions drawn from analysis of the relationship of dimensions found within and without the Great Pyramid:

- A precise definition of the Royal Cubit as it relates to the Earth
- The size and shape of the Earth
- The Mass and Density of the Earth
- The Gravitational Constant
- The Escape Velocity from the Earth to obtain an Open Orbit
- The Escape Velocity from the Earth to obtain escape from the combined Earth's and Sun's gravitational field
- The significance of the location of the Great Pyramid
- The Golden Ratio
- The Mass of the Sun
- The Mass of the Moon
- The Mean distance to the Sun and the Circumference of the Earth's Orbit
- Neutral Points of Gravity between the Earth and the Sun
- The Mean distance to the Moon

- The Orbital Velocity of the Earth
- The Orbital Velocity of the Moon
- The Metonic 19 year cycle of the Moon's orbit of the Earth
- The Lagrange Point (L1) between the Earth and the Moon
- The Speed of Light
- The Orbital Velocity of the Solar System relative to the Center of the Milkyway Galaxy
- The Velocity of the Local Group of Galaxies which includes the Milky Way Galaxy relative to the Universe

'The Great Pyramid of Giza and the entire Giza Plateau may clearly be regarded as a repository of ancient knowledge . . . . It is concluded that the Great Pyramid of Giza and the entire Giza Plateau is of a highly intelligent and fully integrated design. Its construction detail demonstrates extraordinary precision in relaying highly accurate geodetic knowledge of the Earth, astronomy, astrophysics, advanced mathematics and Newtonian mechanics. Since there exist numerous examples of complimentary and corroborating values, which may be interpreted as encoded within the measurements of the Great Pyramid's geometry and the specific placement and alignments of the Pyramids and Sphinx on the Giza Plateau, it is evident, with a very high degree of probability, that the design parameters were expressly intent on conveying this advanced knowledge.'

Also, from Randall Carlson's website, http://sacredgeometryinternational.com/ are found some remarkable facts embedded in the Great Pyramid. Regarding the *scale* chosen, precisely scaling the Earth to the Great, or Phi, Pyramid (for a discussion of the Great Pyramid, see page 181):

- 43,200:   Seconds/Hemisphere @ Equinox'
- 432,000:   Miles/Solar Radius
- 432,000:   Years/Kali Yuga
- 432,000:   Years/Sumerian King's List

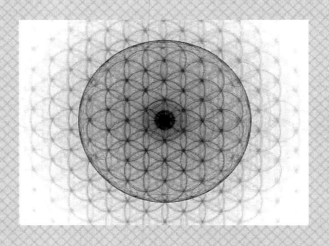

Index

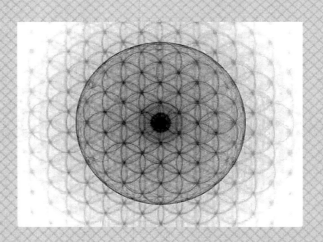

Endnotes

1 'Seeing it all coming alive' is a wonderful quote from the book *The Education of Little Tree*, by Forest Carter. It refers to the times that Little Tree and his Grandfather, both Cherokee Indians living in the Appalachians many decades ago, would walk up to a ridge overlooking where they lived and watch the Sun come up, 'seeing it all come alive.'

2 http://www.theguardian.com/environment/2014/sep/29/earth-lost-50-wildlife-in-40-years-wwf

3 There are different theories about how the Moon was formed, or created, but the impact theory is presently the most accepted for many reasons. But, as the reader will discover, regardless of how the laws of nature . . . physics . . . conspired to accomplish its creation, there were other, deeper, causes at work. Here's one site that covers science's different theories regarding the creation of the Moon: http://www.space.com/19275-moon-formation.html

4 See Graham Hancock's *Fingerprints of the Gods, Supernatural* and *Magicians of the Gods*.

5 A line from the poem *Dover Beach, by* Mathew Arnold.

6 Margaret Mead.

7 http://www.sacred-geometry.com/index.html: 'As far back as Greek Mystery schools 2500 years ago it was taught that there are five perfect 3-dimensional forms - the tetrahedron, hexahedron, octahedron, dodecahedron, and icosahedron . . . collectively known as the Platonic Solids; and that these form the foundation of everything in the physical world. Modern scholars ridiculed this idea until the 1980's, when Professor Robert Moon at the University of Chicago demonstrated that the entire Periodic Table of Elements - literally everything in the physical world - truly is based on these same five geometric forms. In fact, throughout modern physics, chemistry, and biology, the sacred geometric patterns of creation are today being rediscovered. The ancients knew that these patterns were codes symbolic of our own inner realm and that the experience of sacred geometry was essential to the education of the soul. Viewing and contemplating these forms can allow us to gaze directly at the face of deep wisdom and glimpse the inner workings of The Universal Mind.' The Platonic Solids are all derived from this primal, proto-typical, ultimate archetype . . . the Vector Equilibrium, or Stillpoint geometry.

8 From *Cataclysm! Compelling Evidence of a Cosmic Catastrophe in 9500 B.C,* by D.S. Allan and J.B. Delair. See also Graham Hancock's *Fingerprints of the Gods, Supernatural* and *Magicians of the Gods*.

9 http://link.springer.com/article/10.1007%2Fs11538-015-0126-0

10 There is a theory that the sudden appearance of the symbolic art of cave paintings around 35,000 years ago may have been occurred because of the discovery of hallucinogens by human beings, triggering a kind of global transformation of consciousness.

11 From http://dictionary.reference.com/browse/consciousness

12 Tom Waits

13 www.spiraloflight.com/ls_sacred.html

14 http://www.mathematicsmagazine.com/Articles/SacredGeometry.php#.V-r2Ismo2ao

15 Because of the very slightly elliptical shape of the Earth's orbit around the Sun, there are eclipses that are not 'total' at times when the Moon transits the Sun.

16 I have been asked to provide a more visceral description of this transcendent, induced,

other-side-of-the-veil experience, as well as a brief summary of what led to and what prepared me for such an experience. Gurdjieff called what we're attracted to in our lives, our magnetic center. What interests us pulls us into the experiences that define our lives.

I grew up on a cul-de-sac in Northern California playing kick-the-can with many other post-war baby boomers. It was an idyllic time for me. But as I grew older, I became aware that 'my' world was not the same world experienced by billions of people. While attending UC Berkeley, I considered quitting and going to Biafra to become part of the relief effort for the starving Biafrans. I didn't go . . . but never forgot. Instead, I joined VISTA, America's domestic Peace Corp . . . but soon found out that getting a pair of glasses replaced for someone who couldn't afford them, or a trip to the dentist for someone else, or even offering an alternative to young men other than being vacuumed up and shipped to Viet Nam, wasn't going to make a dent in the larger problem. At the age of 19, I first read the quote by Meher Baba included twice already in these pages (and now three) regarding the evolution of consciousness, and my personal path towards unraveling the truth about who we are and why we're here began in earnest.

> 'Before the beginning of all beginnings, the infinite ocean of God was completely self-forgetful. The utter and unrelieved oblivion of the self-forgetful, Infinite Ocean of God in the beyond-beyond state was broken in order that God should consciously know his own fullness of divinity. It was for this sole purpose that consciousness proceeded to evolve. Consciousness itself was latent in the beyond-beyond state of God. Also latent in this same beyond-beyond state of God, was the original whim (lahar) to become conscious. It was this original whim that brought latent consciousness into manifestation (form) for the first time. Slowly and tediously consciousness approaches its apex in the human form, which is the goal of the evolutionary process, and thereby an individual mind gradually differentiates itself from the sea of oblivion . . .'

I immersed myself in Transcendental Meditation, but this was short-lived because for me at the time, I needed more intellectual stimulation. This led directly to spending 4 years in my early and mid-twenties in a Gurdjieff school, studying Man's false personality while trying to 'remember mySelf', along with other fascinating and useful esoteric inquiry. Ultimately though, this was a lesson in learning discernment and to be true to myself amidst the pressure of peers and authority in what was certainly a cult. I left, and in the late 70's purged myself of a lot of past psychological wounding through years of participation in est, learning how to be more honest with myself. While est helped me to heal psychologically and got me back into the world, it also ran its course and I began looking for a way to go deeper. I met with one of the trainers I'd respected after he'd left est to become a Buddhist monk, and I also left work on 'doing' to begin intense work on 'being', attending many Vippassanna 10-day meditation retreats over the next few years, sitting and walking in complete silence while my inner voice learned to be quiet and my awareness expanded. The meditation experience was powerful and invaluable. But my general experience with this world was destroyed when I became involved in the design for the new retreat center the group was establishing. Meditation was one thing, but human beings are quite another and I ended up withdrawing from the project. It broke my heart and had me questioning everything I'd come to trust. This created a vulnerability and confusion and a soul-searching that I'd never experienced before. I began a chaotic relationship with a briefly popular 'crazy-wisdom' teacher while I tried to put the pieces of my life back together and to understand what had happened. But, instead or moving forward, I'd gone from the frying pan into the fire. After a year-and-a-half of constant 'processing', questioning everything I'd ever assumed about myself and life while often betraying my own hard-earned knowing, I was spiritually upside-down. I broke free, and was thrown into the darkest years of my life through the 90's. It took many years of slowly rebuilding my reality, all the while studying sacred geometry, for me to once again feel that I had the strength to push my boundaries and journey into the unknown for the answers I'd been looking for for so many years. In many ways, my life up until this seminal crossroads was a series of coming to hard-earned understandings, only to have the comfort of those understandings turned upside-down, finding myself thrown back into total confusion. And so it is . . . this roller coaster of the evolution of consciousness. Regarding the experience itself, words will never suffice, but here goes:

I had just experienced the most difficult ten years of my life - the decade of the 90's. It was a very dark time for me and any sense of understanding that I thought I'd gained earlier was destroyed, leaving me as spiritually weak and confused as I'd ever been. But by the end of 1999 I'd begun to put my reality back together and was standing on my feet and felt ready once again to *grow*. I reconnected with and began a long relationship with an old friend who was living in Canada. We hadn't seen each other for perhaps 15 years, and because of the courage that often comes from new relationship, she flew down to San Francisco and joined me for what was my second, and her first, experience with what many consider to be the world's most powerful hallucinogen.

A star filled, late Monday night, November 1st, 1999. We fasted all day in preparation and then made the drive to Mill Valley to jump into the void. The guide turned out to be an old friend of hers, and we were permitted to take this journey 'together'. While it was courage and a deep desire to find out more about this miraculous world we inhabit that brought us there, it was all butterflies as the moment approached. We became part of a small group of people - perhaps four - who would smoke this ancient medicine and make the journey into the unknown. The two of us sat up on a bed surrounded by the small group, including the guide who'd prepared the pipe for each of us and was familiar with the territory - but once we inhaled, we were on our own.

The odor of this timeless smoke is indescribable. Smoldering. Decaying. Ancient. Psychedelic. I'd never smelled ancient wisdom before. In hindsight, clearly a visceral element of the mysterious world of the Mayan shamanic ritual catapulting consciousness into the cosmic center.

Someone took the pipe from my hand as I fell softly back onto the bed . . . gone into another world. An atomic explosion went off in my head. I had no choice at all but to totally surrender. The experience gave new meaning to the term 'surrender' as the journey began. I dissolved into the Infinite.

In an instant my inner vision was entirely filled with an indescribable, turbulent array of multi-colored energy, all exploding out of an infinitesimally tiny point at its center – an accurate analogy of the entirety of the manifested Universe exploding out of what science calls the 'singularity' at the moment of the big BANG. For me it was the direct experience of what in later words I would call the face of God. It was immense and vast and primal and archetypal. It was Everything.

A voice emerged out of this unrelenting explosion of energy, challenging me . . . 'You ready for *this?* You think *you're* ready for *this?* I couldn't speak, but my entire being cried 'YES!' . . . and off I went. No longer was I on the outside looking in, but I was *instantly* inside that tiny, still, non-existant *point* at the center of all of it – inside the mind of God. No longer was the Witness observing. *I Am That* - the eternal present moment described so eloquently by the bards, sages, shamans, teachers and scribes of ancient wisdom. It was a direct experience of the Truth of the implicate, *more* real world on the other side of the veil. No experience in my life, before or since, could compare to what happened to me in these brief minutes.

I was awestruck and shaken to my core . . . and it wasn't until the next day that I had the unthinkable thought that perhaps *this* was the reason behind the 'impossible' reality of the Earth/Moon diagram. And so it is.

17 A quote from Graham Hancock's *Magicians of the Gods,* referencing Micrea Eliads *The Sacred and the Profane.*

18 From Stanislav Grof's *When the Impossible Happens*

19 http://gallery.bridgesmathart.org/exhibitions/2014-bridges-conference-short-movie-festival/duganhammock

20 'A statement is called falsifiable if it is possible to conceive an observation or an argument which proves the statement in question to be false.' Wikipedia

21 At that time, 1609, everything discovered scientifically about the solar system was new. Kepler saw a manifestation of the orbits of the planets - expressing pure Platonic geometry - as the expression of God's hand touching the Universe. But then the moons of Jupiter appeared out of the Mystery . . . and they didn't fit the theory. It had to be a *random* phenomenon. It *was.* It *is* . . . for the entire Universe as science has discovered it to be – except *here.* Nowhere is found the perfection of geometric relationships as within this solar system. Not even close. It's hard for us to imagine the influence in daily life by the Church at that time. The reality of the Sun-centered solar system threatened the Church. The reality of the geometry being unique threatened emerging science. The Church won. Both . . . the Church and Science . . . missed the point.

22 Actually, our solar system and the visible stars beyond *was* the known Universe at the time.

23 From the book/film *The Privileged Planet,* 2004.

24 The authors of *The Privileged Planet* only used the first 13 of the 20 factors because fairly reliable variables could be established . . . they wanted to be conservative. One can see, though, that adding any of the other factors puts the chance of our solar system manifesting as it did off the chart. And this is only considering the *normal* way solar systems form as defined by science. Add the phenomena in appendices A, B and C and

another explanation is demanded.

25 Of the perhaps 3,500 planetary 'candidates' to date, 961 have been verified.

26 http://www.foxnews.com/science/2017/02/22/7-new-earth-like-exoplanets-discovered-nasa-announces.html

27 This particular alignment makes possible the Winter Solstice alignment with the Earth, the Sun, and the center of the galaxy . . . the Winter Solstice literally and figuratively signifying the return of the light each year as the nadir represented by the shortest day - and in this case demarcating the nadir of light, and the return of the light - for the 25,920 year long precession of the equinoxes called the Great Year . . . and that since 1987, and through 2018, this alignment has been transiting the galactic center, our local *Stillpoint* – what the Mayan shamans considered the source of their spiritual wisdom.

28 From *A Little Book of Coincidence in the Solar System* by John Martineau.

29 If conscious life somehow requires 'other stranger laws,' then how or why did those laws occur . . . *here?* . . . and not *there?* That is, if life 'required' them *here,* then *how* did this life from the future inspire them to be put in place beyond any stretch of coincidence, ready to 'imbue the host planet with life,' billions of years in the past? The reasoning seems circular.

30 The Sun's diameter is exactly 400 times larger than the Moon's *and* exactly 400 times farther away from the Earth than the Moon. (because of the slightly elliptical nature of the Earth's orbit around the sun, the 'totality' of eclipses vary - 'exactly' as demonstrated by the total eclipse). This particular phenomenon makes possible the total eclipse of the Sun, making possible, in the late 1800's, the discovery of the chemical makeup of the Sun which opened the door to stellar astrophysics and all kinds of discoveries about the Universe itself, and in 1919, Arthur Eddington's verification of Einstein's theory of relativity by verifying that the gravity of the Sun 'bent' light, and all that led to. The *Privileged Planet* is worth watching for this experience alone. We all grew up with the reality that two of the most significant celestial bodies relative to our own existence *just happened* to be exactly the same size in the sky. It was with us every day . . . and invisible to us. None of the other 65 significant moons in our solar system come anywhere close to matching this. I reviewed the film before I began writing this summary and the total eclipse still brought tears - because I was *seeing* the 'impossible' happen.

31 From the book/film *The Privileged Planet,* 2004.

32 The orbital speed of the Moon is 2,288 miles per hour, while the speed of sound it 742 miles per hour, or 3.08 x 742 equals 2,288.

33 'The close number association between the size ratios of the Sun, Moon and Earth, and the orbital characteristics of the Moon, together with the present length of the Earth day, are only applicable to the time that humans have been fully formed. These relationships were not present in the distant past and they will disappear in the distant future. The number sequences which alerted us to the 'message' are clearly meant for the present period.' From *Who Built the Moon?,* by Christopher Alexander and Alan Butler.

34 The term 'total eclipse' as used here means 'perfect' total eclipse. That is, the Moon perfectly covers the face of the Sun. Many years ago, when the Moon was closer to the Sun, total eclipse happened, but the Moon was much larger than the disc of the Sun.

35 https://blogs.scientificamerican.com/life-unbounded/the-solar-eclipse-coincidence/
I am not capable of doing the trigonometry myself and estimates for this window vary, which has been very frustrating. Here's another from someone named Hannah Osborne in Newsweek claiming that the window won't close for another 600 million years: https://www.yahoo.com/news/last-ever-total-solar-eclipse-132530476.html. In the book *The Priviledged Planet*, the authors say that the window has 250 million years left . . . or 5% of the age of the Earth. Either way, humans exist and have evolved within this window to be able to discern the absurd precision and lack of coincidence regarding the total eclipse of the Sun by the Moon.

36 At this point, I'm not sure where I obtained these dates, and other sources vary. Essentially, they represent the time it takes the disk of the Sun to transit the center of the galaxy. From http://alignment2012.com/whatisGA.htm: 'The Galactic Alignment is the alignment of the December solstice sun with the Galactic equator. This alignment occurs as a result of the precession of the equinoxes. Precession is caused by the earth wobbling very slowly on its axis and shifts the position of the equinoxes and solstices one degree every 71.5 years. Because the sun is one-half of a degree wide, it will take the December solstice sun 36 years to precess through the Galactic equator.' Since the location of the exact center is debatable, I am staying with the dates 1987 through 2018.

37 http://thinkprogress.org/politics/2008/04/23/22152/hagee-katrina-mccain/

38 Wikepedia.

39 'Atman project' is Ken Wilbur's term for the journey of the soul through the process of incarnations, or projects, to reach Unity – or Atman.

40 From Arthur Young's *The Reflexive Universe: The Evolution of Consciousnes*.

41 Another possible understanding is that this infinitely evolved consciousness chose to incarnate as the planetary archetypes. That is, our solar system *is* this consciousness.

42 This is the heart of another deep subject - the fact that we must accomplish what we will on our own without higher intervention. Once again, in the spirit of 'as above, so below,' we can see that as much as we may wish to, we cannot live our children's lives – we cannot *do* their lives for them. They have to live their own lives. But we can encourage, guide, teach and suggest - hoping they will find a way to ask and open to our experience.

43 After two decades of certainty regarding the authenticity of alien creation of the crop circles, a twinge of doubt has entered. I have read a number of books on the subject talking about the unnatural way the stalks are bent, dancing light globes appearing before the crop circle appears, the smell of the stalks being heated, the flash-burning of stalks imprinted on rocks in the formation, electromagnetic anomalies, psychic experiences and more. But beyond that, I simply cannot imagine that some of the formations are man made . . . they are phenomenally beautiful and complex. On the other hand, one of the people that I've met along this journey is a crop circle photographer and crop circle maker who tells me that *all* of the formations are man made. I just don't believe it and I'll leave it at that for now. At any rate, it means little in relation to the bigger picture . . . the evolution of consciousness.

44 There is controversy over the interpretation of the binary message with this word. 'Believe' is one interpretation. I do not think that in this case it is meant as 'belief' in one

thing or another, so much as 'do not give up hope.'

45 G. I. Gurdjieff, was an influential early 20th-century mystic, philosopher, spiritual teacher, and composer born in what was then an Armenian region of Russia.

46 From Merriam-Webster.

47 http://www.scientificamerican.com/article/are-virtual-particles-rea/'Quantummeanics allows, and indeed requires, temporary violations of conservation of energy, so one particle can become a pair of heavier particles (the so-called virtual particles), which quickly rejoin into the original particle as if they had never been there.'

48 'Transpersonal', from Merriam-Webster: extending or going beyond the personal or individual . . . especially with esoteric mental experience (such as mysticism and altered states of consciousness) beyond the usual limits of ego and personality.

49 From Graham Hancock's *Supernatural*.

50 See the film *Thrive*, or visit the website www.thrivemovement.com

51 From Stan Grof's *Psychology of the Future*.

52 A wonderful dictionary definition of 'technology' fits perfectly: 'Technology: the system by which a society provides its members with those things needed or desired.' I would stress needed. I want the temple to *be* an opening to the divine. There is no way to find out if this is true without *doing* it. 'The good the truthful the beautiful.'

53 Devil's Tower in anglo . . . Matho Thipila (Bear Lodge) or Ptehe (Brown Buffalo Horn) in Lakota.

54 It is critical here to make the distinction between the *individual* awakenings that have occurred throughout history by many paths and the need for a *global transformation of consciousness*. This attempt to access this higher consciousness cannot simply be the experience of an individual. It must be a *communal* effort, a *communal* attempt – the shared intention of even a small group of people. 'Never doubt the power of a small group of people to change the world. Nothing else ever has.' Margaret Mead.

55 If this experiment ever happens, it will cost money. I cannot think of a more honorable action . . . and an action more filled with integrity . . . than allocating energy usually horded for future survival towards the construction of a possible portal to higher consciousness. Untold trillions are spent for survival, arms, intelligence, obscene homes for the wealthy . . . and recently a new amusement park in Kentucky has opened with a full-scale Noah's Ark costing $100,000,000 (complete with cages displaying dinosaurs). The Mesa Temple would cost a fraction of any of these examples.

56 Max Planck: 'New scientific ideas never spring from a communal body, however organized, but rather from the head of an individually inspired researcher who struggles with his problems in lonely thought and unites all his thought on *one single point* which is his whole world for the moment.'

57 The Axis Mundi: 'Heaven and Earth. As the celestial pole and geographic pole, it expresses a point of connection between sky and earth where the four compass directions meet. At this point travel and correspondence is made between higher and lower realms. Communication from lower realms ascend to higher ones and blessings from higher realms descend to lower ones and be disseminated to all, and functions as the omphalos (navel), the world's point of beginning.'

58 'A green roof or living roof is a roof of a building that is partially or completely covered with vegetation and a growing medium, planted over a waterproofing membrane. It may also include additional layers such as a root barrier and drainage and irrigation systems.' Wikipedia. That is, the roof of the temple will be the living desert.

59 'We have discovered that, though it appears simple to our senses, water – 'humble, and precious, and pure' (St. Francis of Assisi) – knows how to communicate. Its chemical-physical characteristics make it capable of communicating, of recording information and then releasing it . . coherent electromagnetic wave (with higher-order information) managed to inform water by imprinting messages.' Massimo Citro, M.D., *The Basic Code of the Universe.*

60 From *Power vs. Force*, David R. Hawkins, M.D., Ph.D. 2004

61 'Permanent' in our world is relative. Ultimately, the temple is but a sand painting.

62 Locating the site for the construction of the temple is as critical as the temple itself.

63 Simply put, the 'ego' is our sense of self as contrasted with everything else. We live in an ego-centric world – it is 'the organized conscious mediator between the person and reality . . . both in the perception of and adaptation to reality.' https://www.merriam-webster.com/dictionary/ego

64 http://www.thebestschools.org/sheldrake-shermer-god-and-science-opening-statements/

65 http://www.dictionary.com/browse/standard-model

66 http://www.dictionary.com/browse/general-relativity

67 From Sean Carroll's *The Big Picture*, 2016.

68 Ibid.

69 http://www.merriam-webster.com/dictionary/reductionism

70 The fundamental constant of quantum mechanics, expressing the ratio of the energy of one quantum of radiation to the frequency of the radiation and approximately equal to $6.624 \times 10{-}^{27}$erg-seconds . . . http://www.dictionary.com/browse/planck--constant

71 From *The Big Picture*, by Sean Carroll.

72 Alan Guth, responsible for the theory of the inflationary Universe.

73 From *The Big Picture*, by Sean Carroll.

74 Ibid.

75 From a quote by the physicist David Bohm, page 190.

76 From *The Big Picture*, by Sean Carroll.

77 Uri Geller is an illusionist, well known internationally in the 1970's especially, as a magician, television personality, and self-proclaimed psychic. He is known for his trademark television performances of spoon bending. Throughout the years, Geller has used skillful tricks to simulate the effects of psychokinesis and telepathy.

78 Ibid.

79 From Jane Goodall's *Reason to Hope*, 1999.

80 From http://whatis.techtarget.com/definition/superposition: 'Superposition is a principle of quantum theory that describes a challenging concept about the nature and behavior of matter and forces at the sub-atomic level. The principle of superposition claims that while we do not know what the state of any object is, it is actually in all possible states simultaneously, as long as we don't look to check. It is the measurement

itself that causes the object to be limited to a single possibility.'

81 http://www.evolution.berkeley.edu/evosite/evo101/IIIC1aRandom.shtml

82 http://www.pbs.org/wgbh/evolution/library/faq/cat01.html

83 https://www.quora.com/Is-evolution-by-natural-selection-a-random-process a quote from Joshua Engel.

84 http://dictionary.sensagent.com/Emergent%20evolution/en-en/

85 From Jane Goodall's *Reason to Hope*, 1999.

86 From *The Oxford Dictionary*.

87 From *Listen Humanity*, Meher Baba, 1957.

88 It is clear that consciousness has evolved beyond the human level . . . but Baba's words are essentially true.

89 From *God & The New Physics* by Paul Davies, 1983

90 Quotes from Martin Rees 1997 book *Before the Beginning*, found in Rupert Sheldrake's 2012 book *Science Set Free*

91 From *The Privileged Planet*, 2004.

92 From *Discover Magazine*, 2000.

93 Ibid.

94 Ibid.

95 The evidence that this theory is based upon has been brought into question, particularly the redshift of Quasars thought to be much nearer.

96 The 'age' of the Universe is the problem of defining distance in an expanding universe: Two galaxies are near to each other when the universe is only 1 billion years old. The first galaxy emits a pulse of light. The second galaxy does not receive the pulse until the universe is 15 billion years old – the radius of a 30 billion-year sphere; the pulse of light has been traveling for 15 billion years; and the view that from the second galaxy is an image of the first galaxy at 1 billion years old, when it was only 2 billion light years away.

97 There is also evidence that conflicts with this theory of Universe that began at the Big Bang and is expanding. This is called the *static* or *plasma* or *electric* universe theory. The following summary for this theory is taken from Eric Lerner's video series *The Big Bang Never Happened*, Wallace Thornhill and David Talbott's *The Electric Universe*, and Hilton Ratcliffe's *The Static Universe*.

When Einstein presented his Theory of Relativity, it was based upon gravitational forces controlling the Universe. Originally, this modeled a static system. But the equations wouldn't predict a static Universe without the addition of what he called the *cosmological constant* – an arbitrary value shoehorned into the equations. But later, in the 1920's, others provided solutions to Einstein's equation that permitted the Universe to either expand or contract without need of this added value. At the same time, Edwin Hubble made the famous discovery that the redshifts in the light spectrum of distant galaxies, meaning that they were moving away from us – meaning that the Universe was expanding.

But in the 1960's, quasars – very small compact objects containing huge amounts of energy - were discovered in spiral galaxies. Quasars have very high redshifts . . . meaning they should be very far away. But photographic evidence shows quasars close to galaxies with much lower redshifts. Long story short, this meant that the Big Bang theory was based entirely on a misunderstanding of what observed redshifts implied. Plasma theory suggests that new galaxies are born from quasars (proto-galaxies) that are ejected from older galaxies with very high energy – explaining the redshift/distance phenomena, as well as dismissing the expanding Universe for a static, plasma filled Universe – and, unlike the impossibility of proving BB theory in the lab, plasma experiments predict what we observe – a filament filled, plasma filled, static Universe. From astronomer Geoffrey Burbidge: 'What is really happening in these systems is that the centers of the galaxies are where creation is taking place, rather than just in the Big Bang, and so you are looking at all these mini-bangs where matter and energy are literally being created.'

There are many many more facets to this argument for which there is no space here, but a couple of other remarkable aspects need to be mentioned. Big Bang theory predicted the abundance of three light elements: lithium, helium and deuterium – which jived perfectly with the long established value for the density in the Universe. But in the 1980's, with the discovery of dark matter, this prediction was thrown out forever. To compensate, Big Bang theorists – without any basis in fact - simply said that this was an entirely different *kind* of matter – no protons, electrons, neutrons . . . and because of this it didn't count. Yet the theory still stood.

Further, according to the predictions of Big Bang theory, the Universe should be only 8 billion years old . . . but there are stars that are 14 billion years old, their age based on very solid, observed, science. In order to stretch it out to fit established evidence (the age of old stars), BB science had to invent something called *dark energy* . . . invented to *accelerate* the expansion of the Universe. So . . . we now have 70% dark energy, and 28% dark matter . . . leaving only '2% of matter that we can observe through our telescopes here on Earth' – all just to keep the Big Bang theory in place.

Thousands of years ago, the Ptolemaic observers of the movements of the planets invented a complicated series of epicycles (circles moving around other circles) to explain this movement. Layers and layers of artificial constructs heaped upon observations to create a model of the solar system that conformed to the epicycles. It wasn't until Copernicus that it all could be explained by the fact that they all orbited the Sun. This, apparently, is what is happening today in Big Bang Theory.

Bottom line, plasma theory says that the Big Bang never happened, and that the present *stage* of universal evolution is an electric, or a plasma Universe . . . and that we cannot know what stage came before . . . and never will. It is a Universe with no beginning and no end, a Universe continually evolving, at a continually more rapid pace – something evident in our own time on Earth . . . as above, so below.

Either way, to my mind, creation is being generated at the very centers of massive energy systems – and at the center of these centers is the Stillpoint. The beauty in the plasma theory, as well as the implications the Stillpoint phenomena – scientifically speaking – is that they each are derived from *observation* . . . the main pillar of scientific inquiry. The Big Bang theory, on the other hand, presently seems to be held together by glue. The latter is presently accepted dogma. In each theory, the Stillpoint is present . . . as is the torus

98 'Darkness would qualify as a niche subject except for how much of it exists. The current best calculation holds that the cosmos is 4.9% regular matter, 26.8% dark matter, and 68.3% dark energy. In that 4.9% is included all luminous matter contained in billions of galaxies plus a huge amount of non-luminous matter in interstellar dust. So the barest fraction of creation is offering empirical data. Physics has been dealing with the cherry on top of the sundae, the tip of the iceberg, or the grin of the Cheshire Cat after its body has vanished–pick whatever metaphor you like. Most of the universe is at the very least quite exotic.' - http://www.sfgate.com/opinion/chopra/article/Physics-Split-Personality-Is-the-Dark-Side-6406931.php

99 See endnote 100.

100 Perhaps the biggest goal in physics today is to unify the strong nuclear force, weak nuclear force, electromagnetic force, and gravitational force into one unified force, or what physicists call the 'Grand Unified Theory.' It has already been discovered that at high enough energies, electromagnetism and the weak force are the same force, known as the electroweak force. It is theorized that if energies are increased even further, all the known forces will resolve themselves into the same force. According to Einstein's Field theory all the above four forces are different manifestations of a single unified force. He spent the latter part of his life searching for this Grand Unified Theory. His calculations showed that there was only *one force* controlling all phenomena at the moment of origin of the Universe from a Cosmic Egg (the point of singularity).

101 A quote from David Bohm from the Cosmometry website: https://www.udemy.com/fundamentals-of-cosmometry/learn/v4/t/lecture/216131

102 From *Discover Magazine*, November 2002.

103 'The Planck Constant is a physical constant that is the *quantum* of *action* in quantum mechanics. It describes the *proportionality* constant between the energy of a charged atomic oscillator and the frequency of its electromagnetic wave [and is] behavior associated with an *independent* unit – or *particle* - as opposed to an electromagnetic wave and was eventually given the term *photon*. Its relevance is now integral to the field of quantum mechanics, describing the relationship between energy and frequency, commonly known as the Planck relation. In physics, *action* is an attribute of the dynamics of a physical system. It is a mathematical function which takes the trajectory, also called *path* or *history*, of the system as its argument and has a *real number* as its result . . . *the action must be some multiple of a very small quantity* (later to be named the 'quantum of action' and now called Planck Constant). This inherent granularity is counter-intuitive in the everyday world . . . this is because the quanta of action are very, very small in comparison to everyday macroscopic human experience. Hence, the granularity of nature appears smooth to us. Thus, on the macroscopic scale, quantum mechanics and classical physics converge at the *classical limit*. The classical limit is the ability of a physical theory to approximate classical mechanics. The *dimension* of The Planck Constant is the product of *energy* multiplied by *time*, a quantity called *action. The Planck Constant is often defined, therefore, as the elementary quantum of action.'

104 http://math.ucr.edu/home/baez/planck/node2.html

105 These observations taken from a lecture given by Dr. Bob Whitehouse and Dr. Mike Buchele in Berkeley, CA, in 2013 . . . https://vimeo.com/93676817, as well as from Arthur Young's *The Reflexive Universe*.

106 Ibid.

107 Arthur Young's *The Reflexive Universe*.

108 Ibid.

109 Ibid.

110 Ibid.

111 A comment made by Dr. Bob Whitehouse.

112 Ibid.

113 Cosmometry: *http://cosmometry.net/*

114 Ibid.

115 Ibid.

116 See the work of Steve Wilmoth, animated by Dugan Hammock: http://gallery.bridgesmathart.org/exhibitions/2014-bridges-conference-short-movie-festival/duganhammock.

117 'God creating Man in His image.'

118 From *Discover Magazine*, November 2002.

119 Ibid.

120 Ibid.

121 The heavy elements are required for the existance of complex life, and 'are the products of fusion power within stars, subsequently spewed out across the cosmos on the blast waves of supernovae, lacing the interstellar medium with the raw ingredients for planets. To build up enough of these materials, many stars must first live and die, each one contributing to the evolving chemistry of the universe, but how much material is really required to build a planet and how quickly did the universe accrue a sufficient level to do so? http://www.space.com/17441-universe-heavy-metals-planet-formation.html

122 Ibid.

123 From *The Mists of Special Relativity*, by Stephen Earle Robbins, PhD.

124 http://www.iflscience.com/physics/einsteins-spooky-action-distance-confirmed-new-quantum-experiment

125 Ibid.

126 https://www.youtube.com/watch?v=woztlIAYTCU

127 Chreode is a neologism (meaning a 'new word') created by C.H. Waddington (1905 to 1975), a British biologist and geneticist. It is made from the Greek roots chre, meaning 'it is necessary,' and hodos, meaning 'route or path.' nlpmarin.com/chreodes-entelechy-and-human-potentiality/.

128 https://blogs.scientificamerican.com/life-unbounded/the-solar-eclipse-coincidence/

129 From the book *Civilization One*, by Christopher Knight and Alan Butler.

130 See Appendix B, *The Message in Detail*.

131 Hancock feels that the remnants of humanity that survived the cataclysm were taught the rudiments of agriculture, mathematics, astronomy, geometry and megalithic building by envoys from the annihilated advanced civilization, rather than from an extraterrestrial

source. I tend to think it's the latter, but it is impossible to know at this time.

132 From *Science Set Free,* by Rupert Sheldrake.

133 Ibid.

134 Ibid.

135 Ibid.

136 *The Bigger Picture,* Sean Carroll, 2016.

137 https://www.youtube.com/watch?v=bJpyX514cUg

138 'Professor of theoretical physics, best selling author and populizer of science.' mkaku.org For the most part, I take whatever Kaku says with a grain of salt, but in this case I feel that he's correct.

139 http://math.ucr.edu/home/baez/planck/node2.html

140 Very similar to Stephen Hawking's rigid views are those of Richard Dawkins, the famous and well-respected biologist of *The Selfish Gene* fame (support for the Darwinian idea that life is a process of natural selection – that is, due entirely to necessity and chance), and more recently perhaps the most famous atheist in the world. When his book *The God Delusion* came out in 2008, I couldn't wait to read it because I was learning about the huge influence the fundamentalist Christian Right in the United States had on political matters. I couldn't wait to read how the 900 pound gorilla Dawkins was going to tear to pieces the absurdities of this particular belief system . . . as he certainly did. But as I continued reading, I discovered that Dawkins was almost equally as blinded by his own ideology as the fundamentalists . . . by the religion of science. I was impressed by the knowledge the man contains, but was left with Hamlet's words: 'There are more things in heaven and earth, Horatio, Than are dreamt of in your philosophy.'

141 From *Yoga and the Portal,* by Swami Harinanda.

142 http://www.cnet.com/news/physicists-prove-einsteins-spooky-quantum-entanglement/

143 Ibid.

144 According to Dr. Steven Greer, anti-gravity, as well as distant space travel, had been established since 1954. The Carol Rosin Show w/ Steven Greer, 9/15/17.

145 Science is currently rethinking this long accepted figure. New discoveries may place this date as long ago as 750,000 years . . . https://www.youtube.com/watch?v=qkedj59k66A. While this is very interesting and very important, it has little to do with the main focus of this narrative . . . the evolution of consciousness in the Universe. For now, as with other controversial scientific theories . . . say, the Big Bang theory . . . current accepted theories are generally referenced.

146 Possibly due to the discovery of hallucinogens by human beings.

147 Again, see the books written by Graham Hancock - significantly, the story heard by Plato from his ancestor Solon, who got it from the Egyptians, coincided *precisely* with the Younger Dryas geological period 11,600 years ago that ended in cataclysm.

148 See Graham Hancock's *Magicians of the Gods.*

149 Self quote from this book, page 116.

150 http://badarchaeology.wordpress.com/2014/02/15/the-paracas-skulls-aliens-an-unknown-hominid-species-or-cranial-deformation/

151 From *The Forbidden Universe,* by Lynn Picknett and Clive Prince.

Endnotes

152 From *Life Itself,* Francis Crick, 1981.
153 http://www.theatlantic.com/health/archive/2015/01/rethinking-one-of-psychologys-
 most-infamous-experiments/384913/
154 Ibid.
155 Ibid.
156 Ibid.
157 Ibid.
158 *Without Conscience: The Disturbing World of the Psychopaths Among Us,* by Dr. Robert Hare.
 https://www.fastcompany.com/53247/your-boss-psychopath,
 http://www.wanttoknow.info/g/psychopaths-politics-power,
 http://www.bibliotecapleyades.net/sociopolitica/hiddenevil/hiddenevil25.htm
159 The description of the process as well as the floor plan was taken from
 https://en.wikipedia.org/wiki/Milgram_experiment
160 http://www.theatlantic.com/health/archive/2015/01/rethinking-one-of-psychologys-
 most-infamous-experiments/384913/
161 http://www.theatlantic.com/health/archive/2015/01/rethinking-one-of-psychologys-
 most-infamous-experiments/384913/
162 http://www.tomdispatch.com/post/176088/tomgram%3A_rick_shenkman
 %2C_how_we_learned_to_stop_worrying_about_people_and_love_the_bombing/#more
163 http://www.cnn.com/2015/02/11/us/chapel-hill-shooting/
164 lttp//www.democracynow.org/, January 11, 2015
165 A consequence of having enthusiastically taken the red pill some time ago, along with
 being able to share this information, is that normal life has become more and ʳ ̣e
 surreal. Almost 63,000,000 people voted for a man who is clearly a narcissistic, clinically
 ignorant, sociopathic, extraordinarily shallow human being. He is a cartoon character
 who proudly represents those who think that if they have the most toys, win the game of
 life. Many people voted for him precisely because of this . . . if he made so much money,
 there *must* be something very right about him. Many others voted for him because they
 actually believed what he was saying, when it is clear to any healthy common sense that
 the man is a pathological liar and couldn't possibly care less about people affected by the
 issues he used to collect votes. He is a child with his finger on the button. It is also clear
 that many voted for him because the dark inner-nature of the woman he ran against was
 containable no longer, and transparent to much of the country. Long ago now, a small
 town in Texas lost its idiot and I was stunned to watch the coup that put him in the office
 that Abraham Lincoln once held, my long earned sense of reality changed forever. And
 then 9/11 happened and I began an intense process of research into these dark forces that
 now control the world. I thought before this that I had a handle on 'reality' . . . but I had
 absolutely no idea. Most of us, in our busy lives, don't want to be bothered . . . and come
 to our understandings of what goes on in the world through snippets designed for us by
 fewer and fewer media outlets . . . the ownership and control of all we hear and see
 becoming more and more centralized – see the Edward Bernays quote in the Preface. I
 learned that I had to go much deeper to find journalists and writers and filmmakers that I
 could trust. This search led me far beyond where most people cared to go, choosing

instead to accept the con supplied them and get lost in the trivia of trinkets and gadgets . . . while so much of the world was suffering from the actions of those in control. This path inevitably took me into discovering the hidden world of the hideous woman this hideous man was running against, the woman almost 66,000,000 people voted for. I have a good friend where I live, and in a few days he and his wife are celebrating their 50th wedding anniversary. He also chose the red pill and relentlessly charged into the research of the dark on his own, and on a path parallel to my own. He also discovered unthinkable, unimaginable even, truths about the dark reality of this woman . . . but his wife refused to see past the veneer – even as it became paper thin. Just like so many, for whatever reason. I live in the most powerful country the world has ever seen. It has been completely co-opted. There was not even, and hasn't been for some time, even a semblance of a choice when one 'votes' for who will become the most *apparently* powerful person in the world. So, there are two points here. The first is that almost 129,000,000 oblivious human beings had been conned into thinking that there was someone worth their vote. And secondly, the true controllers of this planet, the .1% of the population who by heredity or connections or willingness to sacrifice whatever soul they may once have had, have successfully manipulated the masses into becoming lost in endless, meaningless conversations . . . while they continued to go about their insidious work strangling the planet. The illusion is deeply entrenched . . . how can humanity possibly wake up in time? The core issue exposed here is . . . drum roll . . . a lack of general *consciousness*.

166 http://www.tomdispatch.com/blog/176088/tomgram
%3A_rick_shenkman,_how_we_learned_to_stop_worrying_about_people_and_love_the_bo
mbing/

167 https://books.google.com/books?
id=eMEwSMjnbBMC&pg=PT171&lpg=PT171&dq=Freda+Kirchwey+american+indifferen
ce+during+Korean+War&source=bl&ots=mImcYFVDJ4&sig=w7oXRrXoN3Shmitz3ZeC
vPhUDbU&hl=en&sa=X&ved=oahUKEwjozMevwLTKAhVHiGMKHVcYDasQ6AEI
LzAD#v=onepage&q=Freda%20Kirchwey%20american%20indifference%20during
%20Korean%20War&f=false

168 Ibid.

169 http://www.ncbi.nlm.nih.gov/pmc/articles/PMC3825032/

170 http://www.tomdispatch.com/post/176088/tomgram%3A_rick_shenkman
%2C_how_we_learned_to_stop_worrying_about_people_and_love_the_bombing/#more

171 http://www.prisonexp.org/the-story

172 Ibid.

173 See John Whitehead's *Battlefield America* and *Government of Wolves: The Emerging American Police State*, as well as Radley Balko's *Rise of the Warrior Cop.*

174 http://www.prisonexp.org/the-story

175 Ibid.

176 Ibid.

177 Ibid.

178 Ibid.

179 Ibid.

180 Ibid.

181 Ibid.

182 http://www.sentencingproject.org/template/page.cfm?id=107

183 http://www.apa.org/monitor/2014/10/incarceration.aspx

184 https://hypnosis.edu/articles/truth

185 http://sherryruthanderson.com/the-cultural-creatives/

186 Ibid.

187 From *The Cultural Creatives: How 50 Million People Are Changing the World*, by Paul H. Ray and Sherry Ruth Anderson.

188 https://www.brainpickings.org/2014/08/27/willful-blindness-margaret-heffernan/

189 Ibid.

190 From *Willful Blindness: Why We Ignore the Obvious at Our Peril*, by Margaret Heffernan

191 http://www.brasschecktv.com/videos/out-there-1/willful-blindness.html

192 Ibid.

193 https://en.wikipedia.org/wiki/Aaron_Swartz

194 In Mahayana Buddhism, *bodhisattva* refers to a human being committed to the attainment of enlightenment for the sake of others.

195 http://www.newsfinder.org/site/more/becoming_a_boddhisattva/

196 https://www.studentsforafreetibet.org/about-tibet/self-immolations

197 http://www.dailymail.co.uk/news/article-1082559/The-GM-genocide-Thousands-Indian-farmers-committing-suicide-using-genetically-modified-crops.html

198 http://english.pravda.ru/opinion/columnists/06-02-2014/126767-global_elite-0/

199 http://www.informationclearinghouse.info/article40078.htm   This is another huge subject that can't be covered here. Wars, poverty and oppression wound people irreparably. These people wound their children who wound their children. If the shift we all hope for occurs, the healing and nurturing of our children, providing them with safe and healthy and love-filled lives, is perhaps the most important aspect of the healing that must come.

200 http://www.informationclearinghouse.info/article39165.htm

201 http://williamblum.org/essays/read/overthrowing-other-peoples-governments-the-master-list

202 http://williamblum.org/chapters/rogue-state/united-states-bombings-of-other-countries

203 http://williamblum.org/chapters/killing-hope/us-government-assassination-plots

204 http://williamblum.org/essays/read/suppressing-revolt-and-revolution

205 http://williamblum.org/books/rogue-state, all of the above from Chapter 18 of Blum's book *Rogue State: A Guide to the World's Only Superpower*.

206 Catherine Austin Fitts: https://missingmoney.solari.com/

207 http://fktv.is/ai-being-taught-to-disobey-humans-27543

208 *Common Wealth*, Freddy Silva, 2010.

209 From T.S. Elliot's much longer poem *Little Gidding*.

210 From: http://www.sciencedaily.com/releases/2010/01/100107143909.htm. 'By tuning the system and artificially introducing more quantum uncertainty the researchers observed that the chain of atoms acts like a nanoscale guitar string. Dr. Radu Coldea from Oxford

University, who is the principal author of the paper and drove the international project from its inception a decade ago until the present, explains: "Here the tension comes from the interaction between spins causing them to magnetically resonate. For these interactions we found a series (scale) of resonant notes: The first two notes show a perfect relationship with each other. Their frequencies (pitch) are in the ratio of 1.618..., which is the golden ratio famous from art and architecture." Radu Coldea is convinced that this is no coincidence. "It reflects a beautiful property of the quantum system -- a hidden symmetry. Actually quite a special one called E8 by mathematicians, and this is its first observation in a material," he explains.'

211 The estimated height, including the limestone facing, based upon measurements of Professor Flinders Petrie in 1883, is 484.42'. The radius of the Earth at the equator, scaled at 1/43,200, is 484.62'. Without the limestone, the height equals the *polar* radius of the Earth, or 481.8'. It does appear that the Great Pyramid is, intentionally, a perfect Phi pyramid . . . it is the magic of the *scale* incorporating the *timing* that proves the intentionality of the message. See http://www.thegreatpyramidofgiza.ca/content/

212 Actually, the Great Pyramid's base perimeter relates to the circumference of the Earth at 36 degrees latitude, having to do with the phi proportion and demonstrating even more knowledge regarding the precise measurements of the Earth. See Ekhardt Shmitz's book.

213 In 1990, the geologist Robert Schoch established that the limestone walls of the quarried enclosure, as well as the Sphinx itself – which was carved out of the same bedrock – had been eroded by torrential rain in the distant past. The most recent time that this could have occurred is at the end of the last ice age . . . or 11,600 years ago. Egypt has been dry for at least 8,000 years.

214 A quote from SETI supporter Bernard Oiver found in *The Privileged Planet*.

215 From *The Priviledged Planet*, 2004.

216 As of this writing.

217 *The Cosmic Serpent*, by Jeremy Narby, 1998.

218 http://cropcircleconnector.com/Sorensen/PeterSorensen99.html

219 From *Crossing the Event Horizon by* Nassim Haramein

220 This observation regarding the relationship between primal geometry and our own direct experience of it came from Keith Critchlow at a sacred geometry seminar in 1987.

221 *Common Wealth*, Freddy Silva, 2010.

222 www.cymaticsource.com.

223 From *The Power of Limits*, by Gyorgy Doczi, 1981.

224 Peter Sorensen has told me that he personally knows of many crop circles that are man-made. I am sure that this is true . . . and equally so that many are not.

225 From *Common Wealth*, F. Silva, 2010.

226 John Michell in a personal letter to me, addressing this specifically.

227 When I first saw the inaccuracy of this number . . . 7 x 51.51 = 360.57 . . . not precisely 360, I was as dismayed. Maybe Martin Rees was right when he observed that the orbits of the planets were found to be elliptical . . . demonstrating the 'ugliness' of a randomly created Universe. But I now see a gracefulness perhaps beyond our comprehension, and yet another ever so subtle message. *Seven*. The number never actually 'born', and only

approximated, geometrically speaking.

228 From T.S. Elliot's long poem *Little Gidding*.

229 'Hóka-héy!' means something like 'Let's go!' in Lakota . . . I couldn't find a Lakota translation for 'Today is a good day to die!'

230 This idea of the 'turn' is not a new one. In the ancient Egyptian body of wisdom, Atum, the god of creation whose initiating spark (reminiscent of Meher Baba's 'whim') exploded into manifestation, is not complete without the wisdom obtained from the long journey into this world. From Lynn Picknett and Clive Prince's book *The Forbidden Universe:* 'The creative flow from the god to the material universe is not just a one-way phenomenon. Just as it 'exhales' from Atum, it 'inhales' the life force of individuals, which then travels back to its source. Horus (who occupies a transitional space between the Great Ennedad - the nine gods of the unmanifest world, and the lesser Ennedad - the nine gods of this world), therefore, also represents what Karl Luckert calls the 'turnaround realm,' the point at which the life force can begin the journey back towards Atum. We might need Atum – but he also needs us.'

231 From a section *The Collected Works of Carl Jung*, called *Synchronicity: An Acasual Principle: (1952)*. 'A young woman I was treating had, at a critical moment, a dream in which she was given a golden scarab. While she was telling me this dream, I sat with my back to the closed window. Suddenly I heard a noise behind me, like a gentle tapping. I turned round and saw a flying insect knocking against the window-pane from the outside. I opened the window and caught the creature in the air as it flew in. It was the nearest analogy to a golden scarab one finds in our latitudes, a scarabaeid beetle, the common rose-chafer (*Cetonia aurata*), which, contrary to its usual habits had evidently felt the urge to get into a dark room at this particular moment. I must admit that nothing like it ever happened to me before or since.' Actually, in my case, it has.

232 Gurdjieff's idea of the hierarchal universe: all galaxies, galaxy, star, earth, moon, etc.

233 http://www.bing.com/videos/search?
q=arthur+young+video+discussing+the+turn&FORM=VIRE2#view=detail&mid=2BF7A0

234 From Webster's New World College Dictionary_Copyright © 2010

235 From Stanislav Grof's book *When the Impossible Happens*

236 From Gurdjieff.

237 'A wave that oscillates in place and creates stable nodes of maximum and zero oscillation, produced whenever a wave is confined within boundaries, as in the vibrating string of a musical instrument. Also called *stationary wave*.' Wikepedia

238 http://wn.com/grounded_trailer

239 http://forbiddenknowledgetv.net/wi-fried-27757

240 http://www.brasschecktv.com/videos/science-and-technology/strong-arm-tactics-of-thecorporate-gestapo.html   http://www.globalhealingcenter.com/natural-health/10-shocking-facts-health-dangers-wifi/
https://www.scientificamerican.com/article/who-cares-about-5g-wireless-you-will/

241 From *Common Wealth*, Freddy Silva, 2010.

242 Ibid.

243 Ibid.

Endnotes

244 *Merriam-Webster,* Dictionary.com

245 *Common Wealth,* Freddy Silva, 2010.

246 'Infinity line' in this context is the line where water changes from perfectly horizontal to perfectly vertical . . . imagine a waterfall of a 1/4" layer of slowly moving water over polished black granite, falling over a precise, circular edge of this granite, disappearing almost silently beneath the floor – moving and still at the same time; glistening, shining, reflecting, absorbing.

247 A volume drawing into its center the energy surrounding it.

248 http://wn.com/grounded_trailer

249 *Common Wealth,* Freddy Silva, 2010.

250 To help accomplish this, four crystals will be embedded at floor level at the location of the indentation of each side – located at the intersection of a line drawn 179.5° from each corner of the pyramid's base. This is one of the ways that the abstract is introduced into the physical, creating the geometrical shape required.

251 *MayaCosmogenisis 2012,* John Major Jenkins, 1998.

252 Ibid.

253 Ibid.

254 Ibid.

255 From The Message of the Stargates by Dawn Abel

256 From Gregg Braden's Awakening to Zero Point, 1994.

257 *Common Wealth,* Freddy Silva, 2010.

258 http://slideplayer.com/slide/9560657/

259 From *Common Wealth,* Freddy Silva, 2010.

260 *The Cosmic Serpent,* by Jeremy Narby, 1998.

261 From *Common Wealth,* Freddy Silva, 2010.

262 *MayaCosmogenisis 2012,* John Major Jenkins, 1998.

263 In Graham Hancock's *Magicians of the Gods,* it is suggested that there are as many as 150,000.

264 From *The Cosmic Serpent,* by Jeremy Narby 1998

265 From *Common Wealth,* Freddy Silva, 2010.

266 From Jane Goodall's *Reason to Hope.*

267 http://www.scoop.co.nz/stories/HL0202/S00013.htm

268 Essential wisdom . . . *I Am That,* by Nisargadatta Maharaj

269 And while available today, almost no one know or cares. And there it is.

270 The speed of the shadow on the ground varies due to its path regarding latitude, as well as the angle of the eclipse relative to its angle to the Earth-sphere.

271 *http://biblehub.com/esv/isaiah/60.htm*

272 'Perfect' total eclipse happens only when the Moon is precisely the same size in the sky as the Sun, making the corona visible . . . Annie's 'lifesaver.'

273 From *Teaching a Stone to Talk,* by Annie Dillard, 1982.

274 From Michael Schneider's *A Beginner's Guide to Constructing the Universe,* 2003.

275 From Gregg Braden's Awakening to Zero Point, 1994.

The author makes his living designing architecture and lives in the high desert of the Owens Valley, on the eastern side of the Serra in California.

He's enjoyed a 40-year fascination with sacred geometry. With the help of his son, he's planted 60 trees and built stone circles on the land where he lives.